O. V. S. HEATH, D.Sc., F.R.S., is Professor of Horticulture at the University of Reading.

THE PHYSIOLOGICAL ASPECTS OF PHOTOSYNTHESIS

THE PHYSIOLOGICAL ASPECTS OF PHOTOSYNTHESIS

O. V. S. HEATH

1969
STANFORD UNIVERSITY PRESS
STANFORD, CALIFORNIA

Stanford University Press
Stanford, California

Library of Congress Catalog Card Number: 68-54826

Printed in Great Britain

Preface

THIS book is primarily intended for advanced undergraduate students of botany at a university and for post-graduate research students; I hope, however, that teachers and research workers who have specialized in related fields will also find the work useful. For all such readers I believe that it is more appropriate to adopt a treatment akin to a review, with some critical discussion of evidence and references to the original papers, than to make the unqualified statements of conclusions which are unavoidable in elementary textbooks. The latter method has, however, been used for brevity in the provision of some of the background information in Chapter 1.

The subject of photosynthesis includes, in addition to botany, a great deal of physical- and bio-chemistry and some physics. Recently the exciting advances made in the chemical aspects have led to a relative neglect, both in research and teaching, of the others, which may collectively be called the physiology of photosynthesis: hence this book. Modern theories of the chemistry of photosynthesis are based mainly on experiments with unicellular organisms, especially *Chlorella* spp., and with isolated chloroplasts or parts of chloroplasts. Isolated chloroplasts have the great merit that they do not respire and the chemistry of photosynthesis can therefore be investigated without the uncertainties and confusion introduced by concomitant respiration. Photosynthesis by leaves of higher plants is still, however, a matter of prime importance for the survival of mankind [264, 258] and the complexities that these involve are therefore worthy of attention. Where possible I have drawn on work carried out with higher plants.

In 1957 it was possible to say with conviction that the main problems of the chemistry of photosynthesis had been solved and that only details remained to be elucidated. This state of affairs had come about in under ten years; for the biochemical stages this was largely a result of the brilliant exploitation by Calvin [11] and his co-workers of the new technique of paper chromatography combined with the use of radioactive carbon (supplied as $C^{14}O_2$) and its detection in compounds on the chromatograms by autoradiography. Since then the chemical picture has again changed fundamentally, for the photochemical stages, as a result of developments from some of the last physiological experiments of Emerson before his untimely death in 1959. There are many reviews and semi-popular articles which constantly bring the subject up to date and these should be consulted (for example ref. 320). It is no longer possible to consider the more biochemical aspects of photosynthesis *in vivo* in isolation from other metabolic activities, such as respiration, and they have to a large extent become merged in the wider subject of plant metabolism. The physical chemistry of photosynthesis remains very much a subject in its own right and is developing fast.

One of the functions of the plant physiologist is to assist the biochemist and physical chemist with information as to how the systems studied operate in various environmental conditions, especially those in which the organisms normally exist. Recent work [325] suggests that algae photosynthesizing under low carbon dioxide concentrations such as are found in nature produce a different distribution of compounds from that found by Calvin and others who used concentrations of about three per cent.

Beginning with the work of F. F. Blackman at the end of the last century, a long series of physiological experiments, to study the effects of internal and external factors upon rates of oxygen output and carbon dioxide uptake, has made possible deductions as to the types of physical and chemical processes concerned in the successive stages of photosynthesis (Chapter 9). Such methods have continued to be useful up to the present day, especially when combined with the use of isotopic markers (C^{14} and O^{18}—see Chapter 3). The discovery of the phenomenon of enhancement by Emerson in 1956, which indicated two light reactions in photosynthesis driven by different pigment systems, opened up new fields of investigation which are being actively explored in many laboratories in experiments both with unicellular algae and with isolated chloroplasts, Such work is bringing together the physiological, physical-chemical and biochemical aspects of photosynthesis but it is also extending far beyond the limits set for this book, in which it has been possible to discuss only a selection of the work subsequent to Emerson's discovery (Chapter 9).

The structure of chloroplasts is discussed in Chapter 1, in order that subsequent references may be understood. Some of the remarkable electron micrographs now available are presented; however, the relation of this structure to the processes of photosynthesis remains a matter for speculation. Suggestions that it is possible to see the photosynthetic units (Chapters 1 and 9) are still controversial.

Of the physical aspects of photosynthesis, the diffusion into the plant of carbon dioxide is important and can only be usefully discussed in quantitative terms (Chapter 2). Only elementary mathematics have, however, been used in discussing this topic and the equally important one of interaction of factors (Chapter 4).

More space is devoted to methods (Chapter 3) than is usual in books of this size, partly because I believe that scientific discoveries cannot be assessed or even understood in the absence of information as to how they were made.

A knowledge of the historical development of any branch of science is also necessary if recent developments are to be seen as part of a continuing, though often a tortuous process, and not as final answers. Those who have not already done so are recommended to read the excellent account of the early investigations in the first volume of *Photosynthesis* by Rabinowitch [264]. Subject to considerations of space I have tried to adopt the historical approach in my accounts of more recent work.

The omission of nearly all the chemistry of photosynthesis has helped to keep this book small. Even so, it is necessary only to dip among the 2,000 pages of Rabinowitch's three-volume treatise [264, 265, 266] to realize that any book as small as my own must either be unbalanced or a catalogue written in telegraphic style. I have chosen the former, and have thought it most useful to weight the book in favour of those parts of the subject where my interest is greatest and my knowledge least inadequate. The bias is consequently different from that of many other books. I have also expressed my opinions on such matters as the merits of various methods of investigation. I hope that if the reader holds other views, disagreement will prove a useful stimulant.

Inevitably, in a book such as this, some of the conclusions will be modified or disproved even before it appears and many others will follow them during the next few years. I have tried, therefore, to give less emphasis to the current view than to experimental methods and results, as well as to ways of drawing conclusions from the latter. These are the enduring aspects of science.

I am most grateful to Dr H. Meidner who read the whole typescript, and to Drs R. Hill, J. L. Monteith, B. Orchard, H. L. Penman and J. Prue who read various chapters. Their criticisms and suggestions have been most valuable, but I must take full responsiblity for the errors which the book no doubt contains. I should appreciate correspondence about them, in case a second edition is needed. Finally, I thank my wife; without her support I could not have written the book.

Reading, 1968 O.V.S.H.

Contents

List of Plates

Between pages 38 and 39

(1A and 1B, light micrographs; all others, electron micrographs)

Part 1: The Photosynthetic System

1: Chloroplasts and their Pigments

THE importance of the chloroplasts is indicated by the assertion, based on *in vitro* experiments with isolated chloroplasts, that they are 'autonomous cytoplasmic bodies containing the complete cellular apparatus needed for photosynthesis' [5]. Not only do they contain all the chlorophyll and other pigments which contribute absorbed light energy for the photochemistry but also apparently the whole of the enzyme complement for the Calvin cycle.

A great deal of evidence on the fine structure of chloroplasts is available from electron microscope studies; the structural chemistry of most of the extracted pigments and much of their behaviour *in vitro* in various solvents, both in absorbing light and emitting it again by fluorescence, are now well known. The ability of the pigments to function in photosynthesis, however, apparently depends upon their organization within the chloroplast, together with lipoid substances, proteins and coenzymes, and this organization is still a matter for speculation based on indirect evidence, some of which is mentioned below. The interpretation of *in vivo* absorption and fluorescence spectra is made difficult by the presence of several pigments with overlapping absorption bands, by shifts in the absorption maxima as compared with those of extracted pigments, by selective light scattering and so forth. It can, however, be attempted only on the basis of a knowledge of the structure and behaviour of the individual compounds of the system *in vitro*, both alone and in combination.

A. STRUCTURE OF CHLOROPLASTS

The chloroplasts of higher plants are remarkably uniform in size and shape, being discs or flat elipsoids 3–10 μ (μm in SI symbols) in the longest dimension and about $1\frac{1}{2}$ μ thick. Some early figures of Haberlandt's [133] for number of chloroplasts per unit leaf area are presented on page 199; average numbers per cell in *Ricinus communis* were thirty-six for the palisade and twenty for the spongy mesophyll. More recent figures seem to be available only for the moss *Mnium* [120], in which counting is easier as the 'leaf' is only one cell thick; here the average values were 106 per cell and 9×10^5 per mm^2. The chromoplasts of algae vary greatly in size and shape and generally contain one or more highly refracting pyrenoids: *Chlorella* has a hollow bowl-shaped chloroplast which lines most of the cell wall; *Spyrogyra* has a band-shaped chloroplast arranged in a helix; *Mougeotia* has a huge plate-like chloroplast about 100 μ long; in blue-green algae the chlorophyll is generally stated to be present throughout the cell and apparently occurs in lamellae. In the photosynthetic bacteria the chromatophores are too small to be resolved by the light microscope; they may be as large as 100 nm in diameter (*Rhodospirillum rubrum*) or as small as 30 nm (*Chromatium*).

Towards the end of the nineteenth century, Meyer [225] and Schimper [274] described the structure of chloroplasts as granular and said they saw dark 'grana' embedded in a lighter 'stroma'. This was later attributed to artefacts caused by denaturation of the homogeneous colloidal protoplasm, a view that persisted until the nineteen-thirties when Heitz [154] rediscovered grana; during the next few years they were observed and photographed in chloroplasts (often in living cells) of a great many species of angiosperms and lower plants, though some observers said that they were absent from some species. Heitz [154] stated that the grana were flat discs (Plate 1A) 0·3 to 2 μ in diameter and generally arranged in layers (Plate 1B). Evidence for a laminar structure in chloroplasts was obtained from their birefringence in polarized light by Menke [221, 222] and supported by photographs of chloroplast sections taken with ultra-violet light, which slightly increases the resolving power of the light microscope [222].

In 1940, chloroplasts began to be studied by electron microscopy. Successive improvements in technique have made possible a great elaboration of the above conclusions. Nevertheless, the results obtained with the light microscope, especially of chloroplasts in living cells

(Plates 1A, 1B), provided welcome confirmation that the grana and the laminar structure were not due to the fixation with osmium tetroxide (or other metal-containing compounds) and the complete desiccation used in preparing chloroplasts for electron microscopy. A relatively recent example was the photomicrography of living *Chlorella* cells with light of the 436 nm mercury line [246]; this coincides with the blue absorption peak of chlorophyll *in vivo* and it therefore differentiates the chlorophyll-containing lamellae.

Electron micrographs of thin sections of chloroplasts show that in all species there are more or less parallel lamellae embedded in the stroma. In most higher plants these are packed closely together in places to form the grana which thus appear to be piles of discs, called thylakoids (Plate 2). These can be seen, sometimes spread sideways like a pile

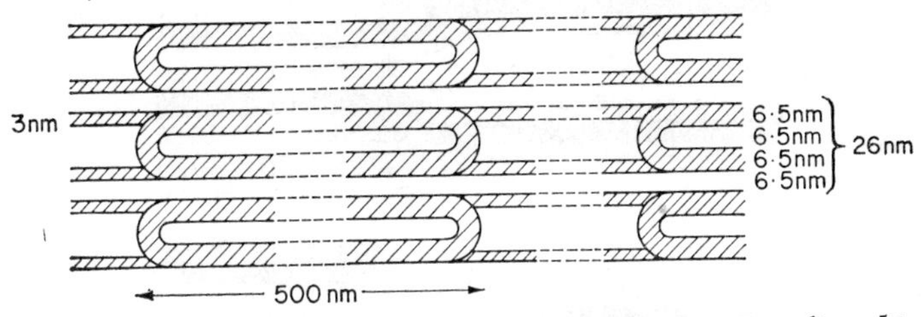

FIG. 1.1 Diagram of grana and lamellae according to Frey-Wyssling. [From 109 after 283.]

of coins that has fallen over, in metal-shadowed preparations of disintegrated chloroplasts (Plate 1C). There appear to be three main interpretations of the structure of grana-containing chloroplasts. According to the first, in the main due to Frey-Wyssling and his school [109], each of the individual discs of the granum consists of a pair of lamellae held together by a U-shaped border to form a flattened sac and each is connected with discs of other grana by two very fine lamellae extending across the stroma. This model, shown diagrammatically in Fig. 1.1, was based on electron micrographs such as Plate 1E; the closed structure of the sacs could be demonstrated by osmotic swelling in hypotonic sucrose solution [108]. All the chlorophyll was believed to be in the grana, a view based on observations that they alone showed fluorescence [109] and reduction of silver nitrate (Molisch reaction) [224, 291].

An alternative interpretation is that of Hodge, McLean and Mercer [170], based on thin sections of maize chloroplasts. These they found to be of two kinds: the chloroplasts of the mesophyll cells had numerous well-developed grana, but in those of the leaf bundle sheaths there were lamellae only (Weier, Stocking and Shumway [315a] find very small and widely scattered grana in the latter—Plates 3A, B). Starch grains formed in these bundle-sheath chloroplasts but, except in old leaves, only occasionally in the mesophyll chloroplasts. The lamellae between the grana of the latter appeared identical in structure with those in the bundle-sheath chloroplasts; both types of chloroplast were green and showed the same fluorescence characteristics. It was concluded that the chlorophyll was probably distributed over the entire lamellar system in each type. In the mesophyll chloro-

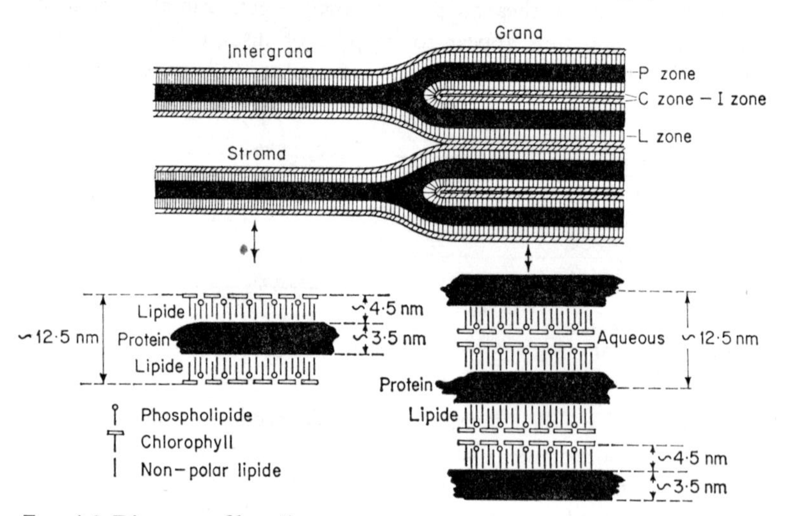

FIG. 1.2 Diagrams of lamellar structure in *Zea mays* chloroplasts according to Hodge. Above: density as seen in osmium-fixed material. Below: interpretation. After Hodge *et al.* [170].

plast sections there appeared to be on average about half the number of lamellae between the grana that there were within the grana (Plate 4). This was attributed to forking of the lamellae, and is not inconsistent with the discs in the grana forming closed sacs which could swell osmotically (Fig. 1.2). This forking of the lamellae could also be seen in the development of grana in an etiolated leaf exposed to light and the fact that lamellae only appear on illumination, concurrently with the formation of chlorophyll, is an indication of the importance of structural organization for photosynthesis.

Thirdly, Heslop-Harrison [156, 157] put forward, first in 1963, an ingenious three-dimensional model of the grana-containing chloroplast of hemp (Plates 5A, B and Fig. 1.3). In this model the lamellae between grana do not form continuous plates around the grana but are fretted and moreover tilted so that they connect the thylakoids of the granum to each other (Plate 6A). Sections of this model in different planes and of different thicknesses could yield some regions with the appearance of forking postulated by Hodge and others resembling the model of Steinmann and Sjöstrand. Obviously much must depend upon the technique of preparation of the sections and the consistency with which the various forms are seen. If this model is correct it has the important consequence that, so far from the discs forming closed sacs,

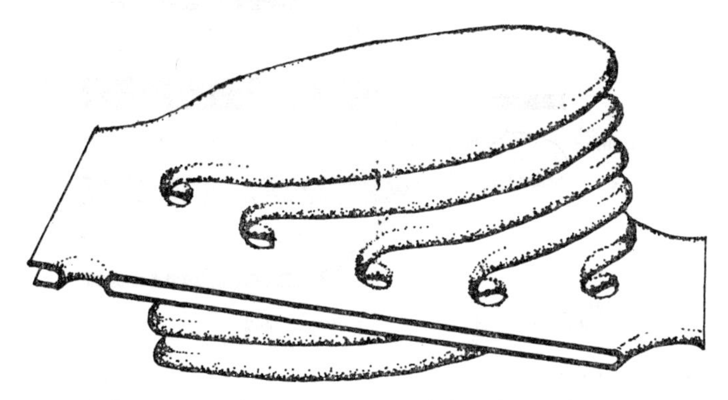

FIG. 1.3 Heslop-Harrison's interpretation of the relation between a stroma lamella and six thylakoids of a granum [157].

the spaces within them are continuous with those of the whole lamellar system of the chloroplast (Plate 4). The swelling of isolated thylakoids in hypotonic solutions must then be attributed to the edges sealing together when fractured. Surface views of grana, from disrupted chloroplasts, seemed to confirm the structure proposed [157] (Plate 7A)

Work on isolated chloroplasts [185, 203] has shown that the chloroplast membrane is easily ruptured and lost; only when this has occurred, as for instance if they have been suspended in distilled water, are the grana easily visible with the light microscope. The proportion of chloroplasts with intact membranes in a suspension can therefore be estimated, and the higher this is the greater the ability of the preparation to fix carbon dioxide. Presumably the loss of the membrane results in the loss of much of the stroma with its enzymes.

In most algae there are apparently no grana but separated lamellae only, though these occur in sets of two to eight with wider spaces between the sets, except in the red algae where they occur singly [117]. The chloroplasts of algae, like those of higher plants (Plate 6B), are bounded by a narrow double membrane. In blue-green algae, however, the lamellae usually appear not to be contained in a chloroplast at all but to spread right around or across the cell [14, 223]; when cell division occurs the lamellae appear to be cut across by the new cell wall [223]. In electron micrography of the blue-green alga *Synechococcus cedorum*, Calvin and Lynch [54] observed round particles about 220 nm

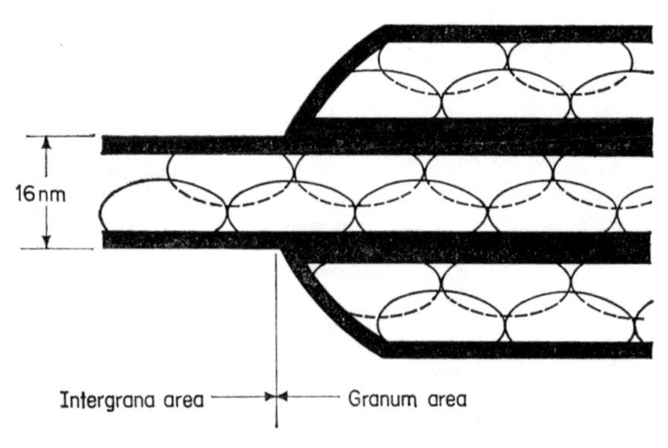

FIG. 1.4 Model for lamellar structure in chloroplast of spinach, after Park and Pon [252].

across and similar sized particles were seen in the solid fraction thrown out by centrifuging at 36,000 g; all the chlorophyll and carotenoids were associated with this fraction and they considered that the particles were analagous to grana.

It was noted as long ago as 1952 [282] that the surfaces of discs from disrupted chloroplasts appeared to be granular and that these granules were too large to be those of the metal used for shadowing (Plate 1D). They were interpreted as macromolecules about 7 nm in diameter making up the lamella on each side of the disc [109]; they did not show in thin sections of lamellae. Park and Pon [252, 253] isolated, from disrupted spinach chloroplasts, portions of lamellae capable of carrying out the Hill reaction more efficiently than whole chloroplasts, perhaps because of the permeability barrier due to the chloroplast membrane surrounding the latter (Plate 6B). Freeze-dried and metal-shadowed preparations of these showed that the lamellar structures totalled 16 nm in thickness but were made up of two layers, each with a maximum thickness of 10 nm. The inner surfaces of the two layers appeared more granular than the outer surfaces and apparently these granules were so packed together as to reduce the total thickness. Park and Pon [252] produced the model shown in Fig. 1.4. The granules were then thought to be oblate spheroids 20 nm in diameter, and later such a granule with its associated membrane was called a quantasome by Calvin on the view that it might be the smallest unit that could carry out the conversion of light energy into chemical energy. Aggregates of seven or eight of these quantasomes were as active as much larger ones in the Hill reaction and, if supplied with the necessary enzymes from the stroma, in fixing carbon dioxide also. Quantasomes can sometimes be found in an extremely regular 'crystalline' array (Plate 7B) and they then measure 16 nm × 13 nm × 10 nm thick; they appear to contain four sub-units [251, 250].

Quantasomes have also been seen by use of the method of freeze-etching [234]. Cells, or portions of tissue about 10^{-1} mm^3, are imbibed with glycerine to prevent formation of ice crystals and frozen very rapidly in liquid propane. The specimen is held at −100°C under high vacuum and a piece chipped off with a microtome knife cooled to about −150°C with liquid nitrogen. The knife is held near to the new surface for about two minutes and some of the ice sublimes onto it, leaving the portions of lower water content standing out in relief; a replica of this 'etched' surface is made by depositing a platinum–carbon film on it from an electric arc and an electron micrograph is prepared. Branton [35] finds that unless membranes are broken exactly at right angles they split in the chipping process so that part of the inner surface of the double layer is exposed. The loss of water content from the watery cytoplasm causes it to shrink below the fractured edge of the outer surface of the membrane, which is thus exposed to view, especially if the etching process is long continued. The method has the advantage that the tissue is not desiccated as in most electron micrography and should give much less shrinkage and distortion. Yeast cells, if thawed quickly at 40°C, can resume active multiplication after going through the glycerine imbibition and freezing process. Quantasomes as seen in freeze-etched preparations measured about 16 nm across and were thought to show sub-divisions into four or five sub-units (Plates 8A, B).

B. THE PIGMENTS OF CHLOROPLASTS IN VITRO

i) *Anatomy of molecules*

Pigments are substances which absorb certain wave lengths of visible light; their colours when viewed in white light are therefore due to the remainder of the visible spectrum, reflected or transmitted as the case may be. The molecules of all organic pigments have a series of regularly alternating double and single bonds, as in the central chain of a carotene molecule (Fig. 1.6), and these are called conjugated double bonds. Although bonds are conventionally shown as lines, each line represents a pair of electrons, so that there are always eight valence electrons around each carbon atom and all these are shared with the neighbouring atoms. Between each pair of carbon atoms is one pair of electrons in a σ bond and these are permanently located. In addition there are pairs of electrons located in π bonds which are mobile over the whole carbon chain; this state of affairs is often called resonance. Extra stability is thereby conferred on the molecule, as is shown for instance by its heat of combustion being lower than would be expected if the double bonds were not conjugated. Such highly mobile paired electrons associated with the conjugated system as a whole rather than with single atoms are easily excited by quanta with energies as low as those in the visible spectrum or near ultra-violet; in this process (called a π–$\pi^\star$ transition) the quantum disappears and the excited electron acquires more energy. Owing to the small energy requirement light is absorbed in the visible part of the spectrum.

(*a*) *Chlorophylls.* In the molecule of chlorophyll *a* (Fig. 1.5) the series of conjugated double bonds forms a closed circuit within the porphyrin nucleus, which consists of a ring of four joined pyrrole rings (I–IV). The π electrons can therefore circulate and have even greater freedom of movement than in carotene, leading to even greater stability.

Each nitrogen atom shares a total of three pairs of electrons with the neighbouring carbon atoms; there is also a pair of electrons in an orbital (page 19) directed towards the magnesium ion. The oxygen atom attached to carbon 9 has four unshared electrons. These non-bonding electrons of the oxygen and those of the nitrogens directed towards the magnesium ion are called *n* electrons; they also are able to absorb energy from and be displaced by quanta of light—a process called an n-$\pi^\star$ transition.

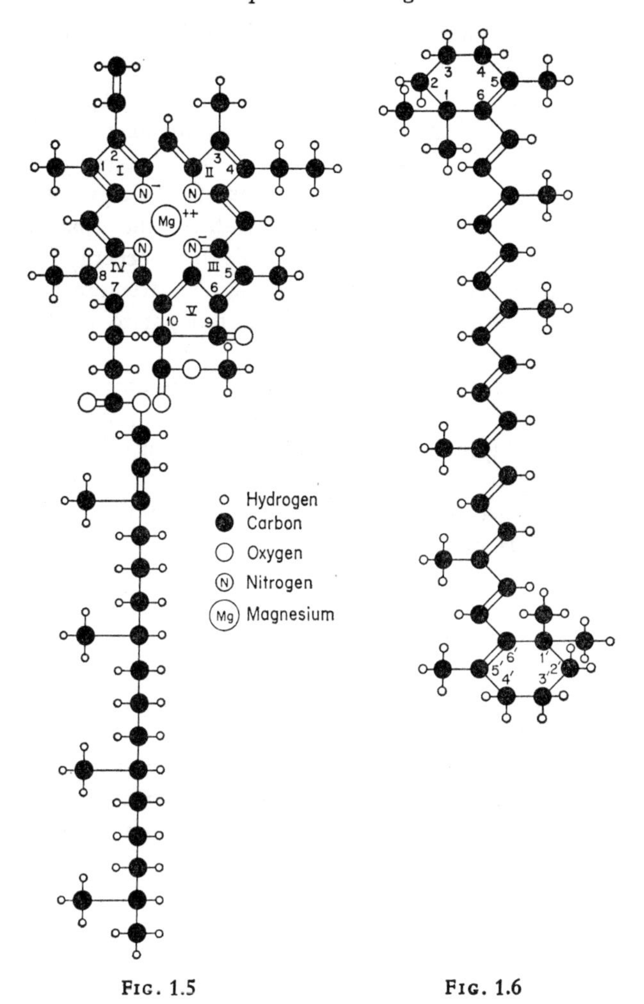

FIG. 1.5 Structural diagram of chlorophyll *a*.

FIG. 1.6 Structural diagram of β-carotene.

Porphyrins are some of the most stable and inert organic molecules known and many examples, believed to be derived from chlorophyll, have been found in crude oil, coal, bituminous rocks and oil shales—some of the latter of Devonian and Cambrian ages and therefore up to 400 million years old [62]. This stability is attributed to the highly condensed ring structure and the resulting resonance, characteristics which also perhaps confer on the chlorophyll molecule its ability to retain the energy absorbed from a quantum of light for unusually long periods and pass it intact to a neighbouring molecule of chlorophyll. In this way energy absorbed from scattered quanta by many chlorophyll molecules can be collected at a reaction centre (Chapter 9) and used for photosynthesis. As chlorophyll *a* is apparently the pigment responsible for the primary process in photosynthesis (the production of reducing power) it is not enough to be able to pass energy from one chlorophyll molecule to another: its molecular structure must also possess a chemically reactive site. This is thought to reside in the 5-membered ring numbered V in Fig. 1.5

The long hydrocarbon chain (phytol) attached to the porphyrin part of the chlorophyll molecule (Fig. 1.5) has lipophilic properties. Its fat solubility is probably important in maintaining the structure and organization of chloroplasts; it accounts for the solubility of chlorophyll in organic solvents such as benzene or acetone and its insolubility in water.

Chlorophyll *a* is present in the chloroplasts or analogous structures of all photosynthetic organisms, except photosynthetic bacteria which have bacterio-chlorophyll *a*. Other pigments are always present also and these differ in different groups. Higher plants and green algae contain chlorophyll *b*, in which there is a $—C\!\left\langle{}^{\displaystyle /\!\!/ O}_{\displaystyle \backslash H}\right.$ group attached to carbon 3 instead of the $—CH_3$ of chlorophyll *a*. Brown algae, diatoms and dinoflagellates contain chlorophyll *c*, but its structure is not fully known. Red algae contain chlorophyll *d* which resembles chlorophyll *a* but has a $—C\!\left\langle{}^{\displaystyle /\!\!/ O}_{\displaystyle \backslash H}\right.$ group attached to carbon 2 instead of $—CH{=}CH_2$. Bacterio-chlorophyll is more reduced than chlorophyll *a*, with two more H atoms at carbons 3 and 4 (which are joined by a single bond) and with $—\underset{\underset{\displaystyle O}{\|}}{C}—CH_3$ attached to carbon 2 instead of $—CH{=}CH_2$.

Green plants and algae also contain small quantities of the more oxidized precursor of chlorophyll *a* known as protochlorophyll *a*, in which carbons 7 and 8 are joined by a double bond and their H atoms are absent. Protochlorophyll is formed in darkness and converted to chlorophyll by exposure to light, which may be exceedingly brief, in the process of greening of etiolated leaves.

(*b*) *Carotenoids.* In addition to the chlorophylls, chloroplasts always contain members of another group of fat-soluble pigments called carotenoids. Practically all these are modifications of the same 40-carbon linear structure, made up of eight isoprene (C_5) units, which may be joined in a ring at one or both ends; the major carotenoid pigment in green plants is β-carotene (Fig. 1.6). α-Carotene has a double bond between carbons 4 and 5, instead of between 5 and 6; carbon 4 has only one hydrogen and carbon 6 has one. The carotenoids are usually classified into (i) carotenes, which are hydrocarbons, and (ii) xanthophylls, which contain oxygen and are more reactive. Five xanthophylls are

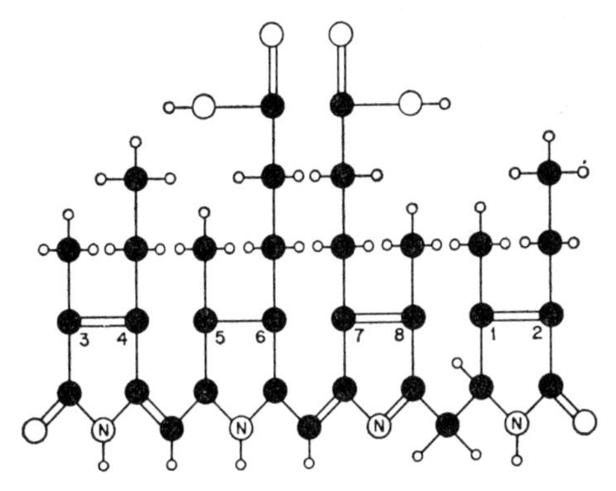

FIG. 1.7 Structural diagram of mesobilivioline (bis-lactam form), the prosthetic group of phycocyanin. After Kamen [186].

common in green leaves; one is lutein which resembles α-carotene but has one hydrogen replaced by —OH both on carbon 3 and 3′. It is thought that only the oxygen-containing xanthophylls pass on the light energy they absorb for use in photosynthesis, that is, act as accessory pigments. The hydrocarbon carotenes may exert a protective function and prevent the destruction of cell contents by photo-oxidation in the presence of light and free oxygen [132, 56]. A carotenoid-free strain of the purple non-sulphur photosynthetic bacterium *Rhodopseudomonas spheroides*, and also another species of the same group (*Rhodospirillum rubrum*) when grown with diphenylamine which suppressed carotenoid formation, grew perfectly well by means of photoreduction, in light under anaerobic conditions. They also grew well heterotrophically in darkness with oxygen present. However, with oxygen and light together, the chlorophyll was destroyed and the cells died.

(*c*) *Phycobilins*. Blue-green and red algae contain different proportions of two types of phycobilin, so called because they resemble bile pigments, namely phycocyanin and phycoerythrin. These are firmly complexed with globulin protein in the living cells and are soluble in dilute salt solutions. The procedures needed to split off the phycobilins

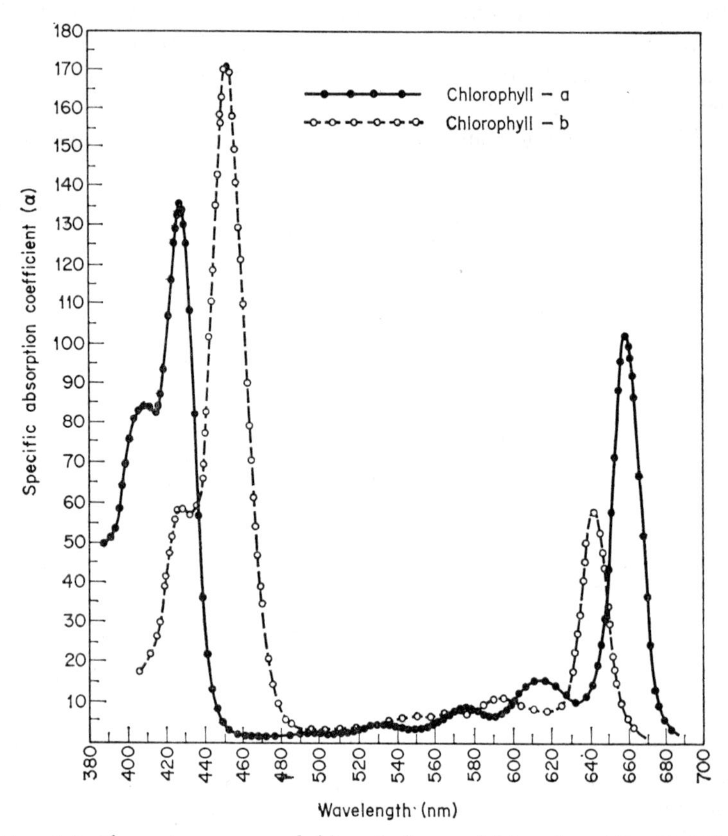

FIG. 1.8 Absorption spectra of chlorophylls *a* and *b* in ether. After Zscheile and Comar [337].

from the protein are drastic and it is possible that their structure is altered in the process. However, they are believed to consist, like porphyrins, of four linked pyrrole rings but arranged in a chain instead of joined in a ring. Fig. 1.7 shows mesobilivioline, which is thought to be the prosthetic (active) group of phycocyanin; that for phycoerythrin is thought to be mesobilirhodin in which there is one less conjugated double bond. No evidence has been found of a metal attached to any of the phycobilins.

ii) *Absorption spectra and fluorescence*

(*a*) *Chlorophylls.* The small difference in structure between chlorophylls *a* and *b* results in considerable differences in their absorption spectra (Fig. 1.8). Chlorophyll *a* in ether has major peaks at 429 and

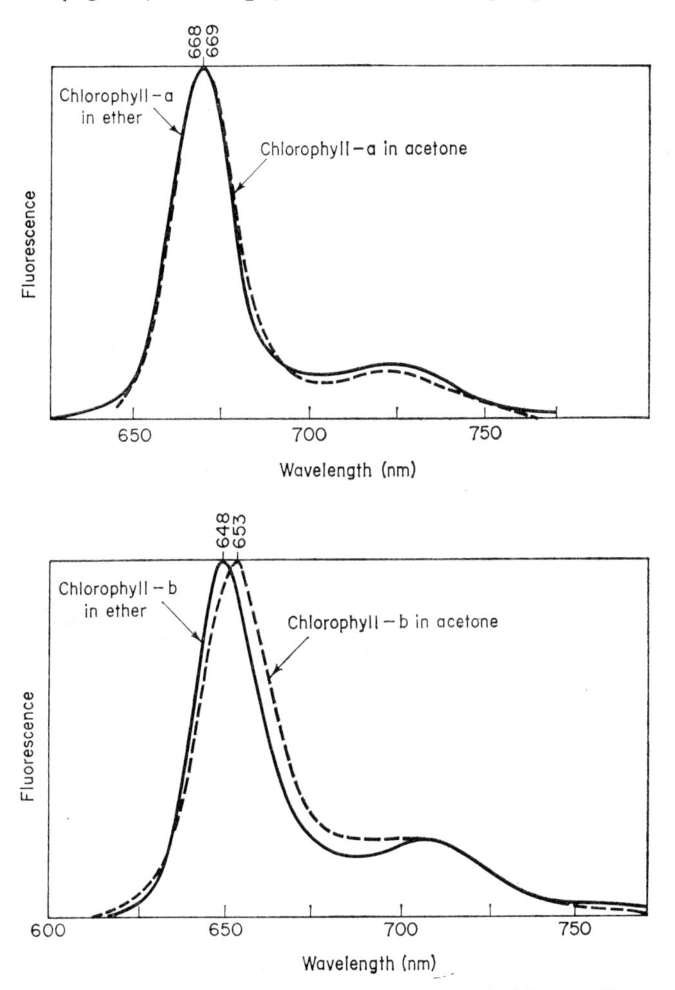

FIG. 1.9 Fluorescence spectra for chlorophyll *a* and chlorophyll *b*. After French [100].

660 nm in the blue and red respectively; for chlorophyll *b* they are at 453 and 643 nm. The former solution appears blue-green to the eye by transmitted light and the latter yellow-green. When viewed at an angle to the light beam, however, they both show a brilliant red fluorescence, somewhat deeper for chlorophyll *a* than for chloro-

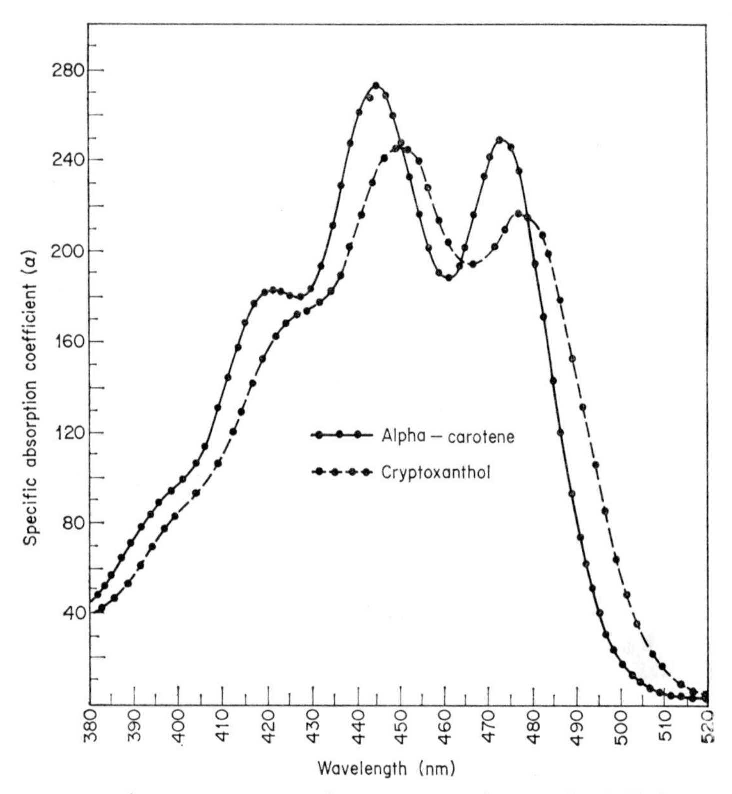

FIG. 1.10 Absorption spectra of α-carotene and a xanthophyll (cryptoxanthol) in hexane solution. After Zscheile *et al.* [338].

phyll *b* and with peaks at slightly longer wave lengths than those for absorption (Fig. 1.9). Although a quantum of blue light absorbed by a chlorophyll molecule contains more energy than one of red light, this extra energy is rapidly lost as heat and the same excited state of the molecule results, with an energy above the ground state of about 167 kJ $mole^{-1}$ (40 kcal $mole^{-1}$) for chlorophyll *a*. This energy may be released in a time of the order of 10^{-8} s and may appear as red fluorescent light; under some conditions up to 33 per cent of the number of incident quanta may be re-radiated from chlorophyll *a* solutions in this way. The fluorescence yield per absorbed quantum is independent of concentration in very dilute solutions, but above 2×10^{-3} M the yield decreases with increasing concentration. This

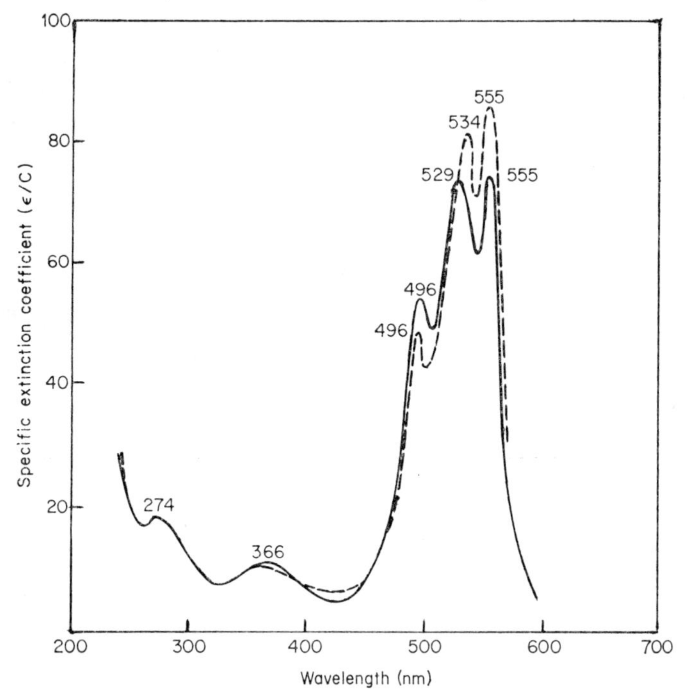

FIG. 1.11 Absorption spectra of phycoerythrin from red algae: solid line, *Ceramium rubrum*; broken line, *Porphyra tenera*. After Svedburg and Katsurai [286].

self-quenching in more concentrated solution is thought to be due to the transfer of energy from one chlorophyll molecule to another.

(*b*) *Carotenoids*. The carotenoids absorb light mainly in the blue-violet, which accounts for their yellow-red colours. As judged from their absorption spectra in organic solvents (Fig. 1.10) there is considerable overlap with the blue absorption bands of the chlorophylls, which confuses the interpretation of photosynthetic efficiencies measured in blue light.

(*c*) *Phycobilins*. The absorption spectra for phycoerythrin from two species of red alga are shown in Fig. 1.11, and for their phycocyanin in

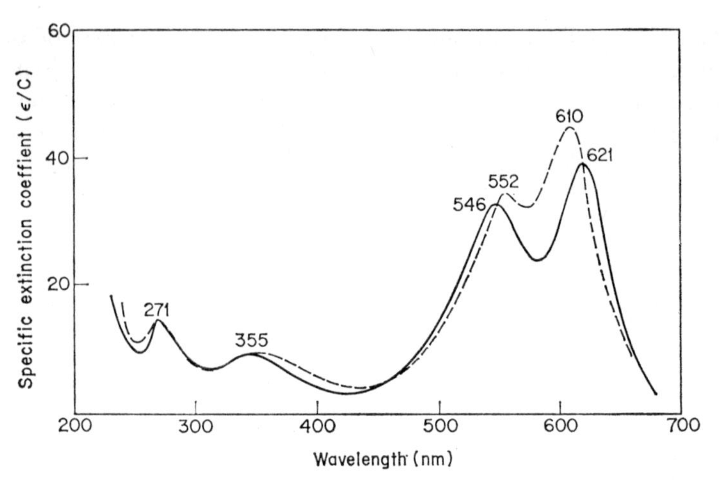

FIG. 1.12 Absorption spectra of phycocyanin from red algae: solid line, *Ceramium rubrum*; broken line, *Porphyra tenera*. After Svedburg and Katsurai [286].

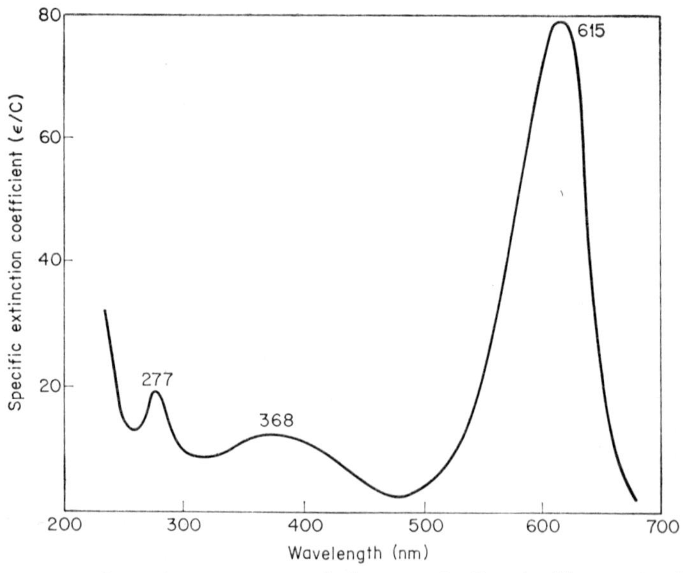

FIG. 1.13 Absorption spectrum of phycocyanin from a blue-green alga, *Aphanizomenon flos-aquae*. After Svedburg and Katsurai [286].

Fig. 1.12. Red algae often grow at considerable depths, where much of the light of long wave length is filtered out by the water, and these pigments do much to fill the gap in the absorption spectrum of chlorophyll *a*. Fig. 1.13 shows the absorption spectrum for phycocyanin from a blue-green alga; this is obviously a somewhat different pigment from that of the red algae.

iii) *Excited states*

The hypotheses for the primary photophysical processes that occur in chloroplasts are nearly all based on results of experimenting with dilute solutions, generally of monomeric solute molecules but sometimes of molecules complexed with the solvent by co-ordinate bonds. These hypotheses are tested mainly by observations of light absorption and fluorescence *in vivo*.

To describe a model system for the various excited states of the chlorophyll molecule which may result from absorbing a quantum of light, electron spin must be mentioned. The regions in space to which electrons are confined are called orbitals and not more than two electrons, which must be of opposite spin (represented as $+\frac{1}{2}$ and $-\frac{1}{2}$), can occupy one orbital. The resultant spin of such a pair of electrons is zero, and if all the electrons in a molecule are paired in this way it has no resultant electronic magnetic moment; it is then said to constitute a singlet system. If, however, the molecule has two unpaired electrons with parallel spin the resultant total spin of unity may be oriented relative to an external magnetic field to give values of $+1$, 0 or -1, giving three different energy levels; this is called a triplet state.

An unexcited molecule in thermal equilibrium with its environment is in its ground state. Its electrons occupy the available orbitals of lowest energy level. A molecule can thus be in a singlet ground state (S_0 in Fig. 1.14), but most organic molecules have even numbers of electrons, all paired, and do not therefore exist in a triplet ground state (T_0). A quantum of red light with an energy content just sufficient to lift a π electron of chlorophyll *a* in ether from one of the highest-energy filled orbitals to one of the lowest-energy vacant orbitals, in a π–$\pi^\star$ transition, would raise the molecule to the first excited singlet state S_1, 180 kJ mole^{-1} (43 kcal mole^{-1}) above the ground state S_0. A quantum of shorter wave length light, of higher energy content, might perhaps raise the molecule to a vibrationally excited level of the second excited singlet state S_2, *a* kcal mole^{-1} above the ground state. This would, however, result first in rapid transfer of vibrational energy (heat) to the solvent, with rate constant k_1, bringing the molecule to the energy level of S_2; secondly, rapid radiationless conversion to the first excited singlet state S_1, with rate constant k_2. In such a radiationless transition the energy is also presumably lost to the solvent as heat. Since this conversion to S_1 usually occurs in less than 10^{-11} s (half-life < 10^{-11} s) the probability of any photo-process such as a chemical

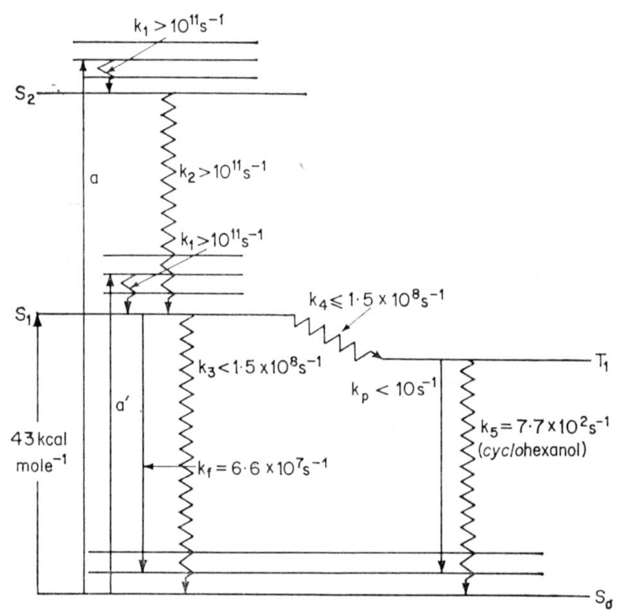

FIG. 1.14 Primary photoprocesses following light absorption by chlorophyll *a* in ether. After Porter [259].

reaction or fluorescence occurring while the molecule is in state S_2 or vibrationally excited is exceedingly remote.

With the thermally equilibrated molecule in state S_1 there are three possibilities. (i) The π electron may drop back to its original orbital (the 'hole' in another system of jargon); the energy may then be emitted as fluorescence with a quantum of red light of rather lower energy content (longer wave length) than that corresponding to S_1; since this always occurs from level S_1 the wave length of the fluorescence is independent of that of the original exciting light. The rate constant for fluorescence is k_f. (ii) There may be a radiationless transition to S_0 with rate constant k_3 but this rarely occurs. (iii) The electron may reverse its direction of spin and the molecule undergo a radiationless transition to the first triplet state T_1, with a rate constant k_4. This state has a rather lower energy level than S_1 and has a much longer lifetime, perhaps because the electron must again reverse its spin if it is to return to its original orbital and the molecule to S_0; its lifetime is of the order of 10^{-3} s and depends almost entirely upon the rate constant k_5 for radiationless transition to S_0. There is also the possibility of radiative transition from T_1 to S_0 (phosphorescence) but this is so slow ($k_p < 10^{-1}{}_{\mathrm{s}}$) that the amount is unobservable except by flash photolysis methods [259]. The long lifetime of the triplet state T_1 allows of chemical reactions; if the singlet state S_1 is also involved in the photochemistry of photosynthesis (as suggested by Franck [97]) the chemical reactant must be complexed with the chlorophyll before excitation occurs, so that it does not depend on a chance collision to capture the excited electron in the 10^{-8} s or so available.

The first excited states S_1 and T_1 have different properties according to whether the electron raised to the lowest-energy vacant orbital was a π electron (π–$\pi^\star$ transition) or an n electron (n–$\pi^\star$ transition). The energy levels of S_1 and T_1 are closer together for the n–$\pi^\star$ state and the probability of radiative transition (fluorescence) is less; the probability of transition from S_1 to T_1 is therefore greater. The π–$\pi^\star$ and n–$\pi^\star$ first singlet states are at very nearly the same energy level in chlorophyll *a*. In a really dry non-polar solvent (benzene) the n–$\pi^\star$ level is slightly the lower and therefore n–$\pi^\star$ transitions predominate; hence the chlorophyll fluoresces very little and more of the molecules cross to the triplet state T_1. In polar solvents and especially those containing water (even as little as 0·01 per cent) the π–$\pi^\star$ energy level is below that for the n–$\pi^\star$ state; π–$\pi^\star$ transitions then predominate and the fluorescence yield is high. The absorption spectrum also differs in the two types of solvent. These changes are attributed to complex formation by the chlorophyll with polar groups of the solvent. Franck [95] suggests that a water molecule becomes attached to two of the nitrogens by hydrogen bonds. This increases the forces holding the n electrons in their ground state and thus raises the n–$\pi^\star$ S_1 level to an energy value above that for the π–$\pi^\star$ transition. The magnesium atom is drawn away from the centre of the ring towards the other two nitrogen atoms, with a consequent change in the charge distribution of the ring system and an alteration of the absorption spectrum.

A donor molecule of chlorophyll in solution, in the S_1 excited state, may transfer its electronic excitation energy to an acceptor molecule in the S_0 state by inductive resonance. The energy gained by the acceptor must equal the energy lost by the donor, and for this to be possible there must be considerable overlap of the fluorescence spectrum of the donor with the absorption spectrum of the acceptor. For this reason it is easier for such transfer to occur from chlorophyll *b* to chlorophyll *a* than from one molecule of chlorophyll *a* to another [314, 65]. In favourable circumstances transfer is possible over distances as great as 10 nm with an efficiency independent of the viscosity of the solvent.

Krasnovsky [198] and other investigators, especially in Russia, have studied extensively the photochemical reactions in which chlorophyll in solution can play a part. These are in the main such as can also be carried out with other fluorescent dye stuffs, whereas for photosynthesis *in vivo* chlorophyll *a* appears to be essential. Perhaps the most relevant is the reduction of oxidizing agents such as quinones by chlorophyll in either the S_1 or T_1 state, in which it is reversibly oxidized and bleached.

It is at least questionable whether the excited states and modes of energy transfer studied in dilute solutions of chlorophyll are relevant to the *in vivo* system; however, problems such as these must be approached from every available angle.

C. *THE PIGMENTS OF CHLOROPLASTS* IN VIVO

i) *Possible organization of pigments*

The gap has narrowed considerably between the size of the chlorophyll molecule and the finest structural detail that has been distinguished in electron micrographs; nevertheless it is still large. The quantasomes (page 9) are about 17 nm in diameter, and if they are divided into four sub-units these are about 9 nm; the porphyrin head of a chlorophyll molecule is 1·5 nm across. However, Moor [234] states that the method of freeze-etching gives a resolution of 2 nm. There appears to be no evidence as yet that the quantasome granules contain chlorophyll: they may simply be macro-molecules of protein, though these are apparently wrapped round with lipid, for extraction with acetone and petroleum ether leaves the quantasomes visible in great relief [250]. Electron micrographs of thin sections of chloroplasts have been of surprisingly limited use in suggesting the organization of the pigment molecules. The apparent dimensions of lamellae differ appreciably with different fixatives, such as osmium tetroxide or potassium permanganate (Plate 5c) [252, 157]; the number and thickness of the electron-dense layers and such features as the continuation of the spaces from the thylakoids into the intergranal lamellae (Plate 4), and whether the latter fork, also vary in preparations from different laboratories. However, such comparisons are usually between different species and may also depend on the plane of the section [156, 157]. Hodge [170] interpreted the lamella (Fig. 1.2) as consisting of a thick central protein layer, which reacted strongly with osmium tetroxide because of its high sulphur content, between two lipid layers thought to contain the pigments, with these layers in turn bordered by extremely thin dense layers (C-zones). Later [169] he stated that the central dense layer (P-zone in Fig. 1.2) was often observed as a doublet made up of two dense zones each about 1·5 nm thick. The diagram was therefore modified, with a space down the centre of the P-zone, to give a pair of structurally asymmetric unit membranes each about 7 nm thick. Park and Pon [252], on the other hand, found thick outer layers to the lamellae which reacted strongly with osmium and no dense inner zone or zones. The interpretation of these and other appearances is discussed in detail by Heslop-Harrison [157].

As long ago as 1936 a scheme for the structure of a chloroplast was put forward on optical and cytochemical evidence [172, 107] which in essentials is still widely accepted. Fig. 1.15 shows a modern development of this scheme (proposed by Calvin [51]) which is discussed further below (page 39); another variant, based on the grana-containing chloroplasts of maize, is that shown in Fig. 1.2. All these schemes have in common the chlorophyll molecules arranged with their phytol tails embedded in a lipid or lipo-protein layer and the porphyrin heads forming a monolayer on a surface between lipid or lipo-protein and protein; in most of them such monolayers coat the outside surface only of the granum discs. In the model proposed by Hodge (Fig. 1.2) the chlorophyll molecules coat these discs on the inside as well as the outside; the structure of the lamellae between grana is the same as within grana except that they are single instead of double and therefore the chlorophylls are on the outside only. It is suggestive that the space within the thylakoids usually appears of rather constant width in electron micrographs, but in chlorophyll-deficient mutants it is much narrower and reduced to the same dimension as the spaces between thylakoids [157, 319].

The area of surface available for the porphyrin heads of the chlorophyll molecules was estimated by Thomas *et al.* [290] for a wide range of organisms. They assumed in their calculations that in grana-containing chloroplasts the chlorophyll was confined to the lamellae within the grana. Although the chlorophyll content of a single plastid might vary from 1 pg in the diatom *Nitzschia*, or 2 pg in spinach and *Elodea densa*, to 340 pg in *Spyrogyra*, the estimated area per

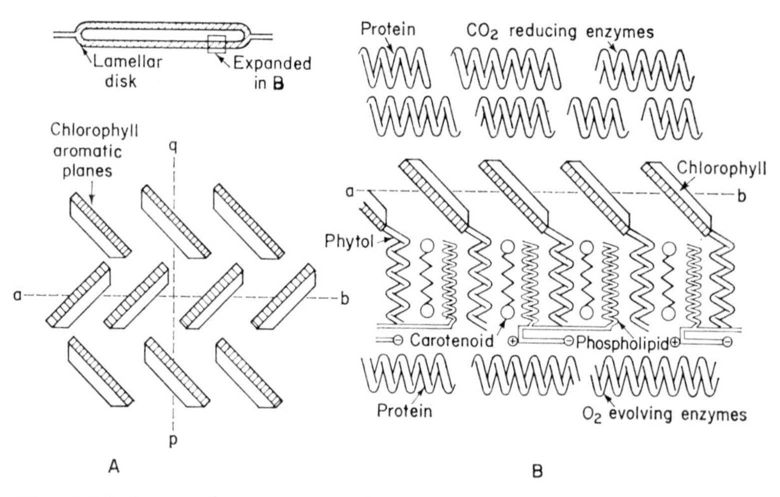

FIG. 1.15 Suggested arrangement of molecules in chloroplast lamellae according to Calvin [51]: *A.* viewed from above the chlorophyll layer; *B.* lamella in vertical section, through ab, viewed from p.

chlorophyll molecule was almost the same (about 2·5 nm²); for ten widely different organisms their estimates ranged from 0·8 to 3·8 with a mean of 2·3. Wolken and Schwertz [334] obtained a similar value of 2·2 nm² for *Euglena*. The area of the porphyrin head of chlorophyll is 2·25 nm². The arrangement they suggested is shown in Fig. 1.16; this would need an average of 2·7 nm² per chlorophyll molecule if these were packed tightly but would allow plenty of room for carotenoids. *In vitro* experiments with chlorophyll monolayers on various liquids have indicated that the porphyrin heads might be tilted at an angle of about 45°, as in Fig. 1.15, and each would then occupy an area of about 1·1 nm². Goedheer [121, 122] considered that the carotenoids probably lay parallel to the lamellae and not at right angles to the porphyrin heads of the chlorophylls as in Fig. 1.16; both he and Frey-Wyssling [109] supposed that these heads formed monolayers on the curved surfaces of macromolecules of lipo-protein; this incomplete orientation would explain the unexpectedly low optical dichroism of chloroplasts. Park and co-workers [252, 253, 251] estimated the molecular make-up

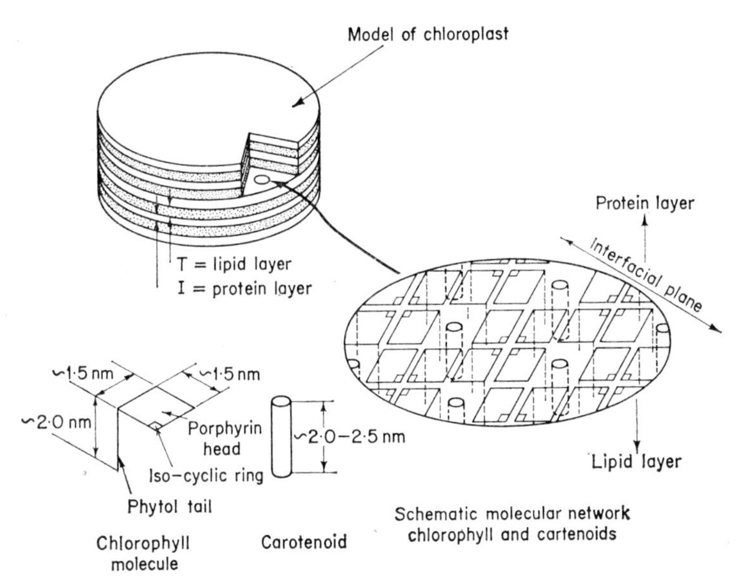

FIG. 1.16 Suggested arrangement of chlorophyll and carotenoid molecules in the chloroplast of *Euglena gracilis* according to Wolken and Schwertz [334].

of quantasomes from studies on disrupted spinach chloroplasts. The ratios of chlorophyll to nitrogen were fairly constant for green lamellar structures from 2,000 nm down to 80 nm in diameter, obtained by centrifuging with increasing multiples of g; the larger structures appeared to have been stripped of grana but the smaller fraction contained grana. This was taken as evidence that chlorophyll was uniformly distributed throughout the lamellar structure of the chloroplast. It was suggested that the usual observation of fluorescence from grana only was due to the greater number of lamellar structures they contained. Small quantities of three transition metals—namely iron, manganese and copper—were found, with the manganese in the lowest concentration. Manganese is needed for oxygen evolution in photosynthesis.

Park and Pon [253] therefore calculated the molecular weight of the smallest unit in the lamella that could possibly carry out photosynthesis, that is, corresponding to one atom of manganese. This molecular weight was estimated as $9{\cdot}6 \times 10^5$. Later the calculations were repeated [251] on the basis of the volumes of the quantasomes as esti-

TABLE 1.1

Composition of quantasomes of Spinacea oleracea. *From Park, R. B. and Biggins, J.* [251]

a. Lipid (in moles per quantasome)		
230 chlorophylls		206,400
160 chlorophyll *a*	143,000	
70 chlorophyll *b*	63,400	
48 carotenoids		27,400
14 β-carotene	7,600	
22 lutein	12,600	
6 violoxanthin	3,600	
6 neoxanthin	3,600	
46 quinone compounds		31,800
16 plastaquinone A	12,000	
8 plastaquinone B	9,000	
4 plastaquinone C	3,000	
8–10 α-tocopherol	3,800	
4 α-tocopherylquinone	2,000	
4 vitamin K	2,000	
116 phospholipids (phosphatidylglycerols)		90,800
144 digalactosyldiglyceride		134,000
346 monogalactosyldiglyceride		268,000
48 sulfolipid		41,000
? sterols		15,000
unidentified lipids		175,600
	Total	990,000
b. Protein		
9,380 nitrogen atoms as protein		928,000
2 manganese atoms		110
12 iron atoms		672
(1 iron atom as cytochrome *b* 1 iron atom as cytochrome *f*)		
6 copper atoms		218
	Total	929,000
Total lipid + protein		1,919,000

mated from measurements on electron micrographs and the effective buoyant density of disrupted lamellar structures in the ultra-centrifuge. The molecular weight for quantasomes was found to be 2×10^6, corresponding to two atoms of manganese. The estimated molecular make-up of a quantasome is shown in Table 1.1. The 10 nm thick membrane is 50 per cent lipid and 50 per cent protein; hence allowing for the difference in density (1·0 : 1·4), about 6·5 nm of the membrane thickness is lipid, which is consistent with a double lipid layer.

ii) *Absorption spectra*

Although chlorophylls *a* and *b* in solution absorb very little light between 500 and 600 nm (Fig. 1.8), leaves absorb something like 70 per cent of the incident light over this range (Fig. 1.17). This is a matter of some importance for the plant, as sunlight provides most energy in the middle part of the visible spectrum with a peak value at about 500 nm. It may be attributed in part to the usually high chlorophyll content of leaves which to some extent compensates for the inefficient absorption per chlorophyll molecule, and in part to internal reflections within the leaf which lengthen the paths for the predominantly green light which passes through a chloroplast without being absorbed the first time. If this were the whole explanation, however, absorption would be expected to be almost complete at the red and blue absorption peaks of chlorophyll. The very great relative increase in absorption in the middle wave lengths may be partly attributed to differential scattering by small particles and by the chloroplast pigments (see below). Another difference from the absorption spectra for the chlorophylls in solution is that the red absorption peak is at a longer wave length—about 680 nm. This can be attributed to a different state of the chlorophyll and probably to its association with protein.

Fig. 1.18 shows the absorption spectrum for a spinach leaf (as in Fig. 1.17) together with those for the isolated chloroplasts, disintegrated chloroplasts and a crude methanol extract, all containing the same quantity of pigment per unit cross-sectional area of light beam. As the system was simplified there was a progressive reduction of absorption in the central region of the spectrum and a sharpening of the red absorption peak. The last was at the same wave length (680 nm) except for the leaf extract, where it showed a shift to about 665 nm.

Methods of estimating light absorption, by measuring the difference between incident intensity and the amounts transmitted, reflected and scattered, are discussed in Chapter 3. When such methods are used with narrow wave bands of incident light for determining absorption spectra, the results obtained represent the actual absorption by the

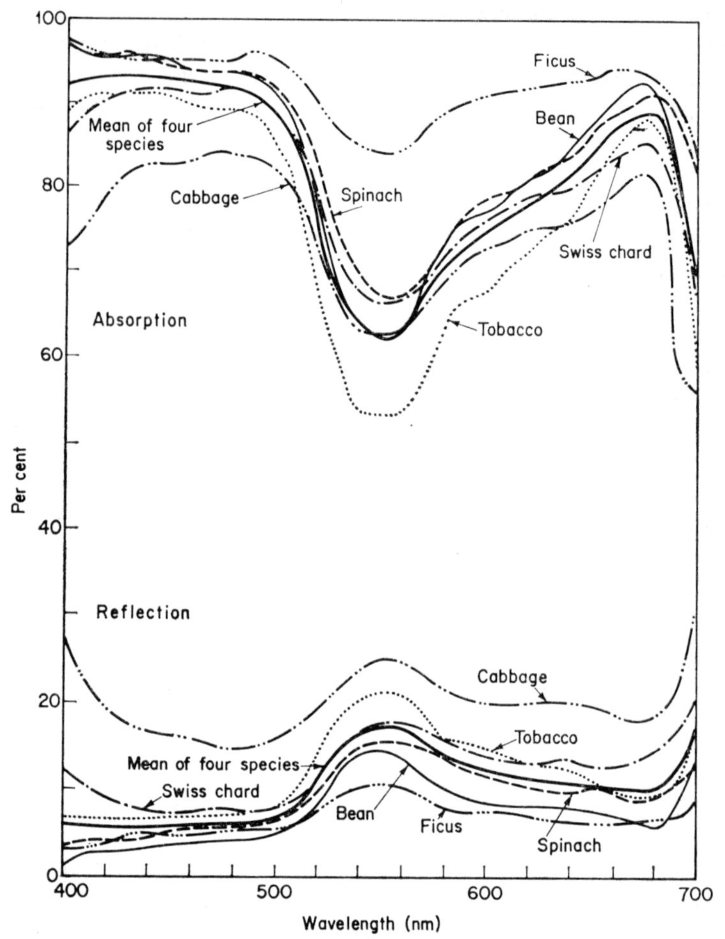

FIG. 1.17 Absorption and reflection spectra for leaves of six species and mean curves for four of them (bean, spinach, Swiss chard and tobacco). For these four species average absorptions were: 400–500 nm, 92 per cent; 500–600 nm, 71%; 600–700 nm, 84%. After Moss and Loomis [237].

object used—leaf, cell suspension or suspension of isolated chloroplasts; but they are exceedingly difficult to interpret in terms of the individual pigments. The difficulties are greatest for leaves. Light that has entered the leaf encounters heterogeneities of many kinds. First, it is reflected and refracted by cell walls, especially in a leaf of a land plant with air-filled intercellular spaces; secondly, it is scattered by the many particles of different sizes and refractive indices within the cells. Its

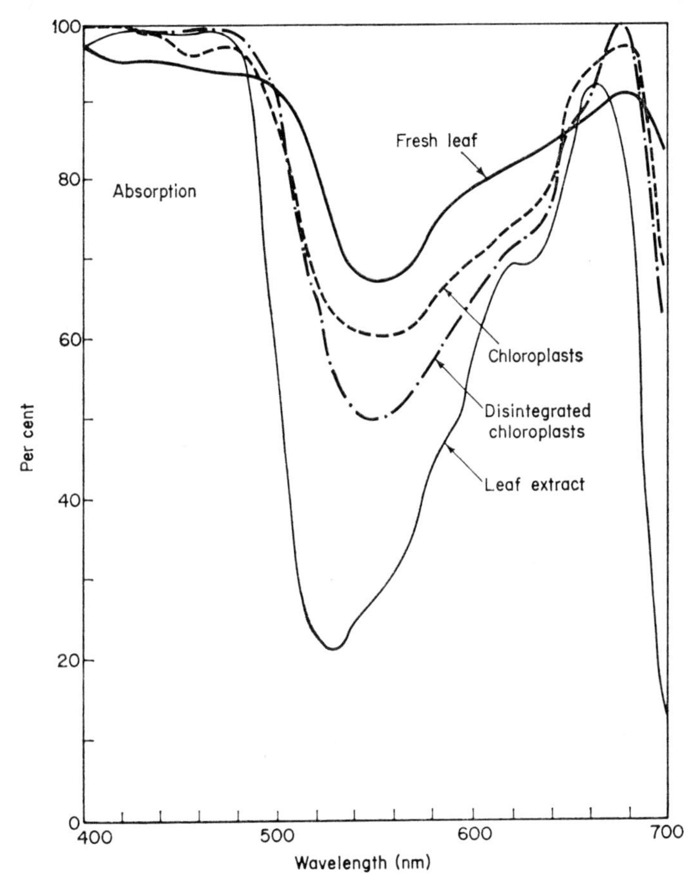

FIG. 1.18 Absorption spectra for spinach leaf, isolated chloroplasts, disintegrated chloroplasts and crude methanol extract—all with the equivalent quantities of pigments. After Moss and Loomis [237].

paths within the leaf are therefore of various and unknown lengths; some light may pass entirely between chloroplasts, while some may pass through several chloroplasts and even through the same ones more than once. The uncertainties are fewer for suspensions of unicellular algae or chloroplasts but even so they are considerable. It is a property of light that when it meets any discontinuity of refractive index, part of it is scattered. The scattering at the surface of algal cells, due to the difference in refractive index between their walls and the water, can be almost eliminated by suspending them in concentrated protein solution of refractive index similar to that of the walls [10]. Scattering within the cells may be more important because the particles involved are smaller and because of the presence of pigments. With very small particles, of diameter less than the wave length of light, the amount of scattering is inversely proportional to the fourth power of the wave length (Rayleigh scattering); this highly selective scattering especially increases the mean path lengths for the shorter wave length light. For larger colourless particles there is less variation with wave length in amount of scattering, but the refractive indices of pigments change rapidly in the neighbourhood of their absorption bands (anomalous dispersion) and this results in marked changes in the amount of scattering in those regions of the spectrum. This selective scattering may be very large, owing to the high concentrations of pigments, and may then be expected to modify considerably the absorption spectra obtained: even if the scattered light emerging from the cells were measured and allowed for, errors of unknown magnitude would be introduced by the different path lengths within the cells for light of different wave lengths.

Absorption spectra for suspensions of *Chlorella* and of spinach chloroplasts are shown in Fig. 1.19 together with the proportion of the incident light scattered. These curves are remarkably similar in spite of the differences in plant material and in the angles at which the scattered light was measured. Maximal scattering occurred on the long wave length sides and minimal scattering on the short wave length sides of the absorption peaks. In this the scattering curves resemble a curve for refractive index plotted as a function of wave length within an absorption band of a pigment (anomalous dispersion curve) [201]. Not all cell suspensions, even of *Chlorella*, gave this type of relation between absorption peaks and scattering maxima and minima [202]: for some suspensions the scattering minima coincided with the absorption maxima; hence the predominant effect on amount of scattered light was the attenuation of the incident light by absorption and apart from this there was little effect of wave length. The two contrasting types of relation between scattering and absorption are attributed partly to differences in cell size as related to the phase shift in light passing through the cell relative to that passing round it [202].

Butler [50a] states that selective scattering does not introduce important errors in absorption spectra unless they are measured with a spectrophotometer admitting a very narrow angle of light from the cell suspension. Multiple non-selective scattering in dense suspensions results in longer path lengths for wave lengths of lower absorbance, as suggested above for leaves, and he has attempted to correct for these differences in effective path length. For dilute suspensions, light passing between the cells raises the measured value of transmitted light of all

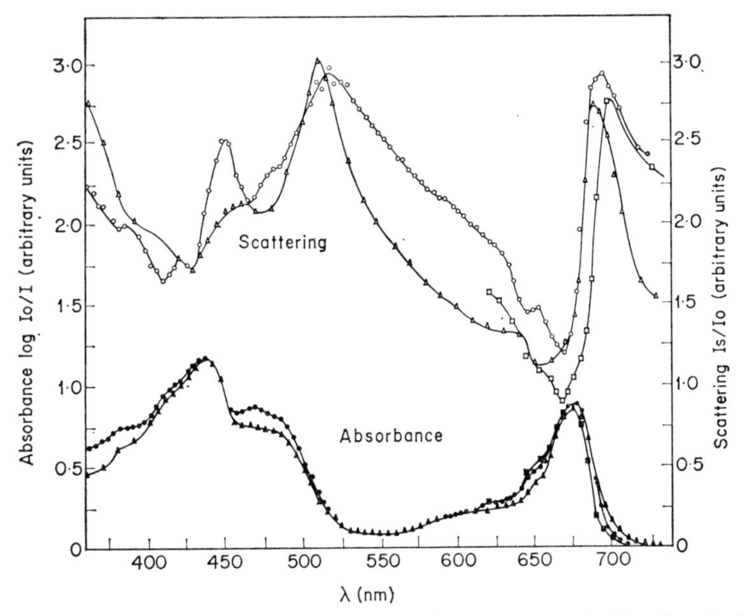

FIG. 1.19 Scattering and absorption spectra for suspensions of *Chlorella* and of spinach chloroplasts. Triangles—*Chlorella* cells, scattered light measured at 90° ± 15°; squares—*Chlorella* cells (different investigators), scattered light measured at 22–85°; circles—spinach chloroplasts, scattered light measured at −5–49°. After Latimer [202].

wave lengths and so effectively reduces the absorption peaks ('sieve effect' [257*a*]). This error is greatest for widely scattered particles of high light absorption, such as chloroplasts. Single *Euglena* chloroplasts may absorb up to 80 per cent of light of 675 nm wave length [334*a*], as estimated with a microspectrophotometer from the transmission of a beam of light much narrower than the chloroplast diameter—perhaps the best way of obtaining absorption spectra.

In the absorption spectra of leaves, algal cells or chloroplasts it is difficult to detect the contributions of the various pigments (Figs. 1.17 to 1.19) and, although the red absorption peak for chlorophyll *a* usually shows clearly, the effects of the other pigments are seen only as shoulders or changes of slope on the sides of the main peak. Giese and French [98] designed a derivative spectrophotometer which automatically plotted against wave length the first derivative or slope of

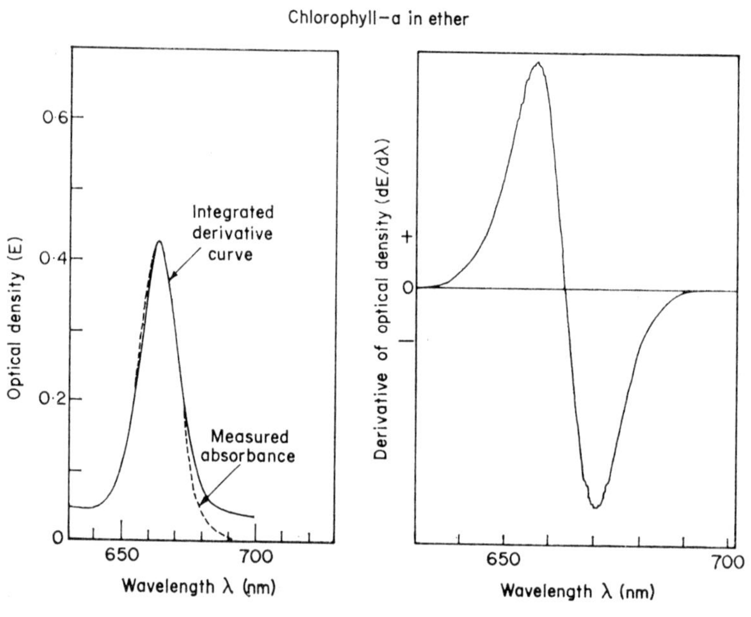

FIG. 1.20 Broken line, left-hand graph: absorption spectrum for chlorophyll *a* in ether, obtained with a conventional spectrophotometer; solid line, right-hand graph: first derivative, obtained with derivative spectrophotometer; solid line, left-hand graph: integrated derivative curve. After French [98].

the absorption curve. This derivative curve could, if desired, be integrated to give a curve equivalent to the absorption spectrum; where the absorption peak gave a normal (Gaussian) curve the derivative was also symmetrical, as for the ether solution of chlorophyll *a* shown in Fig. 1.20. However, with an asymmetrical absorption peak, or one with shoulders as in that for *Chlorella* (Fig. 1.19), the derivative curve showed up the irregularities with much greater sensitivity (Fig. 1.21). A graphical computor made it possible to fit first derivatives of up to five hypothetical normal probability curves which could be modified by trial and error so that when summed they gave a good approximation to the measured curve (Fig. 1.21). In this way absorption peaks were estimated as occurring at 672, 683 and 694 nm for chlorophyll *a* with one at 653 attributed to chlorophyll *b*. The different absorption

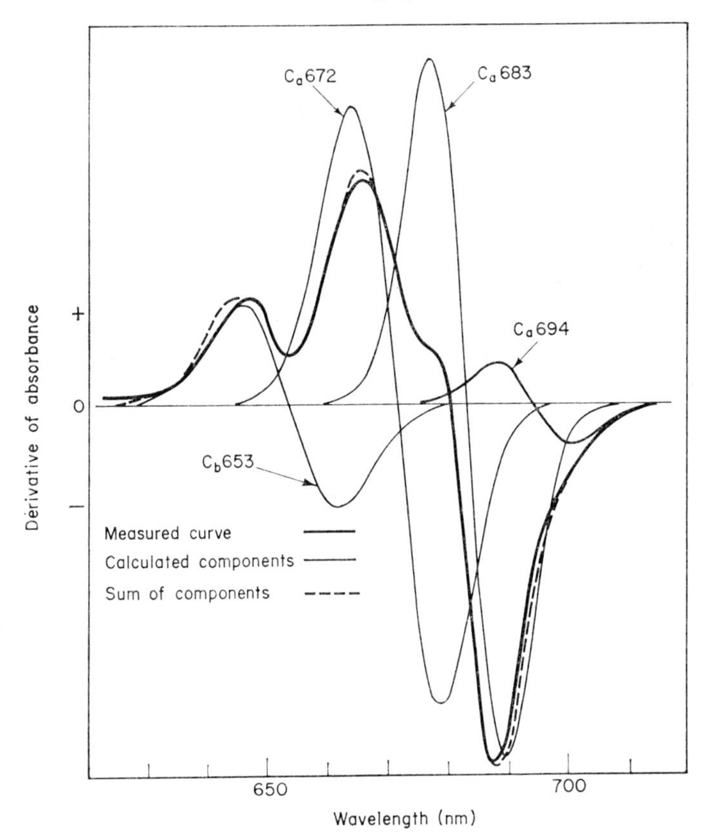

FIG. 1.21 Measured derivative absorption spectrum for suspension of *Chlorella*, calculated derivative curves for four hypothetical components and the sum of these four for comparison with the measured curve. After Brown, J. S. [48].

maxima for chlorophyll *a in vivo* may represent the pigment in combination with different proteins. Two of these forms of chlorophyll *a* are apparently formed in succession early during the process of greening of etiolated leaves (Fig. 1.22). In dark-grown leaves the protochlorophyll has an absorption maximum at 648 nm. The chlorophyll

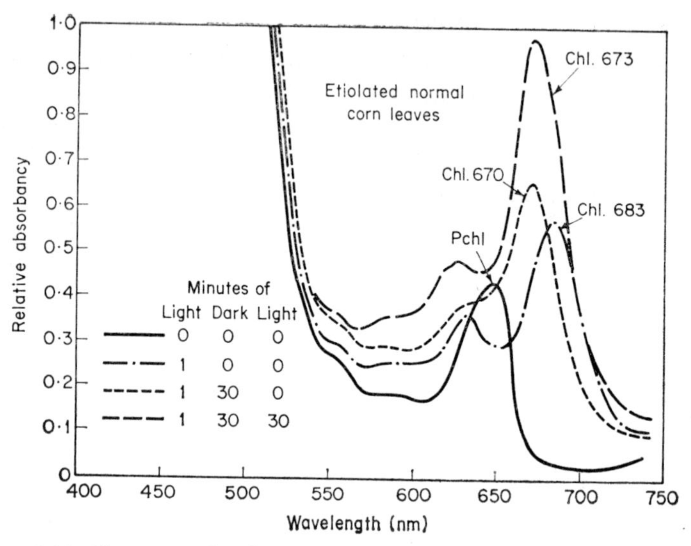

FIG. 1.22 Changes in the absorption maximum of chlorophyll in greening maize leaves. After Smith *et al.* [278].

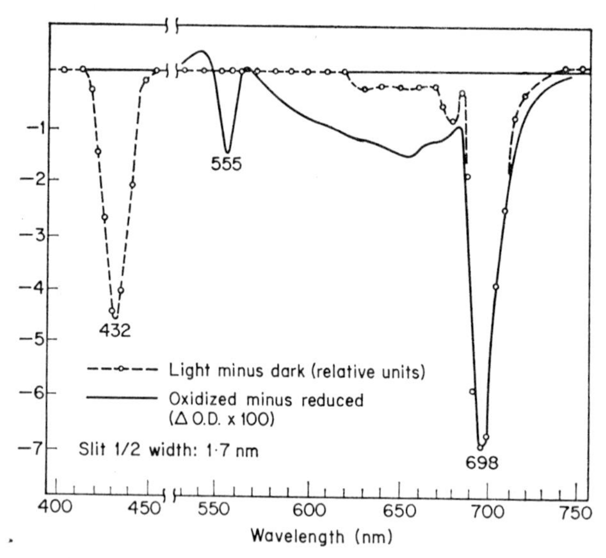

FIG. 1.23 Effects of actinic light or darkness and of oxidation or reduction on absorption by the pigment P700, in a suspension of chloroplast fragments extracted with acetone. Solid line, oxidized minus reduced difference spectrum; broken line, light minus dark difference spectrum. After Kok [195].

formed on first exposure to light has a peak at about 683 nm; in darkness this changes to 670 nm with no increase in the amount of chlorophyll and on further illumination the peak again moves towards longer wave lengths with the increased formation of chlorophyll [278] ultimately reaching a value approaching 680 nm.

Absorption changes caused by light are not confined to greening etiolated leaves. Kok [193, 194, 195] obtained evidence from difference spectra for the presence in algae and chloroplasts of a pigment (P700), probably a form of chlorophyll *a*, with an absorption band at about 700 nm; this was apparently rapidly bleached or broken down by such far-red light, or by flashes of intense white light from a mercury arc, for its absorption fell rapidly. The pigment, as judged by the absorption changes at 700 nm, could be regenerated by white or red light, or in darkness (Fig. 1.23). Such changes could also be produced in chloroplast particles by oxidation and reduction (Fig. 1.23); apparently, therefore, the effect of far-red or intense white light was to oxidize the pigment and in darkness or red light it was reduced once more. Chlorophyll *a* was about 300 to 500 times as abundant as P700 and the bleaching of the latter caused by even weak far-red illumination suggested that energy transfer from the bulk chlorophyll *a* must be responsible.

iii) *Fluorescence and energy transfer*

The inaccuracies in measuring fluorescence are probably even greater than those for absorption. Not only must an attempt be made to separate, by means of filters, the fluorescent light to be measured from the scattered light emerging from the cell (some of which will be at the same or almost the same wave lengths—Figs. 1.19 and 1.24) but there will be additional errors due to reabsorption of the emitted light within the cell. These last can, however, be minimized by using tissues of very low chlorophyll content. Fig. 1.24 shows that a so-called chlorophyll-free leaf gives a fluorescence spectrum much resembling that for chlorophyll *a* in ether (Fig. 1.9) though with a slight shift towards longer wave length such as occurs for the *in vivo* absorption spectrum. In the normal leaf, reabsorption near the absorption maximum for chlorophyll *a* has greatly reduced the fluorescence peak and shifted it to an even longer wave length; at the same time the far-red fluorescence peak at 740 nm, where chlorophyll scarcely absorbs at all, is much higher because of the greater chlorophyll content. Scattering and internal reflection within a leaf, which increase reabsorption by lengthening the light paths, can be reduced by infiltration of the intercellular air spaces with water [304]: the same method can be used to improve absorption spectra.

Duysens [64, 65, 66] showed in 1951 that whether *Chlorella* was illuminated with light of 420 nm wave length, absorbed mainly by chlorophyll *a*, or with light of 480 nm absorbed mainly by chlorophyll

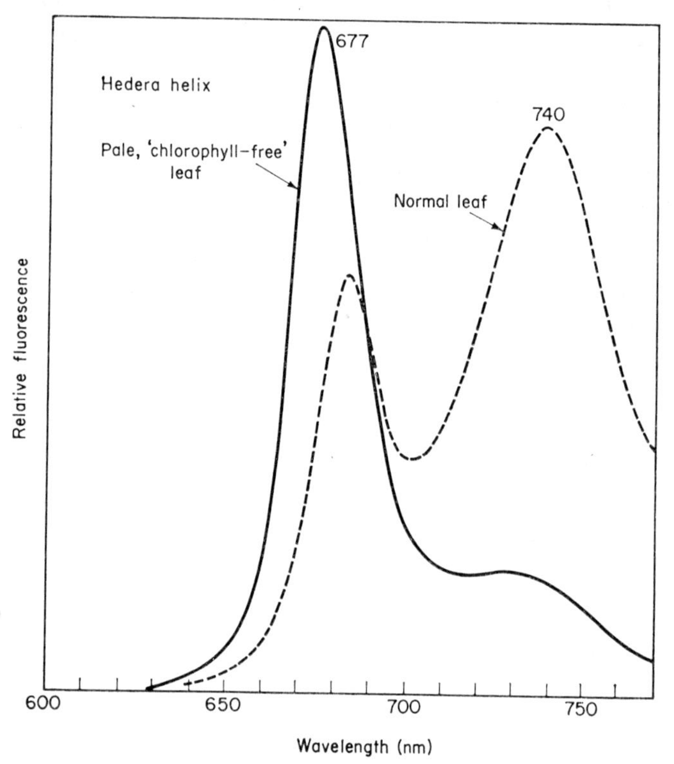

FIG. 1.24 Fluorescence spectra of normal and 'chlorophyll-free' parts of a variegated ivy leaf (*Hedera helix*), with the same incident light (436 nm line from a mercury arc) and sensitivity of the recording instrument. After Virgin [305].

b, the fluorescence spectrum obtained was that of chlorophyll *a* and chlorophyll *b* did not fluoresce. The action spectrum (see page 221) for production of fluorescence of chlorophyll *a* showed that light absorbed by chlorophyll *b* was as efficient as that absorbed by chlorophyll *a* itself, but that absorbed by carotenoids was only 40–50 per cent efficient. However, Dutton *et al.* [63] had found that the diatom *Nitzschia closterium* gave the same yield of chlorophyll *a* fluorescence whether illuminated with blue-green light predominently absorbed by the carotenoid, fucoxanthin, or with red light absorbed by chlorophyll *a* only. The phycobilin pigments of blue-green and red algae are strongly fluorescent and light absorption by these does cause them to fluoresce but it also causes the fluorescence of chlorophyll *a*. Duysens [64, 65] and

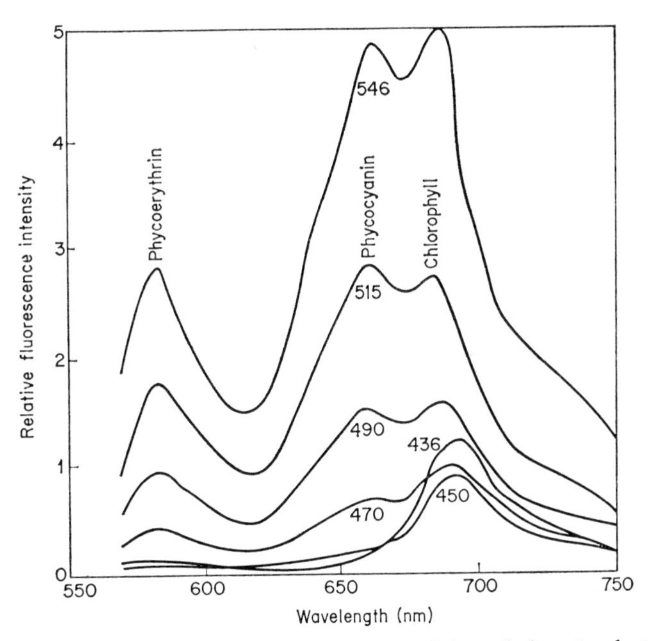

FIG. 1.25 Fluorescence spectra of a suspension of the red alga, *Porphyridium*, obtained with equal numbers of incident quanta of the wave lengths shown by each curve. After French and Young [105].

French and Young [105] found that if the phycoerythrin was excited, with light of 546 nm wave length, the fluorescence spectrum showed peaks characteristic of phycoerythrin, phycocyanin, and chlorophyll *a*; however, light of 436 nm absorbed almost entirely by chlorophyll gave chlorophyll *a* fluorescence only (Fig. 1.25). Apparently there was a one-way transfer of energy, in the direction of longer wave length fluorescence, from phycoerythrin to phycocyanin and thence to chlorophyll *a*. The remarkable result was obtained that fluorescence of the chlorophyll *a* was produced with much greater efficiency by light absorbed by the phycobilins than if it was directly absorbed by the chlorophyll. This resembled a finding of Haxo and Blinks [137] that in the red algae the phycobilins were more efficient than chlorophyll *a* in the use of light for photosynthesis.

On the basis of the similarity between the action spectra for photosynthesis and for fluorescence of chlorophyll *a* in a wide range of species, Duysens [64, 65, 66] proposed that only chlorophyll *a* was active in photosynthesis and that the other pigments only assisted in so far as they transferred their absorbed energy to it. The curious results

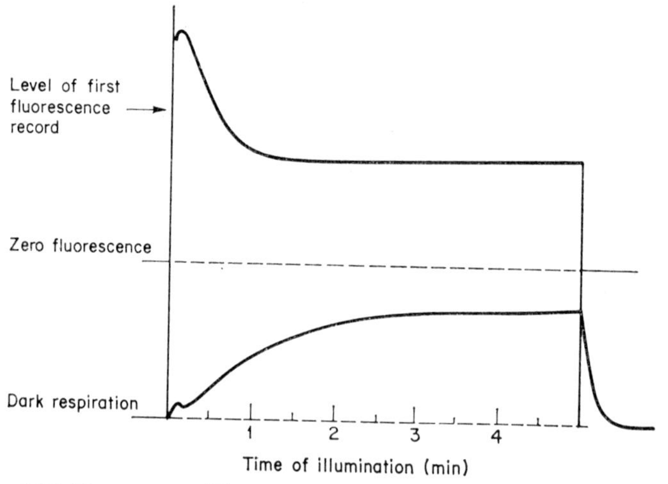

FIG. 1.26 Time curves of fluorescence yield (upper curve) and carbon dioxide uptake (lower curve) following bright illumination of wheat seedlings in normal air, after a 30 minute dark period. After McAlister and Myers [206].

for red algae were explained by supposing that a large part of their chlorophyll *a* was in a form inactive in photosynthesis, non-fluorescent and not receiving energy from phycobilins.

It should be noted that fluorescence yields for chlorophyll *in vivo* are very small compared with the yield of up to 30 per cent that can be obtained in solution (page 17) and have not been found to exceed 2·5 per cent; this would be consistent with the conversion of a large part of the excitation to the triplet state. In solution, chlorophyll gives a fluorescence intensity proportional to incident light but *in vivo* the relation is affected by the rate of photosynthesis. In general, fluorescence decreases as photosynthesis increases and *vice versa* but the relations are not simple. An example is shown in Fig. 1.26. This serves both to

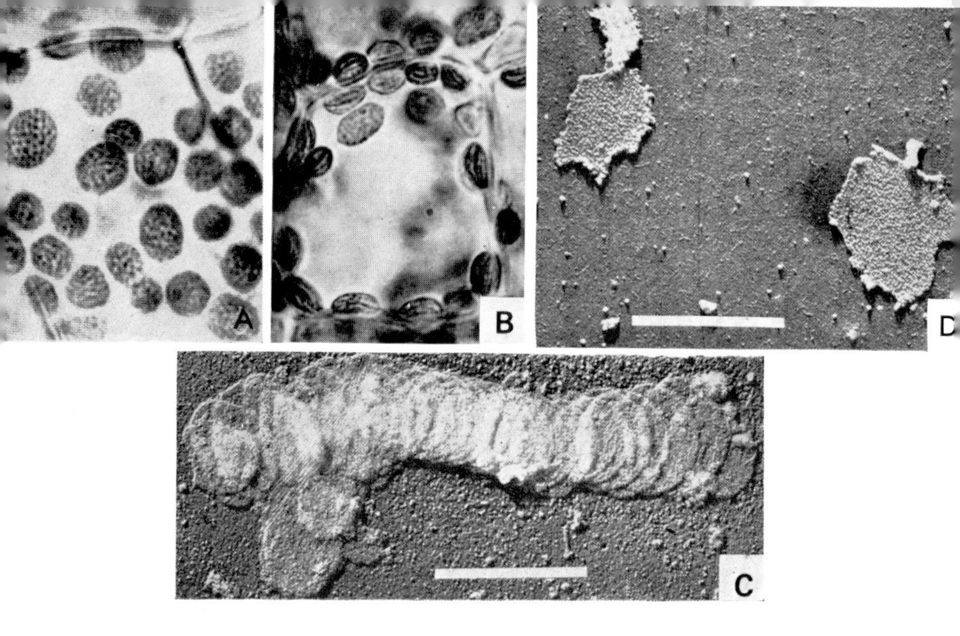

PLATE 1A. Chloroplasts with grana in cells of intact living leaf of *Aponogeton ulvaceum* × *fenestrum*. (Reproduction from half-tone copy with inevitable loss of detail.) From Heitz [154].

PLATE 1B. Chloroplasts with grana in intact cells of leaf of *Todea superba*, in 4 per cent sucrose solution. (Reproduction from half-tone copy with inevitable loss of detail.) From Heitz [154].

PLATE 1C. *Aspidistra*. Granum isolated in distilled water and shadowed. (Reproduction from half-tone copy with inevitable loss of detail.) × 21,000. From Steinmann [282].

PLATE 1D. *Spyrogyra*. Chloroplast fragments obtained by sonic disintegration. Fixed osmium tetroxide and shadowed. (Reproduction from half-tone copy with inevitable loss of detail.) × 21,000. From Steinmann [282].

PLATE 1E. *Aspidistra elatior*. Electron micrograph of a thin section of a chloroplast. Osmium tetroxide fixation. × 62,000. From Steinmann and Sjöstrand [283].

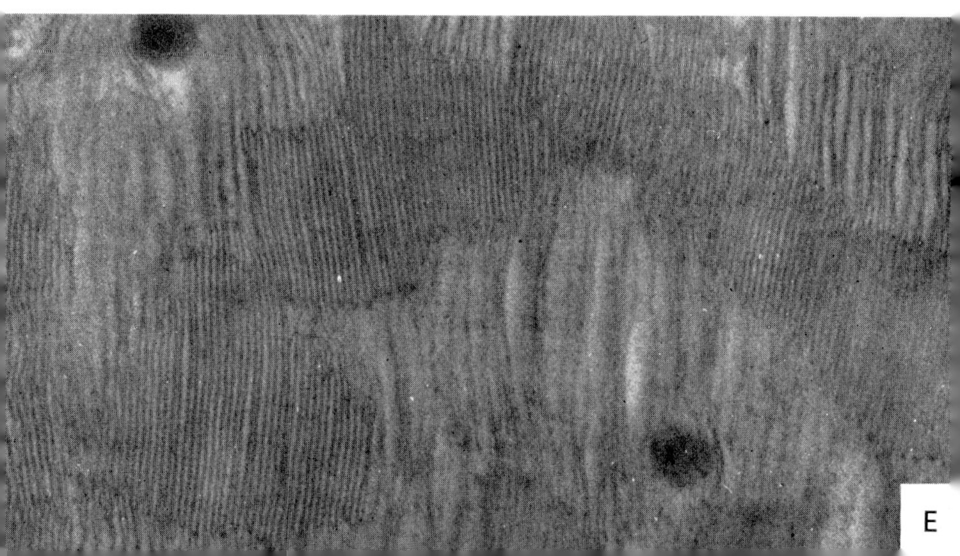

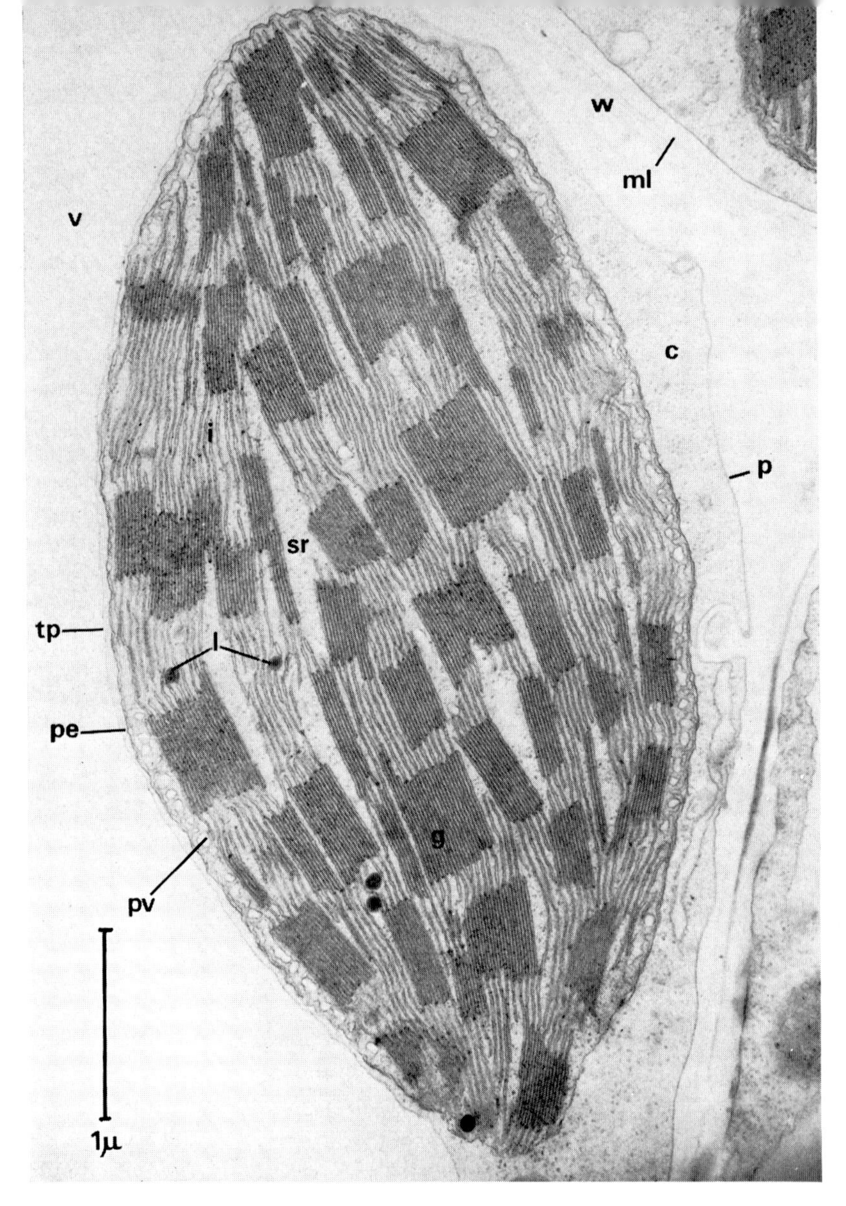

PLATE 2. *Zea mays.* Electron micrograph of a thin section with a chloroplast in the corner of a mesophyll cell. Leaf fixed in gluteraldehyde and post-fixed in osmium tetroxide; section stained with lead citrate. The cytoplasm has retracted from the cell walls during preparation. g, granum; i, intergranal lamellae; sr, stroma; l, lipid droplets; pv, peripheral vesicles in stroma; pe, plastid envelope (2 membranes); tp, tonoplast; p, plasmalemma; c, cytoplasm; v, vacuole; w, cell wall; ml, middle lamella. × 26,000. (*Courtesy of A. D. Greenwood and K. Skinner.*)

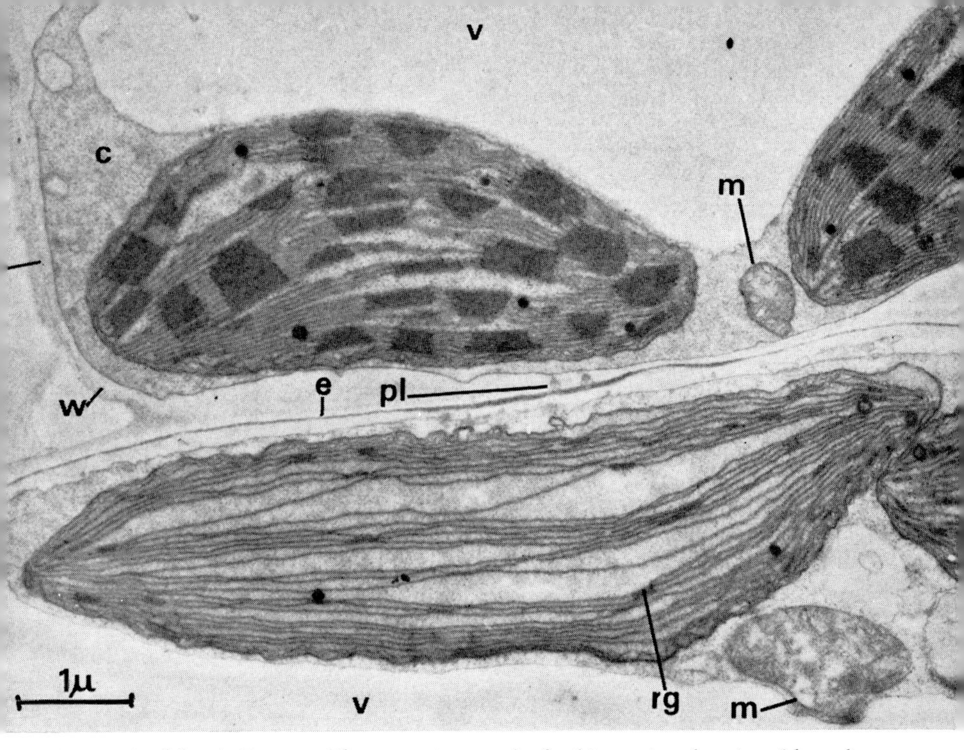

PLATE 3A (*above*). *Zea mays.* Electron micrograph of a thin section showing chloroplasts in contiguous cells of mesophyll (above) and bundle-sheath (below). Fixation and symbols as for Plate 2; also: m, mitochondrion; pl, plasmodesmata; e, electron-dense middle lamella between mesophyll and bundle-sheath; rg, rudimentary granum. × 16,000. (*Courtesy of A. D. Greenwood and K. Skinner.*)

PLATE 3B (*below*). *Zea mays.* Electron micrograph of part of a bundle-sheath chloroplast. Fixation and symbols as for Plate 3A; also: s, starch grain (lies in stroma and has no surrounding membrane). × 68,000. (*Courtesy of A. D. Greenwood and K. Skinner.*)

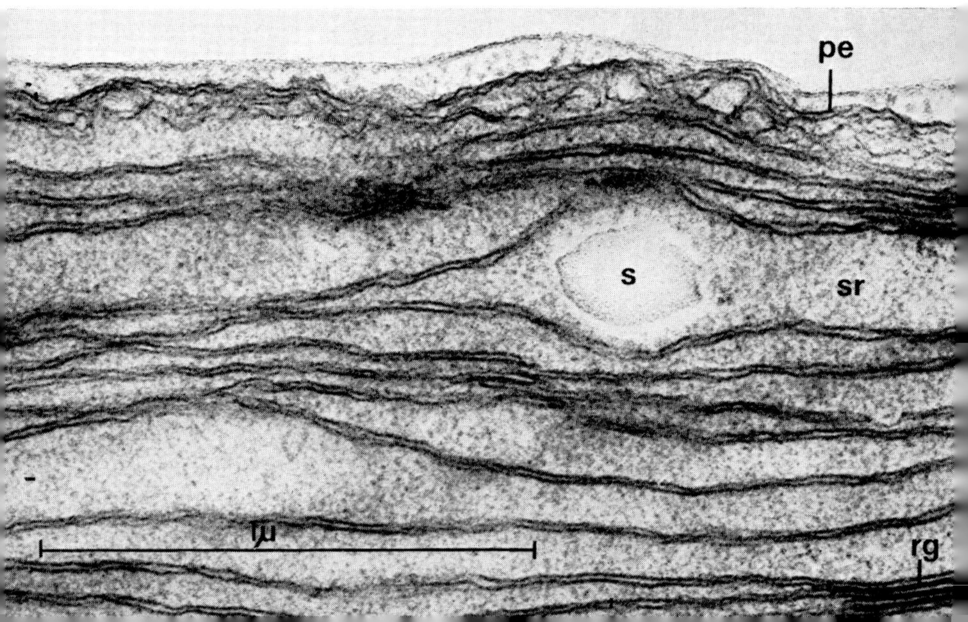

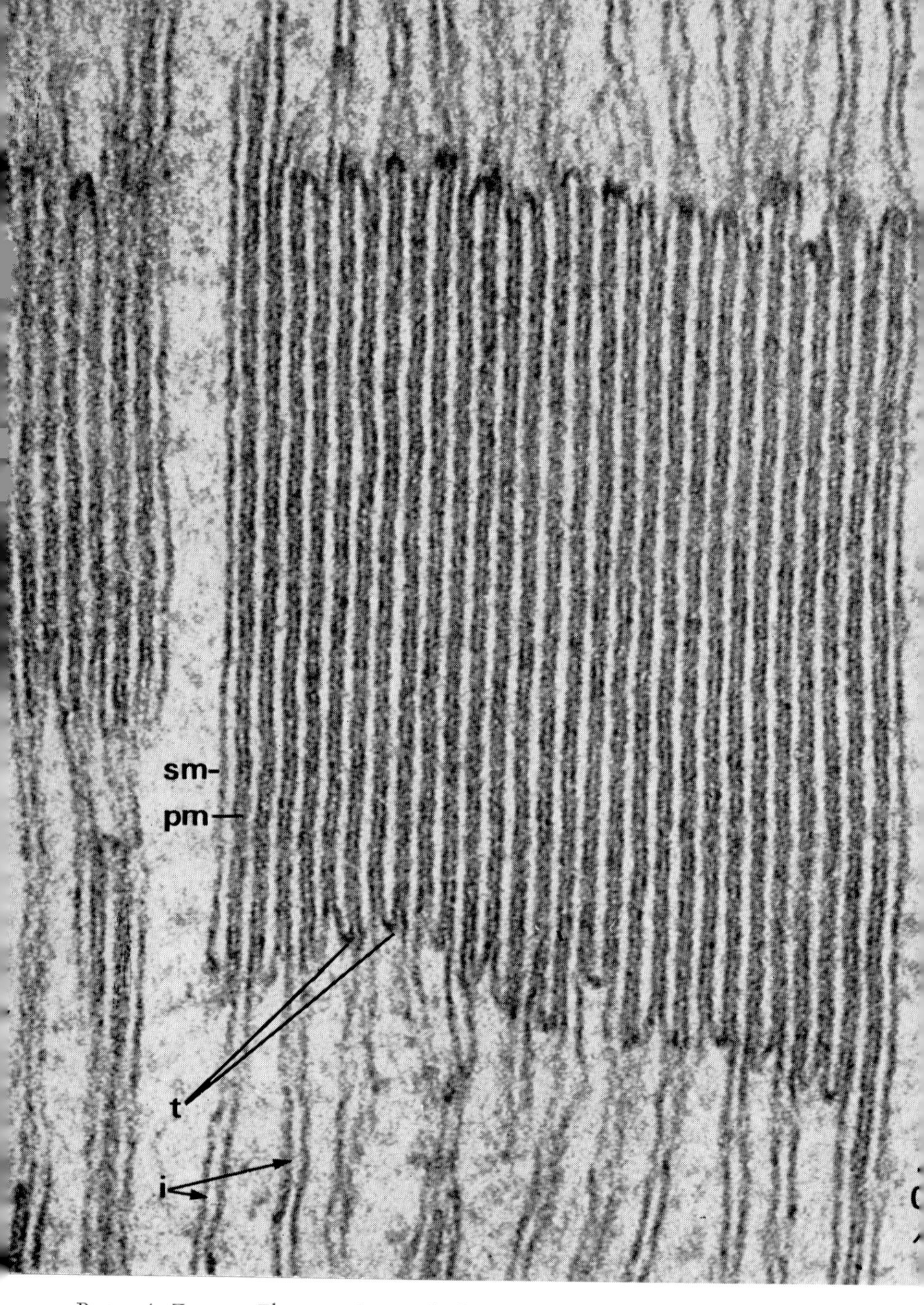

PLATE 4. *Zea mays*. Electron micrograph of a granum in a mesophyll chloroplast. Fixation as for Plate 2. i, intergranal lamellae; t, edges of lamellae (thylakoids) terminating at boundary of granum; sm, single membrane of outermost thylakoid of granum; pm, paired membrane formed by adjacent thylakoids. × 197,000. (*Courtesy of A. D. Greenwood and K. Skinner.*)

PLATE 5, A, B. Models to illustrate postulated mode of growth of the membrane system in a chloroplast of *Cannabis sativa*. In A, successive stages in building up a granum are shown (1 to 4). In B, the complexity matches that of part of actual chloroplast. From Heslop-Harrison [157].

PLATE 5C (*below*). *Cannabis sativa*. Sections of grana with different fixation. (a) Potassium permanganate; (b) gluteraldehyde fixation followed by normal osmication; (c) gluteraldehyde fixation with reduced osmication and uranyl acetate post-staining. × *c*. 201,000. From Heslop-Harrison [157].

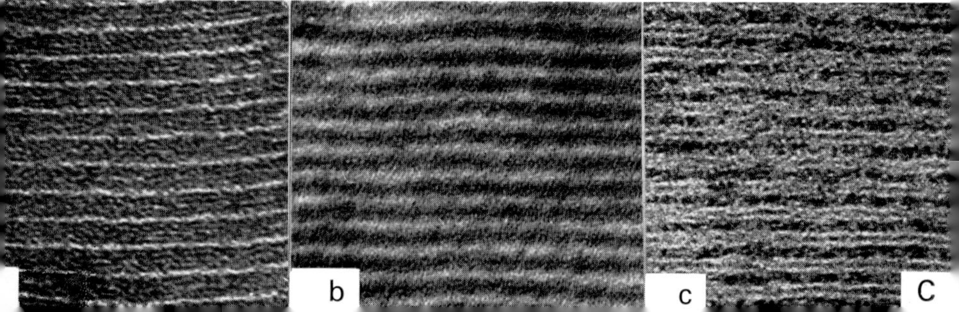

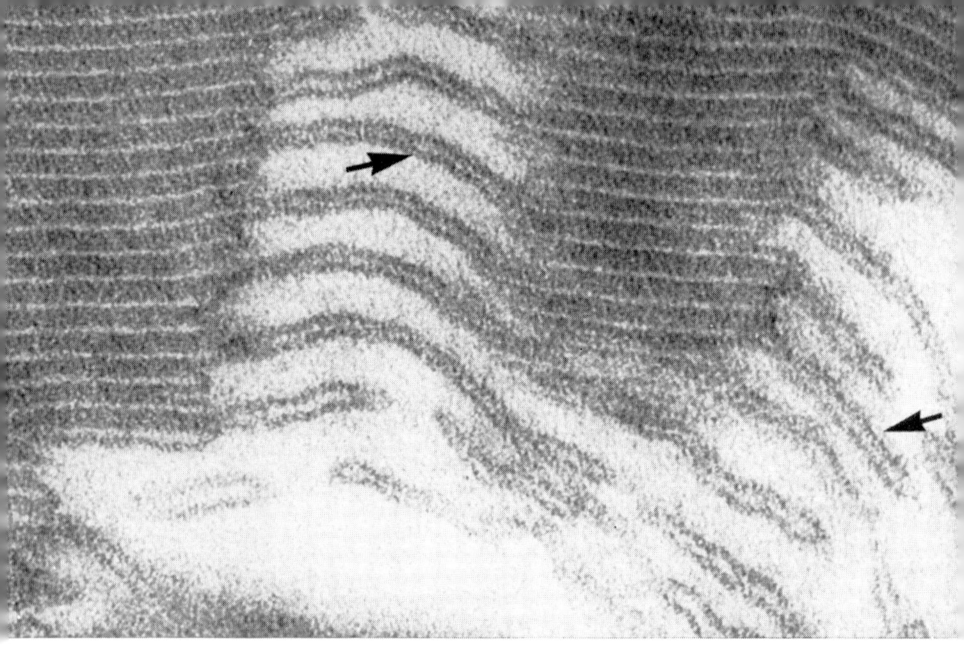

PLATE 6A. *Cannabis sativa*. Electron micrograph of oblique section through a granum. Arrows indicate two parts of the same inclined stroma lamella. (Compare Figure 1.3.) Fixed in permanganate. × *c*. 185,000. From Heslop-Harrison [157].

PLATE 6B. *Cannabis sativa*. Electron micrograph of chloroplast, sectioned *in situ*. Symbols as in Plates 2 and 3. Permanganate fixation. × *c*. 57,000. From Heslop-Harrison [157].

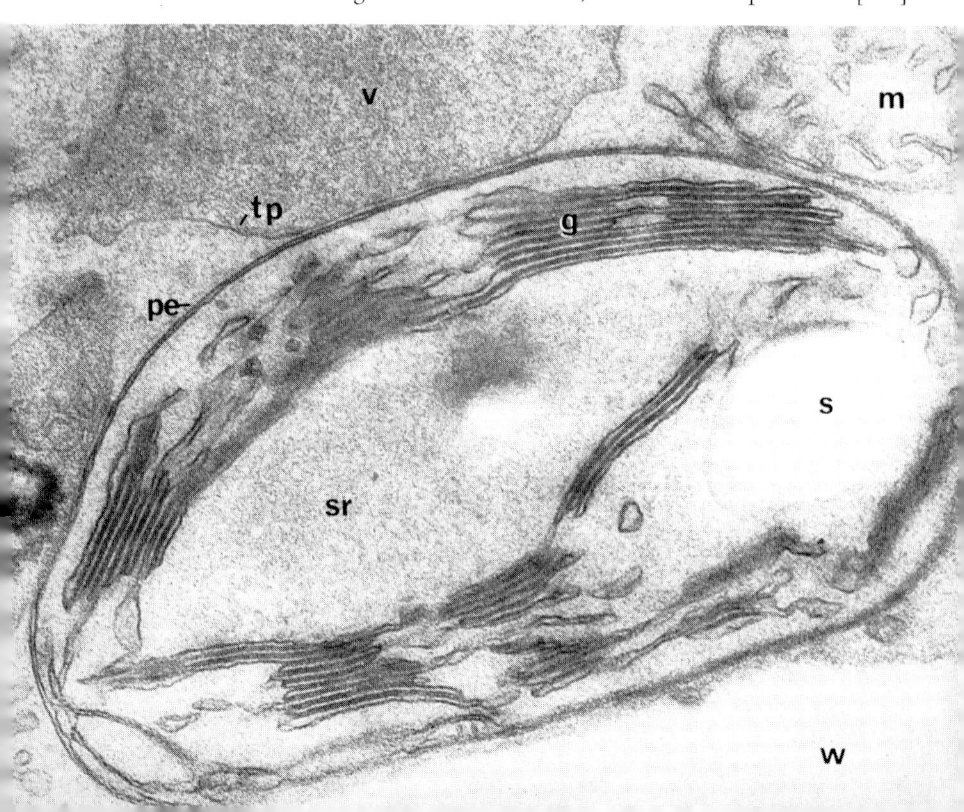

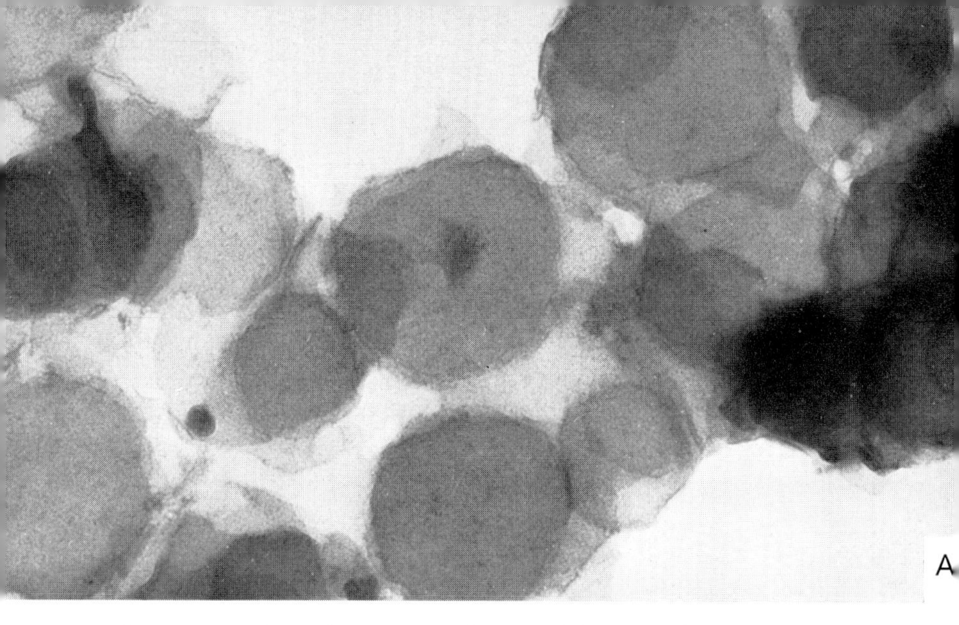

PLATE 7A. Fragment of membrane system of a young chloroplast of Swiss Chard, showing fretted stroma lamellae and various stages of thylakoid growth. Fixed in permanganate. × *c*. 63,000. From Heslop-Harrison [157].

PLATE 7B. Shadowed preparation showing 'paracrystalline' array of quantasomes. Circle shows one quantasome with sub-units. × 340,000. From Park [250].

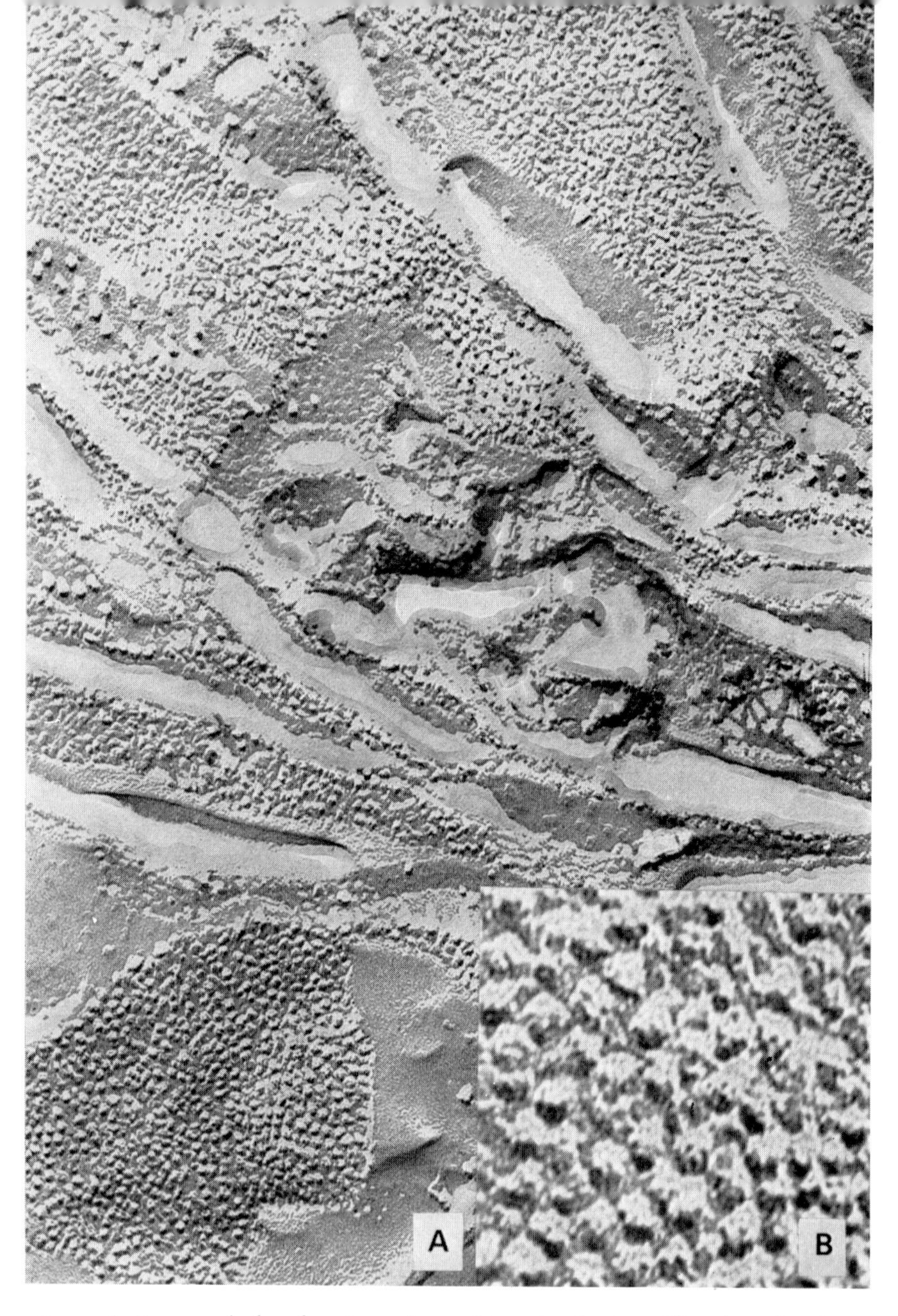

PLATE 8. Freeze-etched surfaces in a tobacco chloroplast, broken obliquely to the plane of the lamellae, showing quantasomes on the grana. A, × 81,000. B, × 288,000. From Moor [234].

illustrate the well-known Kautsky effect, in which the fluorescence of a leaf suddenly illuminated rises rapidly to a maximum and then decays to a steady value, and to give an example of induction phenomena in photosynthesis. Changes in the fluorescence yield of chlorophyll *a* under altered conditions of photosynthesis have been used to infer the proportion of the pigment in the excited singlet state.

Detailed discussion of the mechanism of energy transfer between pigment molecules in the chloroplast lies beyond the scope of this book but some mention must be made of the words that are so freely used in the literature of photosynthesis. In the early 1950s it was general to refer to such transfer as due to inductive resonance. It was considered to be a relatively 'slow' type of transfer such as could take place between the weakly coupled molecules in a solution, for instance from chlorophyll *b* to chlorophyll *a* [259] (see page 22). More recently, with the increased interest in and knowledge of solid-state physics, one school of thought prefers to consider the chlorophyll molecules to be arranged in a two-dimensional molecular crystal lattice and to behave somewhat like a semi-conductor. This would imply that neighbouring molecules are so strongly coupled that their orbitals fuse. Migration of the exciton, described as an electron and a positively charged 'hole' travelling together through the lattice [187], may then be so rapid that it cannot be associated with a single molecule at any one time. The possibility of the chloroplast being such a solid-state system was supported by Arnold and Sherwood's [4] finding that if dried chloroplasts were illuminated at room temperature and then heated at temperatures rising to 140°C in darkness, they emitted light. This would occur in a semi-conductor if some of the excited electrons were trapped in faults in the crystal lattice and were later released, by absorption of the infra-red quanta, to drop back into the holes. Similar effects were observed, but for technical reasons not measured, with fresh *Chlorella* suspensions and leaves. It was suggested that the same mechanism was concerned in the delayed emission of very small amounts of light when green tissues were darkened at normal temperatures after illumination [285].

These ideas have been pursued much further by Calvin and his school [51]. The suggested arrangement of chlorophyll molecules is indicated in Fig. 1.15. The porphoryn heads, oriented at 45° to the plane of the lamella (a–b in Fig. 1.15 (A)) are supposed tilted in two directions relative to the two other axes also (Fig. 1.15(B)), so that their edges are always presented to the faces of neighbouring molecules. The exposed hydrogens (exposed positive charge) round the edges of the porphoryn heads (Fig. 1.5) would repel each other if the edges were adjacent (as in Figs. 1.2 and 1.16) and the arrangement of molecules would thus be unstable. With the arrangement proposed, these hydrogens would tend to be buried in the π-electron clouds of the neighbouring heads, giving maximum attraction between molecules and hence stability. This layer of tilted aromatic heads is supposed to provide uniform horizontal conductivity in the lamellae; on either side of it are supposed sites for trapping electrons and holes respectively, thus separating the reducing and oxidizing parts of the exciton. The chemical identity of these traps is a matter for speculation but the production of free (unpaired) electrons in illuminated chloroplasts has been demonstrated and measured by the method of electron spin resonance. This method [52] involves irradiating cells or chloroplasts at a constant wave length (about 3 cm) in a magnetic field of variable strength. The spinning unpaired electrons (page 19) orient themselves either with or against the field; the energy difference between the two positions depends upon the strength of the field and corresponds to a frequency. When the appropriate field strength is reached the electrons absorb the radiation. The number of unpaired electrons can thus be estimated from the strength of absorption. The interpretation of these measurements for such a complex system as the chloroplast must be a matter of great difficulty, though the method has proved most valuable in elucidating the numbers and locations of unpaired electrons in simpler organic molecules [52].

There are apparently theoretical difficulties in accounting for the separation of the electron and hole of the exciton, essential for photosynthesis, with the low energy levels available from light absorption by chlorophyll [187]. Further, Porter [259] has pointed out that although the concentration of chlorophyll in the granum is high (about 10^{-1} M) its absorption spectrum is like that of amorphous chlorophyll or even dilute solutions and quite unlike that of crystalline chlorophyll. He suggests therefore that the lifetime of the excitation of an individual molecule in the first excited singlet state must be longer than 10^{-13} s, but may be not much longer in view of the favourable conditions for transfer given by the overlap of the fluorescence and absorption spectra. As the observed lifetime of the singlet state of the chlorophyll in the chloroplast is 10^{-9} s there could be transfers of the energy from a single quantum through nearly 10^{-4} molecules. This is of the right order of magnitude for a photosynthetic unit (Chapter 9).

2: The Diffusion Paths

A. THE EXTERNAL PATH

FOR photosynthesis to occur, carbon dioxide must reach the chloroplast. A small quantity will come from respiration in neighbouring mitochondria (chloroplasts do not, apparently, respire) but this obviously cannot result in any net gain of carbon for the plant; the main supply must come from the environment—either the carbon dioxide dissolved in the water bathing the submerged leaves of a water plant or that present in the air surrounding the leaves of a land plant. The photosynthesizing plant will therefore tend to deplete the supply in its immediate neighbourhood and thus give rise to a concentration gradient along which carbon dioxide will diffuse owing to the random thermal movements of its molecules—these result in a net movement from a region of higher to one of lower concentration. In completely still water or air this gradient would in theory ultimately extend to infinity before the rate of supply became steady, but with a large volume supplying an isolated plant the change of rate would soon become negligible. Moreover it is impossible to obtain really still air in the presence of a transpiring plant (see page 51) and even in water there are probably nearly always at least small convective movements due to small local differences of temperature. In spite of such unavoidable small movements, however, in air or water as still as can be obtained experimentally the carbon dioxide supply to the plant is much slower than when the medium is actively stirred; the diffusive flow towards the plant can therefore be considered as opposed by a *resistance* due to the molecules of the medium (air or water) impeding those of the carbon dioxide.

The resistance to diffusion of carbon dioxide is about 10,000 times as great for a given length of path in water as in air, i.e. with the same concentration difference between the two ends of the path the rate of diffusion in water will be 10^{-4} times that in air. This high resistance gives rise to one of the major engineering problems in the proposed bulk culture of *Chlorella* for food, for even with high concentrations of carbon dioxide in the liquid culture medium the latter has to be kept in a state of extreme turbulence if high yields are to be obtained. It can be demonstrated by a simple experiment by the bubble-counting method (page 83) with the Canadian pondweed (*Elodea canadensis*). If a cut shoot of this plant is placed upside down in a large, brightly illuminated vessel of unstirred distilled water, which is in equilibrium with ordinary air (that is, containing about 300 p.p.m. by volume of dissolved carbon dioxide, for the absorption coefficient is about unity at room temperature), bubbles of gas will be seen to emerge from the cut end of the stem. These are largely of oxygen, produced by photosynthesis in the leaves; this gas builds up a pressure in the intercellular spaces and is forced out through the stem. The rate at which bubbles emerge gives a measure of rate of photosynthesis. If the number of bubbles is counted in say successive half-minute periods from the time the shoot is first placed in the vessel, it will be found that the rate of bubbling falls considerably, until at last a nearly steady rate is reached indicating that an effectively constant gradient of carbon dioxide has been established. If the water is now gently stirred, when stirring ceases the rate of bubbling returns to its initial value because a concentration of 300 p.p.m. carbon dioxide is once more close to the surfaces of the leaves. This experiment is discussed in the following chapter and shown to be in some respects less simple and straightforward than it at first seems, but the demonstration of high diffusive resistance is not invalidated.

The very different rates of diffusion of carbon dioxide in air and in water can be expressed by its 'diffusion coefficients' for these two fluids. The diffusion coefficient K, for a given substance diffusing in a given medium, may be defined as the mass M in g of the substance passing across 1 cm² in the diffusion path per second per unit density gradient:

$$K = \frac{M}{At} \cdot \frac{l}{(\rho_2 - \rho_1)} \qquad (2.1)$$

where A is the cross-sectional area of the path, l is its length, ρ_2 and ρ_1 are the densities of the substance at its two ends and t is the time. Dimensionally (2.1) can be written:

$$K = \frac{\text{g}}{\text{cm}^2\,\text{s}} \cdot \frac{\text{cm}^3\,\text{cm}}{\text{g}} = \text{cm}^2\,\text{s}^{-1} \quad \text{or} \quad L^2T^{-1}$$

An alternative definition of K, which may be derived from (2.1), is the *volume* in cm^3 of a diffusing gas (measured at unit pressure but at the temperature of the experiment) passing across 1 cm^2 in the diffusion path per second per unit *partial pressure* gradient:

$$K = \frac{P_T V}{At} \cdot \frac{l}{P_2 - P_1} \tag{2.2}$$

where P_T is the total pressure for the experiment, $P_T V/1$ is the volume at unit pressure, P_2 and P_1 are the partial pressures at the two ends of the path and the other symbols are as before. This gives the same value for K as (2.1) and dimensionally can be written:

$$K = \frac{\text{bar cm}^3}{\text{cm}^2\,\text{s}} \cdot \frac{\text{cm}}{\text{bar}} = \text{cm}^2\,\text{s}^{-1} \quad \text{or} \quad L^2T^{-1}$$

The volume may be measured at N.T.P. but then the partial pressure gradient must also be corrected to 273°K and the final result is the same.

The diffusion coefficient for carbon dioxide in air at 20°C and 1 bar pressure is $0{\cdot}160$ cm^2 s^{-1}; the corresponding value in water would therefore be $0{\cdot}160 \times 10^{-4}$ cm^2 s^{-1}.

Equation (2.2) may be rearranged:

$$\frac{P_T V}{t} = K\,A.\frac{(P_2 - P_1)}{l} \equiv \frac{dq}{dt} \tag{2.3}$$

This states that the rate of diffusive flow, which could also be written dq/dt for the rate at an instant, is directly proportional to the cross-sectional area of the path and to the difference in concentration (partial pressure) between its two ends, and inversely proportional to its length (Fick's Law). The rate dq/dt is in terms of cm^3 s^{-1} at unit pressure and the temperature of the experiment.

By analogy with Ohm's Law for electricity ($R = E/C$) we may consider dq/dt as the current, P as the potential difference (where $P = P_2 - P_1$) and $\dfrac{P}{dq/dt}$ as the resistance r to diffusion. Hence from (2.3):

$$r = \frac{l}{KA} \tag{2.4}$$

Dimensionally, $r = \text{cm}.\dfrac{\text{s}}{\text{cm}^2\,\text{cm}^2} = \text{s cm}^{-3}$ or $L^{-3}T$

The reciprocal of r is the conductance s and is thus in terms of cm^3 s^{-1}.

In the systems we shall consider, such as a leaf assimilating carbon dioxide from a large volume of still air, the various diffusion paths concerned are seldom of uniform cross-sectional area (as implied in equation 2.3) and while some involve diffusion of carbon dioxide in air, in others it is diffusing in water. In order to be able to summate resistances in series and to compare the various components it is convenient to express each of the resistances as an equivalent air path. This is the length of a tube of uniform *unit* cross-sectional area, which, when filled with air and with the same concentration difference maintained between its two ends as between the two ends of the actual diffusion path, would allow the same rate of diffusive flow as occurs in that path. This measure of resistance is in purely geometrical terms, independent of the substance diffusing:

$$R = \frac{KP}{dq/dt} = \frac{l}{A} \text{ cm}^{-1} \qquad (2.5)$$

It will thus give comparable values for either carbon dioxide or water vapour diffusing in air; for carbon dioxide diffusing in water, equation (2.5) must be multiplied by the ratio of the diffusion coefficients in air and in water ($\times 10^4$) to make the values comparable with those for diffusion in air. We shall use such geometrical measures of resistance from now on and this will be implied by the symbol R, with or without suffix.

For many purposes it is convenient to consider the effective length (L_{eff}) of a tube of air, of uniform cross-sectional area equal to that of an absorbing or evaporating surface, which similarly would allow the same rate of diffusion for the same concentration difference as that across the system. This is *not* a measure of the resistance (except for resistance per cm^2 of the surface), for the rate of diffusion will also be affected by the area. Thus $R = L_{eff}/A$ cm^{-1} and

$$L_{eff} = RA \text{ cm} \qquad (2.6)$$

The letter L, with or without suffix, will imply an effective length or a resistance per cm^2 of surface.

The resistance to diffusion of carbon dioxide up to the outer surface of the plant will be called the *external* resistance (R_{ext}), and we have seen that if the medium is in motion the effect is as though this external resistance had been reduced. Actually, of course, this reduction in resistance is due to a shortening of the diffusion path, for the distance from the plant at which the concentration of carbon dioxide is effectively that in the main bulk of the medium is less. The *internal* paths, for the diffusion of carbon dioxide from the plant surface to the chloroplasts, may be partly in air and partly in water, as in a leaf of a land plant, or may be entirely in water, as in a water plant. In either case it is almost always assumed that the medium is motionless but the possibility exists that in the final aqueous path up to the chloroplast within the photosynthesizing cell the passage of the carbon dioxide may be accelerated by cytoplasmic movements.

A large area of a photosynthesizing crop which covers the ground, such as a pasture, may to a first approximation be considered as one large flat leaf absorbing carbon dioxide. In perfectly still air the gradient should extend to infinity. In fact, however, some rather indirect estimation, based on observed evaporation rates from an open water surface under different weather conditions, indicates that when 'zero' wind velocity is measured at a height of 2 m above the ground the external resistance per cm^2 is only $1{\cdot}3\ cm^{-1}$ [255, 228]. There is of course some internal resistance to be added to this, as we shall see shortly, but the rate of photosynthesis will be *as if* a concentration of 300 p.p.m. carbon dioxide were maintained in really still air at a distance of 1·3 cm above each cm^2 of foliage-covered land. This resistance is reduced as the wind velocity measured at 2 m height increases from zero, becoming $0{\cdot}38\ cm^{-1}$ with an 8 $km\ h^{-1}$ wind and $0{\cdot}16\ cm^{-1}$ with one of 24 $km\ h^{-1}$. Over the range 3 to 11 $km\ h^{-1}$ comparable values have been derived [231] by a more direct method, from observations of profiles of wind speed, carbon dioxide concentration, water vapour content and temperature above a crop of field beans. These results all suggest that, at least for the uppermost layers of a field crop, the external resistance is seldom important. Within a dense crop less air movement occurs and hence resistances are much higher [231] but in spite of this the considerable quantities of carbon dioxide produced by respiration of the roots and micro-organisms in the soil (of the order of $1{\cdot}0\ mg\ cm^{-2}\ day^{-1}$) move rapidly up through the crop. This can occur because the continuous production in the soil results in a large potential difference building up. It is estimated that the soil normally supplies about one-fifth of the carbon dioxide assimilated by a vigorously growing crop [231].

In really still air an isolated leaf would have a much larger supply of carbon dioxide available than an equal area of sward. For the latter the rate of supply would be proportional to the area (equation 2.3) and the lines of diffusive flow would be parallel to each other and normal to the land surface. For the isolated leaf the flow lines would converge—theoretically from 'semi-infinite space' if we consider one leaf surface only. Theory, based by Stefan on an analogy with electrostatics, suggests that the rate of diffusion dq/dt should then be directly proportional to the diameter rather than to the area. For a flush circular

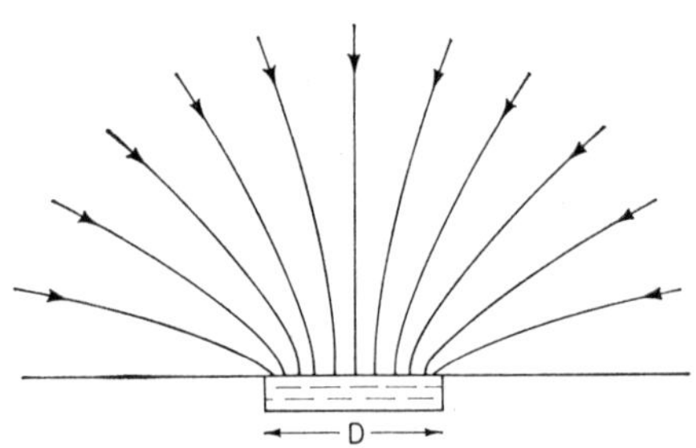

FIG. 2.1 Theoretical lines of net diffusive flow of carbon dioxide, in still air, converging on a flush circular absorbing surface of diameter *D*.

absorbing surface, such as a widely flanged dish of caustic soda filled to the brim (Fig. 2.1):

$$\frac{dq}{dt} = 2KPD \tag{2.7}$$

where *D* is the diameter and *P* is the difference in partial pressure of carbon dioxide at the absorbing surface and at infinity. It is inconvenient to deal with gradients extending to infinity, even in theoretical treatment, and a finite effective length may be arrived at as follows:

For diffusion from infinity to a disc, the resistance is given by:

$$R_d = \frac{KP}{dq/dt} = \frac{1}{2D} \quad \text{(from 2.7)} \tag{2.8}$$

Similarly for diffusion from one end of a tube to the other, the resistance is:

$$R_t = \frac{l}{A} = \frac{4l}{\pi D^2} \quad \text{(from 2.5)} \tag{2.9}$$

Thus, if we consider diffusion from infinity to the mouth of a flanged tube of circular cross-section and thence down the tube to an absorbing surface at the bottom (Fig. 2.2), the total resistance is:

$$R_d + R_t = \frac{1}{2D} + \frac{4l}{\pi D^2} = \frac{4}{\pi D^2}\left(l + \frac{\pi D}{8}\right) \tag{2.10}$$

This is the same as the resistance of a tube of length $l + \pi D/8$; a concentration difference of P between the two ends of such a tube would give the same rate of diffusion as that in the system we are considering. The effective length (page 44) is therefore $l + \pi D/8$ and $\pi D/8$ is the so-called end correction for the tube.

Now consider a tube connecting two semi-infinite volumes of air, with diffusion in at one end, through the tube and out at the other (Fig. 2.3). The sum of the two end corrections would then be $\pi D/4$ and the effective length $l + \pi D/4$. If now the septum were made so

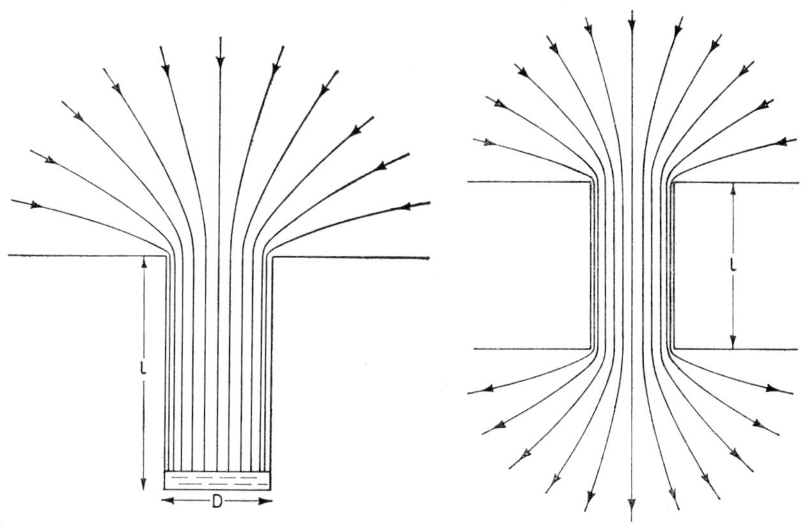

FIG. 2.2 Theoretical lines of net diffusive flow of carbon dioxide, in still air, to the mouth of a flanged circular tube, diameter D and length l, and thence down the tube to an absorbing surface at the bottom.

FIG. 2.3 Theoretical lines of net diffusive flow of carbon dioxide towards, through and away from a circular tube, diameter D and length l, which pierces a partition between two large volumes of still air.

thin that l became practically zero (Fig. 2.4), the effective length for the circular aperture would be $\pi D/4$. Similarly the effective length for diffusion up to the flush dish of caustic soda (Fig. 2.1) is given by $\pi D/8 = 0{\cdot}39D$. Thus, with the hemisphere of semi-infinite space to draw on for its carbon dioxide, in perfectly still air the isolated leaf should assimilate at the same rate as it would if surrounded by an extensive area of crop with the unreduced carbon dioxide concentration (say 300 p.p.m.) only about two-fifths of its diameter away. For a leaf of 10 cm diameter this effective length would therefore be 4 cm, or 0·2 cm for one of 0·5 cm diameter. For an isolated submerged leaf of a water plant, 0·5 cm in diameter, the corresponding effective length in terms of air path would be no less than 20 m. However, in natural waters dissolved carbonates and bicarbonates will release more carbon dioxide into solution as it is removed and so greatly reduce this theoretical effective length.

If assimilation by a circular isolated leaf is proportional to its diameter it must also be proportional to its circumference (πD). It is easy to

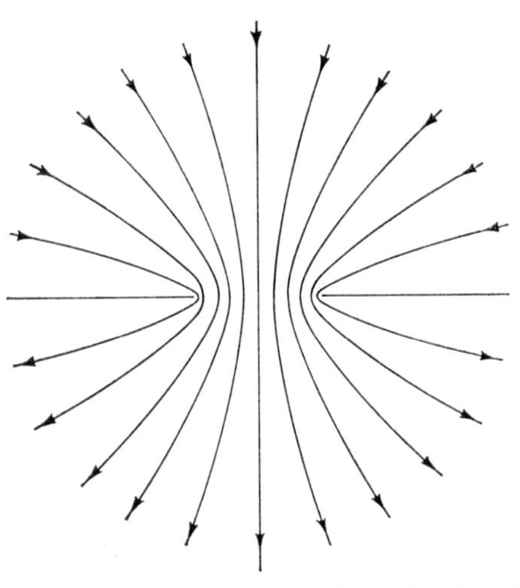

FIG. 2.4 Theoretical lines of net diffusive flow of carbon dioxide towards and away from a circular pore, diameter *D*, in an exceedingly thin septum between two large volumes of still air.

see that most of the absorption of carbon dioxide and loss of water vapour in still air will take place at the edge of the leaf if we consider as a model water loss from a circular dish (Fig. 2.5). For a small unit volume of air at (1) over the centre of the dish water vapour will both enter and leave at the sides at about the same rate, but for a similar unit volume at the same height above the edge of the dish at (2) there will be a net loss on the side remote from the centre; to maintain the same partial pressure of water vapour as in (1) the unit volume over the edge must be closer to the dish as (3). In this way contour surfaces of equal partial pressure of water vapour could (in theory) be mapped out over the water surface and they would lie much closer together at the edge than over the middle (Fig. 2.6). These contour surfaces are the so-called diffusion shells, which of course have no more discrete existence in Nature than contour lines on a hillside; they are half ellipsoid in shape, with the foci of the ellipses on the edges of the dish, and they approach nearer to a hemispherical shape the further they are from the surface. The flow surfaces (hyperboloids) which cross them at right angles, just as water flows downhill at right angles to the altitude contours, must bend outwards except over the centre of the dish, and diffusive flow at the edges will be the most rapid because the contours are closest together and the gradient steepest there. Note that the first half ellipsoidal shell will be a plane surface coincident with

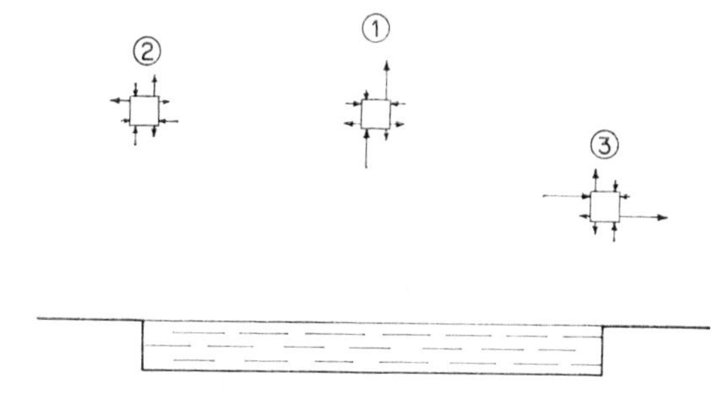

FIG. 2.5 Diagram of vertical and lateral components of diffusion of water vapour in still air over a circular water surface in a flanged dish. (1) Small unit volume of air over the centre, in a vertical vapour pressure gradient but with air at the same vapour pressure at the sides. (2) Small unit volume of air at the same height as (1) but over the edge of the dish, with drier air both above and towards the outside; vapour pressure in (2) is less than in (1). (3) Small unit volume of air at such a height above the edge of the dish that the vapour pressure is the same as in (1). Lengths of arrows indicate, very approximately, relative rates of diffusion.

the water surface and having P_s = saturation pressure of water vapour, or for diffusion of carbon dioxide to a perfect absorbing surface P_0 = zero pressure. The flow surfaces must cross this at right angles and therefore normal to the surface.

If the above theory is even approximately correct, small isolated leaves in still air should lose water faster per unit area than large ones, or considering diffusion in the opposite direction, they should assimilate carbon dioxide more rapidly. Similarly, if it also applies to other shapes, then much-dissected or narrow leaves, having a greater perimeter, should both transpire and assimilate faster than entire and more nearly isodiametric leaves of the same area.

We may now consider how nearly the theory applies in practice. The end correction $\pi D/8 = 0{\cdot}39D$ has been confirmed experimentally for the acoustics of flanged organ pipes; for a pipe with no flange the end correction is found to be $0{\cdot}31D$ and hence for a circular leaf with only one surface assimilating carbon dioxide, but with diffusive flow past the leaf edges, the resistance might be expected to be similarly reduced. It has not been found possible to confirm the theoretical treatment by experiments on diffusion itself. Brown and Escombe [46] attempted to measure the rates of diffusion of carbon dioxide through

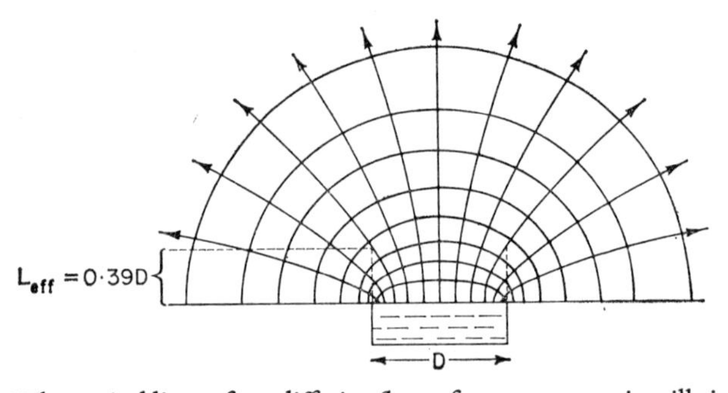

FIG. 2.6 Theoretical lines of net diffusive flow of water vapour, in still air, from a circular water surface in a flanged dish and contours of equal water vapour pressure. The diagram represents a median section, the 'flow lines' are sections of hyperboloidal flow surfaces crossing at right angles a series of contour surfaces (half ellipsoids). L_{eff} is the effective length, that is, the length of a tube of the same diameter D as the dish which would offer the same resistance to diffusion as that for unrestricted flow from the water surface to infinity.

circular apertures of various sizes and concluded that, as predicted by Stefan, the rate was proportional to the diameter ($dq/dt \propto D^{1{\cdot}0}$) and not to its square. Penman (unpublished) re-examined their data and found that these indicated a power of D of about 1·2 to 1·3—hardly good enough to confirm the theory though clearly less than 2·0.

The failure of various experimental attempts to confirm the so-called diameter law for diffusion in still air cannot be attributed to the space available being less than semi-infinite, for the experimentally observed rates are always higher than theory demands. It may of course be that this law does not apply to diffusion; more probably the failure has been due to the use of experimental systems involving the evaporation of water. Diffusion of vapour away from a dish of water should in theory obey equation (2.7), with of course the appropriate diffusion coefficient K for water vapour (0·257 $cm^2\ s^{-1}$ at 20°C), and be as in Fig. 2.6. However, the mere fact of evaporation must give rise to mass movements of air. The air close to the surface becomes saturated with water vapour, which lowers the density, and cooled by the evaporation which has the opposite effect. Moreover, most evaporation occurs at the edges of the dish and the extra cooling there will both reduce the amount of water vapour the air can hold at saturation and have more effect in increasing the density. Vertical convection currents and lateral mass movement are therefore to be expected and really still air cannot be obtained. Air currents of 10 cm s^{-1} have been measured over water in 'still' air [263]. For air as still as obtainable indoors the effective length for a circular water surface up to 10 cm diameter was found by Penman and Schofield [255] to be given by $L_{eff} = 0{\cdot}4D^{0{\cdot}6}$ instead of $0{\cdot}39D^{1{\cdot}0}$. Thus for a dish or leaf of 10 cm diameter $L_{eff} =$ 1·59 cm instead of 3·93 cm. Slight draughts would further reduce the effective length. Somewhat similar air movements probably occurred during Brown and Escombe's experiments in which they allowed atmospheric carbon dioxide to diffuse through a hole in a diaphragm covering the mouth of a tube and then down to the surface of a solution of caustic alkali at the bottom. The water vapour content of the air in the tube must have caused convection currents, with moist air rising out of the tube through the hole and dry air moving in to replace it.

Although an isolated leaf photosynthesizing in really still air is not apparently to be found in Nature, the theoretical treatment is important as giving an upper limit for the external resistance (R_{ext}); the theory is also useful in considering the *internal* resistances where, as noted, the medium is usually assumed to be stationary.

The importance of wind in reducing R_{ext} for a continuous area of crop has already been mentioned; this effect will be even greater for more or less isolated leaves, as on scattered plants in a garden bed or in a semi-desert situation, or on a tree or shrub that stands up clear of surrounding vegetation.

In theory [181], when wind blows over a flat surface such as a leaf, streamline flow occurs in a layer close to the surface (the boundary layer) and here the ordinary diffusion coefficient K applies. This layer is thought to be about 1 mm thick with a wind speed (u) some distance from the leaf of 2 m s^{-1} or 7·2 km h^{-1}. Beyond this boundary layer the flow is turbulent and here the rapid mixing results in the carbon dioxide being transported much faster. The coefficient of eddy diffusion increases with height, with wind speed and with the roughness of the surface; it was estimated [254] as 10^3 for $u = 2$ m s^{-1} a few cm above a wheat crop (compared with 0·16 for K). The contribution of the turbulent layer to the external resistance is thus negligible in comparison with that of the boundary layer. We may therefore consider the ambient carbon dioxide concentration as being maintained constant at the outer surface of the boundary layer and the thickness of the latter as being L_{eff}, at least for parts of the leaf remote from the edges.

Unfortunately the thickness of the boundary layer is difficult to determine. It depends on wind speed in a rather ill-defined way and

TABLE 2.1

Effective lengths (cm) for evaporation from circular free water surfaces, as influenced by size and wind velocity. (In brackets are the resistances $R = (L_{eff}/A)$ cm^{-1}.) From Table 2 Milthorpe, F. L. [228].

Diameter cm	Area cm^2	Wind velocity			
		'Still' air	1 m s^{-1} 3·5 km h^{-1}	3 m s^{-1} 10·6 km h^{-1}	5 m s^{-1} 17·7 km h^{-1}
2·0	3·15	0·60 (0·191)	0·09 (0·029)	0·05 (0·016)	0·04 (0·012)
10·0	78·5	1·59 (0·020)	0·18 (0·0023)	0·10 (0·0013)	0·07 (0·0009)
20·0	314·1	2·39 (0·0076)	0·25 (0·0008)	0·13 (0·0004)	0·10 (0·0003)

it increases from zero thickness at the leading edge to a constant value further back so that the mean value will depend on the leaf size. It may also be reduced by leaf flutter. For these reasons, Penman and Schofield [255] made an empirical approach, using data for evaporation from a circular disc in a wind tunnel [263]. They estimate that $L_{eff} = 0{\cdot}89D^{0{\cdot}44}/u^{0{\cdot}56}$. Here L_{eff} includes both the turbulent component and the boundary layers.

Making use of this expression for moving air and $L_{eff} = 0{\cdot}4D^{0{\cdot}6}$ for 'still' air., Milthorpe [228] has calculated L_{eff} for evaporation from a circular free water surface as shown in Table 2.1.

Transpiration rates for leaves of the sizes shown in the table should be inversely proportional approximately, to the resistances and rates per unit area to the effective lengths. The very important effects of a slight breeze and the much smaller effects of further increases of wind speed are clearly shown. For a leaf, the effects of wind speed may be expected to be partly offset by stomatal closure. It may be noted that over a wide range of wind speeds and sizes of evaporating surface the effective length approximates to 1 mm, suggesting that the boundary layer is frequently of this order.

B. THE EXTERNAL AND INTERNAL PATHS

For the purposes of the discussion of the external path the leaf has been treated as a flat uniformly absorbing surface. This is a reasonable approximation for a submerged leaf of a water plant, such as *Elodea canadensis* (Fig. 2.7); here the whole diffusion path may be considered

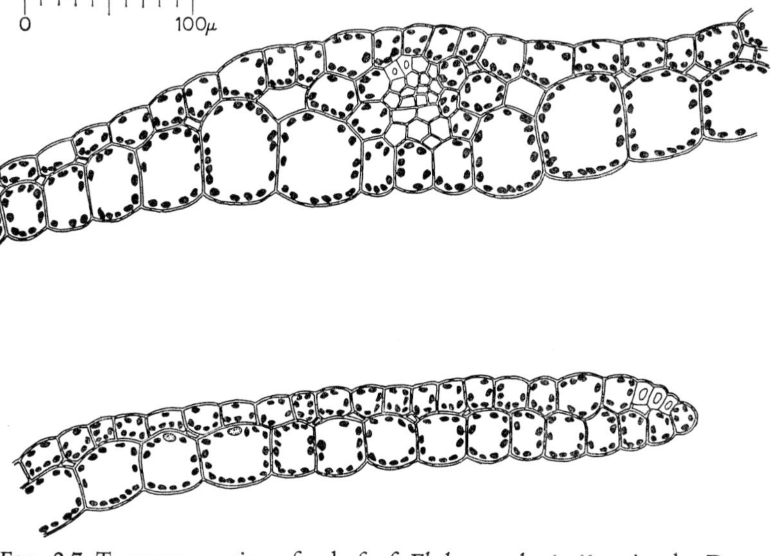

FIG. 2.7 Transverse section of a leaf of *Elodea canadensis*. Drawing by Dr R. M. Grill.

as aqueous and dissolved carbon dioxide passes through the water-imbibed and scarcely cuticularized cellulose outer walls of the cells to the many closely adjacent chloroplasts. If the distance to the nearer of these is taken as 2·5 μ the effective length in terms of air will be 2·5 cm —this is appreciable, though of course negligible if added to the theoretical 20 m external effective length for really still pure water (page 48). This internal resistance for diffusion to the peripheral chloroplasts, and even more to those further in, may well be reduced by movements of the cytoplasm; in *Elodea*, under the influence of a bright light, such movements actually carry the chloroplasts round the cell.

If a shoot of *Elodea* is exposed to the air it quickly dries out and the leaves of land plants need a more or less impervious coating to prevent this from happening. This is provided by the cuticle—a layer of wax-like cutin which covers the outer cellulose walls of the epidermis and interpenetrates their fibrils—and often by external layers of waxes which interpenetrate the cutin. The thickness of the cuticle and the

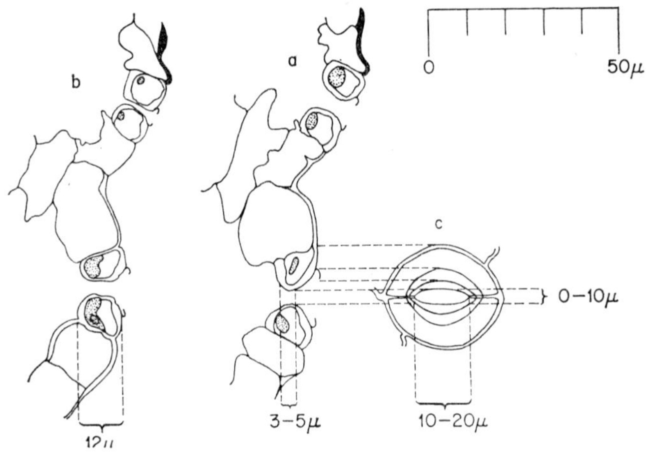

FIG. 2.8 Stomata in the lower epidermis of a leaf of *Pelargoneum zonale*. The dimensions are average values. *a*, Camera lucida drawing of two open stomata in transverse section, near centre of aperture. *b*, The same, near one end of the elliptical aperture. *c*, Diagram of surface view of one of the stomata. The width of the pore would only reach 10 μ under reduced carbon dioxide concentration; its length changes only slightly with changing aperture and 20 μ would be for a very large stoma. From Heath [139].

types and amounts of wax differ greatly for leaves of different species; they are generally thicker on the upper than on the lower surface of any one leaf. Estimates of effective lengths for their resistance per cm^2 to the passage of water vapour [228] vary from 1·1 cm (in *Antirrhinum*) to 50 cm (in *Rhododendron*); comparable estimates for the passage of carbon dioxide do not seem to be available but there is little doubt that in many species the resistance is considerable and that most of the carbon dioxide entering the leaf does so through the stomatal pores. Each of these is bordered by a pair of specialized guard cells which enlarge, diminish or close the elliptical aperture by changing their shape (Fig. 2.8); such changes are brought about by alterations in the difference of turgor between the guard cells and the neighbouring epidermal cells—they occur in response to a variety of external stimuli but especially changes of light intensity or carbon dioxide concentration [142, 143, 147]. Here we shall be concerned with the changing apertures (and resistances to diffusion) of the pores rather than the mode of action of the guard cells.

In many land species, especially of trees, the stomata are confined to the lower leaf surface (hypostomatous leaves); in these the upper surface, unless so heavily cuticularized and wax coated as to have infinite resistance, must again behave as a uniform absorbing surface, though a very much less efficient one than the leaf of *Elodea*. If we omit any consideration of diffusion past the edges of such a hypostomatous leaf to the lower surface, the model for uptake through the upper epidermis (cuticular uptake) is provided by Fig. 2.2 and equation (2.10). In these, D is the diameter of the leaf (assumed circular), $\pi D/8$ represents the upper limit for the external effective length (in completely still air) and l is made up of the effective lengths for the wax plus cuticle and for the aqueous diffusion path through the epidermis to the chloroplasts of the underlying palisade parenchyma. It may be noted that in many dicotyledonous leaves a few chloroplasts do occur in the ordinary epidermal cells, especially when grown under shade conditions; as far as I am aware this is not so in monocotyledonous aerial leaves. Thus for most of the carbon dioxide assimilated through the cuticle the resistance of a very considerable aqueous path (say 15 μ of water, equivalent to 15 cm of air) is added to that of the wax plus cuticle and l must be large.

In contrast with cuticular uptake, the carbon dioxide entering through the stomatal pores can diffuse *in air* to all parts of the leaf in the intercellular space system and so reach the surfaces of the actual assimilating cells (Fig. 2.9); the final aqueous diffusion path through the wet cell wall and up to the chloroplast is thus kept as short as possible. The approximate magnitudes and effective lengths for these various resistances are considered later—first we may discuss how to calculate them.

For carbon dioxide diffusing to a *single* isolated stomatal pore in a heavily cuticularized leaf, the lines of diffusive flow and contours of equal partial pressure over the pore would be as indicated in Fig. 2.6

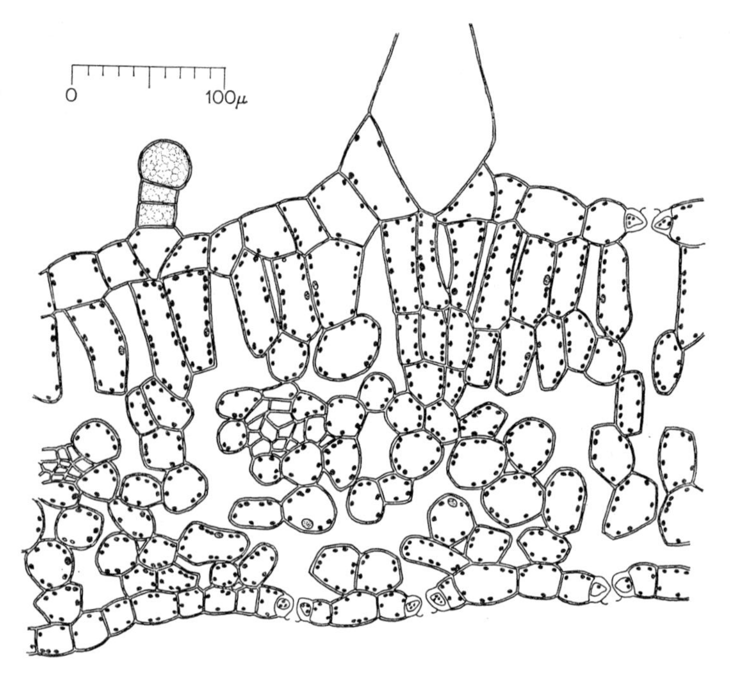

FIG. 2.9 Transverse section of a leaf of *Pelargonium zonale*. Drawing by Dr R. M. Grill.

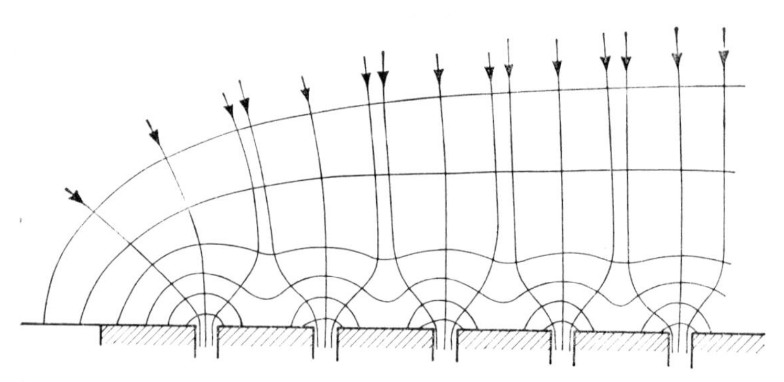

FIG. 2.10 Theoretical lines of net diffusive flow of carbon dioxide, in still air, towards part of a single leaf and into the stomatal pores which pierce its upper surface; also contours of equal partial pressure of carbon dioxide. No account is taken of diffusive flow past the edge of the leaf or to the lower surface and this model would be appropriate for a water-lily leaf floating on water.

but with the gradients and directions of flow reversed. With a number of stomata in the leaf surface, the carbon dioxide diffusing towards the leaf from semi-infinite space would be shared between them. The lines of diffusive flow would thus converge towards the leaf as if the whole surface were absorbing and then separate to pass into the individual pores (Fig. 2.10)—this would occur nearer to the leaf surface with more numerous or larger stomata. In the limit, when the whole leaf surface was occupied by the pores, the diagram would again be as in Fig. 2.6, but this would represent the leaf as a whole instead of a single stoma.

We see that for stomatal assimilation the external resistance has two components: the first is the equivalent air path for the leaf as a whole and in series with this are the external resistances for all the stomata in parallel. We may now consider a simplified model for stomatal assimilation (Fig. 2.11). A circular absorbing surface of caustic alkali is at the bottom of a tube of uniform cross-sectional area A (diameter D). Part way up the tube is a diaphragm of thickness l pierced by n circular holes each of cross-sectional area a (diameter d). The perforated diaphragm represents the leaf epidermis pierced by stomatal pores of similar number and dimensions; the absorbing surface represents the chloroplasts and is assumed as efficient as they are; the lengths L_2 and L_3 are such that $L_2/A = R_{int}$ is the equivalent air path for the overall resistance to diffusion in the intercellular spaces and $L_3/A = R_{aq}$ is that for the aqueous diffusion paths to the chloroplasts; L_1 is the effective length for the external resistance of the leaf as a whole, which is supposed circular and of area A. Note that in the leaf the total internally exposed surface of assimilating cells is very large, estimated as amounting to about six times the area of the leaf; on the other hand the cross-sectional area of the intercellular spaces totals

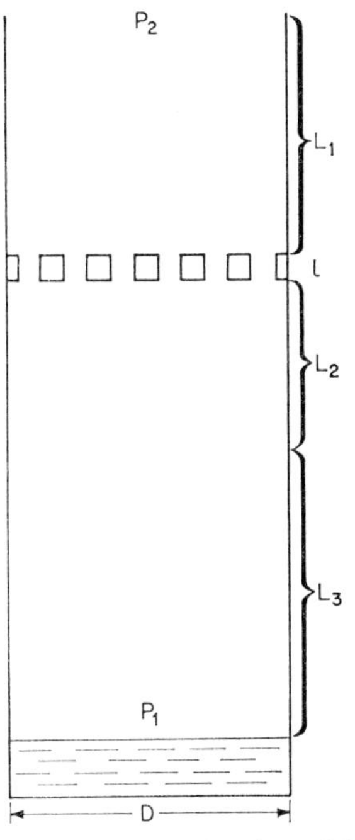

FIG. 2.11 Simplified model for stomatal assimilation of carbon dioxide. For explanation see text. Modified from H. L. Penman (unpublished).

about one-third of the leaf area. For these reasons L_2 and L_3 are not effective lengths for the intercellular spaces and aqueous paths though L_2/A and L_3/A are the equivalent air paths. If the carbon dioxide concentration difference ($P_2 - P_1 = P$) between the mouth of the tube and the absorbing surface is the same as that between the air remote from the leaf and the chloroplasts, the rate of uptake should be the same.

The total resistance to diffusion in the model is made up of six components in series, which are shown below, together with their equivalents for the leaf, assuming perfectly still air.

Model	*Leaf*
$\frac{L_1}{A} = R$ up to diaphragm	$\frac{1}{2D} = R_d$ for leaf as a whole
$\frac{1}{n} \times \frac{1}{2d} = R$ into n pores	$\frac{1}{n} \times \frac{1}{2d} = R_d$ for n stomata
$\frac{1}{n} \times \frac{l}{a} = R$ through n pores	$\frac{1}{n} \times \frac{l}{a} = R_t$ for n stomata
$\frac{1}{n} \times \frac{1}{2d} = R$ out of n pores	$\frac{1}{n} \times \frac{1}{2d} = R_d$ for n stomata
$\frac{L_2}{A} = R$ through 1st part of tube	$\frac{0{\cdot}3L_2}{0{\cdot}3A} = R_t$ for intercellular spaces
$\frac{L_3}{A} = R$ through 2nd part of tube	$\frac{6L_3}{6A} = R_t$ for hydrodiffusion

The chemical reactions in the chloroplast take time to complete and may also, therefore, be considered as opposed by resistances but these are not a part of the diffusion paths. They will result in the concentration of carbon dioxide at the chloroplast surface (P_1) being greater than zero and in the model may be considered to be the result of the caustic alkali not being a perfect absorber, again resulting in P_1 exceeding zero.

We may now sum the six diffusive resistance components for the leaf:

$$R_{\text{Leaf}} = \frac{1}{2D} + \frac{1}{n}\left(\frac{l}{a} + \frac{1}{d}\right) + \frac{L_2}{A} + \frac{L_3}{A}$$

$$= \frac{1}{A}\left[\frac{A}{2D} + \frac{A}{na}\left(l + \frac{a}{d}\right) + L_2 + L_3\right]$$

$$= \frac{1}{A}\left[\frac{\pi D}{8} + \frac{A}{na}\left(l + \frac{\pi d}{4}\right) + L_2 \quad L_3\right] \qquad (2.11)$$

Here the terms within square brackets make up the total effective length: $\pi D/8 = 0{\cdot}39D$ is the theoretical effective length for a disc in perfectly still air (page 47), A/na is the reciprocal of the proportion of the leaf area occupied by pores, L_2 is three times the effective length of the diffusion path through the intercellular spaces and L_3 is one-sixth of that (in terms of air) of the hydrodiffusion path up to the chloroplast. It will be noted that the sum of the two end corrections used for a stomatal pore ($\pi d/4$) is that for perfectly still air in two hemispheres of semi-infinite space. As far as stillness of the air is concerned, it can be assumed to be motionless within the leaf, and on the outer surface the first contour or shell of equal concentration characteristic of the leaf as a whole will be well within the boundary layer at all ordinary wind speeds; the stomatal component is thus independent of wind velocity. The fact that the path between the first shell just referred to and the interior of the leaf is shorter than that between one end of infinity and the other will only reduce the effective length of the stoma by some 10 per cent [255] as much of the resistance is due to l: this correction is perhaps just worth making. On the other hand $\pi D/8$ is a large term and much greater accuracy can be obtained by inserting a value based on the empirical investigations referred to earlier (pages 51, 52).

We may use equation (2.11), with the appropriate adjustments, to estimate the total resistance for a typical mesophytic leaf, such as that of *P. zonale* (Fig. 2.9); hence we may calculate a rate of photosynthesis for conditions of high light intensity and adequate temperature, when carbon dioxide supply to the chloroplasts may be assumed to be the main rate-limiting factor (Chapter 4).

The following approximate data are for the lower surface of a large leaf of *Pelargonium* with wide open stomata: $D = 10$ cm; $A = 78{\cdot}5\text{cm}^2$; $n = 1{\cdot}5 \times 10^4 \times 78{\cdot}5$; $d = 1 \times 10^{-3}$ cm; $a = 7{\cdot}85 \times 10^{-7}$ cm^2; $l = 1{\cdot}2 \times 10^{-3}$ cm; $L_2 = 3 \times 1{\cdot}5 \times 10^{-2}$ cm, $L_3 = \frac{1}{6} \times 2{\cdot}4$ cm.

The value taken for the mean stomatal diameter ($d = 10\ \mu$) is very approximate and based [255] on the length and breadth of the elliptical pores as seen in fixed preparations made by Lloyd's method [204]—it may well be an overestimate [140] and is probably such as might only be expected under conditions of carbon dioxide deficiency, for example in very still air with active photosynthesis proceeding. In arriving at L_2, the mean distance from the inside of the lower epidermis to the middle of the palisade layer (about 150 μ) has been used. For L_3 the path through the cell wall to the chloroplast has been taken as

$2{\cdot}4\ \mu$ of water, equivalent to 2·4 cm of air. If light and temperature are at low levels, thus slowing the chemical reactions, the chemical resistance will be high and the concentration of carbon dioxide at the chloroplast will greatly exceed zero. Even for favourable levels of light and temperature a chemical resistance equal to that of the hydrodiffusion path has been suggested by Maskell; this we shall use for the purposes of this chapter although some other workers [110] consider that the concentration at the chloroplast is zero in high light. In this strain of *P. zonale* the upper stomata were usually less than one-tenth as numerous as those on the lower surface, and for simplicity the assimilation through the upper surface will be ignored. If we use, in the first instance, $0{\cdot}4D^{0{\cdot}6}$ instead of $\pi D/8$ for 'still' air (page 51) and reduce $(l + \pi d/4)$ by 10 per cent we obtain:

$$R_{\text{leaf}} = \frac{1}{78{\cdot}5}\left[1{\cdot}59 + \frac{100}{1{\cdot}18}(0{\cdot}0012 + 0{\cdot}00079)\frac{9}{10} + 0{\cdot}045 + 0{\cdot}4 + 0{\cdot}4\right]$$

$$= \frac{1}{78{\cdot}5}\left[\underset{\text{Leaf as a whole}}{1{\cdot}59} + \underset{\text{Stomata}}{0{\cdot}152} + \underset{\text{Intercellular spaces}}{0{\cdot}045} + \underset{\text{Aqueous path}}{0{\cdot}4} + \underset{\text{Chemical}}{0{\cdot}4}\right]$$

$$= \frac{1}{78{\cdot}5} \times 2{\cdot}59\ \text{cm}^{-1}$$

Note that the stomatal pores contribute only 6 per cent of the total resistance in 'still' air and assimilation would therefore be 94 per cent of what could in theory be achieved if the epidermis were entirely absent. This is because their dimensions are very small compared with the leaf as a whole, which in 'still' air contributes most to the resistance. In a 3·5 km h^{-1} breeze the external effective length would be reduced to 0·18 cm (Table 2.1), the total to 1·18 cm and the stomatal contribution would now be 13 per cent even if the wind caused no closure. If, however, the stomata closed to have a mean diameter of $3\ \mu$, the effective length for the stomatal term would become 1·22 cm or 54 per cent of the total, which would now be 2·25 cm. In a wind of 17·7 km h^{-1} the external contribution would be 0·07 cm and further closure to $1\ \mu$ would then increase the stomatal contribution to 91 per cent of the total (Table 2.2).

At very small pore widths, comparable with the mean free path of the diffusing molecules, the diffusion coefficient K decreases; therefore the stomatal resistance in fact rises more steeply with the approach of closure and becomes an even greater proportion of the total resistance than the figures in Table 2.2 imply. This effect becomes important at pore widths of the order of 0·1 μ, which of course correspond to considerably larger mean diameters.

These effects of wind and of stomatal closure may be compared with the corresponding effects for resistances to transpiration (Table 2.2). In calculating the effective lengths for transpiration there is no chemical resistance to be considered; it is usually assumed that the resistance of the mesophyll cell walls is negligible and since much of the water

TABLE 2.2

Relative contributions of the stomata to total resistances for carbon dioxide assimilation and transpiration by a Pelargonium leaf 10 *cm diameter under different wind velocities. The tabulated figures are:*

Stomatal L_{eff}/*Total* L_{eff} = *Relative resistance due to stomata*

Wind speed:		*'Still' air*	1 *m* s^{-1} 3·5 *km* h^{-1}	5 *m* s^{-1} 17·7 *km* h^{-1}
Mean stomatal diameter				
10 μ	Assimn.	0·152/2·59=0·06	0·152/1·18 =0·13	0·152/1·07 =0·14
	Transpn.	0·152/1·74=0·09	0·152/0·332=0·46	0·152/0·222=0·68
3 μ	Assimn.	1·22/3·66 =0·33	1·22/2·25 =0·54	1·22/2·14 =0·57
	Transpn.	1·22/2·81 =0·43	1·22/1·40 =0·87	1·22/1·29 =0·95
1 μ	Assimn.	9·76/12·2 =0·80	9·76/10·8 =0·90	9·76/10·7 =0·91
	Transpn.	9·76/11·4 =0·86	9·76/9·94 =0·98	9·76/9·83 =0·99

vapour will come from cells close to the inner mouths of the stomata the intercellular space resistance is also probably small enough to be ignored [228]. With these three components absent, the stomatal contribution will obviously be a larger percentage of the total, especially if the external component is much reduced by wind when the stomatal pores will provide almost the only resistance for transpiration. Thus comparison of the ratios of the total effective lengths from Table 2.2 shows that stomatal closure from 10 μ to 3 μ in 'still' air would reduce assimilation by 29 per cent but transpiration by 38 per cent; with a breeze of 3·5 km h^{-1} the corresponding reductions due to closure would be by 48 and 76 per cent; in a 17·7 km h^{-1} wind they would be 50 and 83 per cent. With further closure to 1 μ the stomata would provide the main resistance for diffusion in either process—even in 'still' air the reductions as compared with the rates for 10 μ would be 79 and 86 per cent while in a 17·7 km h^{-1} wind they would be 90 and 98 per cent. Thus the values in Table 2.2 show that the relative effect of a given stomatal closure always depends on the magnitude of the other resistances—it is greater in wind than in still air, for transpiration than for assimilation and greatest for transpiration in wind. This preferential control of transpiration, although it arises from the unavoidable high internal resistances to assimilation, is an excellent arrangement for the survival of the plant. The table shows that even if the stomata remained wide open (10 μ) the presence of a cuticularized epidermis pierced by pores would greatly reduce the risks of excessive water loss (especially in wind) without very much reducing assimilation. The partial stomatal closure which occurs in response to higher carbon dioxide concentration and to water loss—both brought about by wind—reinforces this protective function and with the reduction of the external effective length to about 0·1 cm in wind the stomatal resistance to transpiration exerts almost complete control.

If we treat the suggested effective length of 0·4 cm for the chemical resistance as representing the extension of the diffusion path to a point where the concentration of carbon dioxide is zero (under high light intensity and favourable temperature), we can calculate rates of assimilation corresponding to the various total effective lengths shown in Table 2.2. Thus for $d = 10\ \mu$ and 'still' air

$$\frac{dq}{dt} = \frac{KPA}{L_{\text{eff}}} = 0{\cdot}160 \times 3 \times 10^{-4} \times 1 \times 3{,}600/2{\cdot}59$$

(see equation 2.3)

$$= 0{\cdot}067 \text{ cm}^3 \text{ cm}^{-2} \text{ h}^{-1}$$

(The 1 has been inserted for A as the rate is required per unit area, and 3,600 to give the rate per hour.) The results of this calculation are given in Table 2.3.

An experimental value to compare with these is 0·061 $cm^3\ cm^{-2}\ h^{-1}$ for a *Pelargonium* leaf exposed to a mean concentration of 250 p.p.m. carbon dioxide in moving air at 27°C in a leaf chamber under fairly high light intensity (19,400 lux of tungsten filament light, or approximately 80 J m^{-2} s^{-1} in the visible region of the spectrum); this would be equivalent to about 0·074 cm^3 cm^{-2} h^{-1} in air of 300 p.p.m. carbon dioxide content.

If rates of assimilation and transpiration are measured simultaneously for the same leaf, it is possible to obtain experimental estimates of the total effective lengths for the two processes, from $dq/dt = KPA/L_{eff}$ in each case. Subtracting one from the other gives an estimate of the extra internal resistances (per cm^2 of leaf surface) for assimilation, that is intercellular spaces plus hydrodiffusion paths plus chemical resistance. Further, an experimental estimate of the external effective length can be obtained from the rate of water loss from a model evaporating

TABLE 2.3

Calculated assimilation rates in cm^3 cm^{-2} h^{-1}, for ordinary air (300 p.p.m. carbon dioxide) at 20°C and high light intensity, corresponding to values of Total L_{eff} given in Table 2.2

Wind speed	*'Still' air*	1 *m* s^{-1} 3·5 *km* h^{-1}	5 *m* s^{-1} 17·7 *km* h^{-1}
Mean stomatal diameter			
10 μ	0·067	0·147	0·161
3 μ	0·048	0·077	0·081
1 μ	0·014	0·016	0·016

surface, of the same shape as the leaf and placed under the same conditions. This value can then be subtracted from the total effective length for transpiration to give an estimate of that for the stomata. This method was employed by Gaastra [110] whose values may be compared with those in Table 2.3. He used leaves of turnip in a leaf chamber and measured the carbon dioxide uptake from a rapid flow of ordinary air (300 p.p.m. carbon dioxide content) with an infra-red gas analyser. Transpiration was measured by an electrical resistance method. He assumed that the intercellular space resistance was small, and similar for both transpiration and assimilation—it was therefore included in his estimates of stomatal resistance. He also assumed zero concentration at the chloroplasts so that the three extra internal resistances mentioned above were all included in the hydrodiffusion resistance or mesophyll resistance. His estimate of the external effective length for a leaf of 69 cm^2 was 0·24 cm, i.e. slightly more than the value in Table 2.1 for a 10 cm diameter surface (78·5 cm^2) in a 3·5 km h^{-1} breeze. His lowest stomatal effective length was 0·39 cm, which for *Pelargonium* in our Table 2.2 would correspond to a mean diameter of 6 μ, and he found considerable variation in the mesophyll resistance, between 0·28 and 1·4 cm for different leaves. (This range would include our value of 0·845 cm for intercellular spaces + hydrodiffusion paths + chemical resistance.) Using these values he estimated maximum rates of diffusion of carbon dioxide, in high light and at 20°C, of about 0·075 cm^3 cm^{-2} h^{-1} with a mesophyll resistance of 1·4 and about 0·180 cm^3 cm^{-2} h^{-1} with one of 0·28. The importance of such variations in these internal resistances is clearly demonstrated, whether they are in fact mainly chemical or in the hydrodiffusion path.

It may be noted that the range of values of external effective length estimated for a field crop, namely between 1·3 cm for 'zero' wind speed and 0·16 cm for a 24 km h^{-1} wind (page 45), is not very different from that for an isolated leaf of 10 cm diameter (Table 2.1). The analysis put forward for the latter's photosynthesis and transpiration should therefore apply approximately to the upper layers of a field crop having similar mesophytic leaves: at the very least it helps in thinking about the problem and in the design of equipment for field measurements.

The type of leaf can in fact have an important bearing on the efficiency of the diffusion processes. If such large variations can exist between the values of the mesophyll resistance for different individual leaves of turnip, even greater differences may be expected between species. A consideration of equation (2.11) and the values inserted in it (page 60) will show what large differences in the total effective length can be caused by variation in the number and size of stomatal pores. The number of pores may be as low as 2 × 10^3 per cm^2 (wheat) or 1·6 × 10^3 (*Tradescantia zebrina*) and may be as high as 6·6 × 10^4 (*Cucumis*). The dimensions of the pores also vary greatly between species: the long axis of the ellipse from 5 to 40 μ and the short axis at maximal opening from 2 to 10 μ. Widths as large as 10 μ are not normally seen, however, unless the leaf has been in nearly carbon dioxide-free air in a bright light—under such circumstances the apertures of wheat stomata, for instance, may measure as much as 40 × 10 μ whereas in ordinary air they seldom exceed 2 or 3 μ in width.

A further complication that affects the diffusion systems of leaves of various species is that in some the stomata are not uniformly scattered over the leaf surface but arranged in rows (wheat), more or less isolated patches (*Begonia*; Fig. 5.2), sunk in pits or grooves (many xerophytes) etc. More elaborate models than that of Fig. 2.11 could be devised on similar lines to suit such leaves.

This chapter has been mainly concerned with diffusion *paths* (resistances) rather than with potentials but both are equally important for rate of diffusion: $dq/dt \propto P/L_{eff}$. We have treated P_1 as zero at some point beyond the chloroplast surface. If diffusion of carbon dioxide from the external air to this point represented the whole system, then however large the resistances en route, as long as they were not infinite, a leaf photosynthesizing in a limited volume of air should in time take up all the carbon dioxide and reduce the external concentration to zero. Experiment shows that this does not normally happen [111, 141, 150] and that a steady-state low concentration of carbon dioxide is brought about and maintained in the external air. Apparently some of the carbon dioxide of respiration from the mitochondria escapes into the intercellular spaces instead of being assimilated. This complicates the question of the potential and hence will have some effect on estimates of resistances where these are based on experiment rather than on measurement and calculation. The phenomenon itself may be supposed due to the resistance in the diffusion paths from the respiring mitochondria to the chloroplasts and its relation to that in the paths from the mitochondria to the intercellular spaces. It will be discussed further in Chapters 5 to 7.

Part 2: The Physiology of Photosynthesis

3: Methods of Investigation

PLANT physiology seeks to measure and explain the responses of living plants or parts of living plants to the physical and chemical factors of the environment. The ultimate object of the study is to explain the behaviour of the plant as an intact and integrated individual —or even that of a population such as a field of crop plants, though here physiology grades into ecology, especially if biological factors of the environment are considered as such rather than in terms of their physical and chemical effects. In order to approach this object it is necessary to study also the responses of organs (especially leaves in the case of photosynthesis), individual cells or even parts of cells such as isolated chloroplasts, and it becomes appropriate to consider also factors of the internal environment to which cells or their components are exposed. As this analysis of the plant into its components proceeds to smaller and smaller units it grades imperceptibly into biochemistry and physical chemistry. The apparent simplification due to studying smaller parts of the complex organism is accompanied by increasing doubt as to whether the phenomena observed occur in the intact plant (or population). On the other hand as larger sub-units or even the whole plant or field crop are studied it becomes necessary to make more and more simplifying assumptions, if complete empiricism is to be avoided, and these again introduce uncertainties. Investigation should therefore be carried out at all levels of organization to provide as many cross-checks as possible. Here I shall limit physiology to the study of *living* plants, organs and cells, leaving (almost entirely) the reactions in isolated chloroplasts and smaller units to biochemistry.

Plant physiology is essentially an experimental science and as such needs methods of changing the factors of the environment by known amounts and measuring the plant responses. In measuring amount of photosynthesis or the rates of the process (kinetic studies) there are six objects for quantitative estimation, as is indicated by the overall equation

$$CO_2 + 2H_2O + \text{light energy} \rightarrow \frac{1}{n}(CH_2O)_n + H_2O + O_2 + \text{chemical energy} \quad (3.1)$$

Five of these have been used as the basis for such estimation, *viz.*: uptake of carbon dioxide, output of oxygen, gain of dry matter or estimation of chemical products, total energy fixed, light energy absorbed (minus heat emitted). The sixth is the *incorporation of water* in the photosynthetic reaction. It would be impossible to make use of this without isotopic markers because of the great abundance of water in plant tissues (often as much as 80–90 per cent) and because some water appears on both sides of the equation. Marking some of the oxygen in water, as the isotope O^{18}, has shown that this is not incorporated in the carbohydrate formed but is evolved as molecular oxygen. On the other hand, half the hydrogen from the water used does appear in the carbohydrate and it might be thought that marking this as tritium in $H^3{}_2O$ would enable newly photosynthesized carbohydrate to be distinguished. However, marking either the oxygen or the hydrogen would be of no avail for quantitative purposes unless the proportion of marked water molecules in the plant were known, and this might well take so long to equilibrate that exchange and other reactions (e.g. hydrolyses) would distribute the markers too widely. Further, there is considerable discrimination against tritium because it has an atomic weight three times that of hydrogen.

In considering methods of measurement, a distinction should be drawn between the *sensitivity* of a method and its *accuracy*. The former is the smallest *change* in the quantity measured that can be detected and this information is usually provided in publications describing new methods. This does not tell us the accuracy with which the quantity itself may be determined as this depends on all the various sources of error in the method. It can only properly be assessed by statistical methods and expressed in the appropriate statistical terms. The most suitable procedure would be to state the confidence limits within which the true mean (or slope of the line, etc.) might be expected to lie with a certain probability (say 20 to 1). Such information is rarely given. It is, however, useful to know the sensitivity, for obviously this sets a limit to the accuracy and it may also for some purposes be useful to be able to detect a change, such as the beginning of photosynthesis when a plant is first illuminated, even if its magnitude cannot be measured very accurately.

A. CORRECTION FOR RESPIRATION RATE

If respiration continues in the light, the five measures listed above will all give the *net* rate or amount of photosynthesis. Thus the carbon dioxide intake observed will be the difference between the amount taken up in photosynthesis and that evolved in respiration, the converse will hold for oxygen, the dry matter increase will be a net increase and so will the increase in chemical energy, the heat emitted will include that due to respiration. High rates of photosynthesis may be ten to thirty times the (dark) respiration rate and under such circumstances any effect of respiration on the measured rate is often deemed unlikely to be important and is ignored. This may for some purposes be the best procedure whatever the rate; for instance, where the results are to be correlated with growth it is obviously the net rate of photosynthesis that is of interest. When the interest centres on the photosynthesis itself we should, ideally, measure production of reducing power which is the primary photosynthetic process. As we shall see in Chapter 5, there is some reason to think that actual photosynthetic *production* of oxygen (not net production) may give a good measure of this, though it does not include possible adenosine triphosphate (ATP) production in cyclic phosphorylation in which no oxygen is produced, if indeed this occurs *in vivo*.

Before the use of isotopic tracers, indirect evidence was brought forward by various workers between 1886 and 1952 to suggest that respiration was inhibited, accelerated or (in most cases) unaffected by light. This last was generally assumed to be true and in most assimilation experiments the net or 'apparent' rate of photosynthesis observed was corrected to the gross, 'true' or 'real' value by the addition of the mean dark respiration rate as measured before and after the light period. This assumption could not be adequately checked experimentally until the availability of isotopes of oxygen and carbon made it possible to measure simultaneously both the production and the consumption by the plant of the same gas—molecular oxygen or carbon dioxide. Using O^{18}, A. H. Brown [42] in 1953 found that the respiratory oxygen uptake of *Chlorella*, certain other algae and several higher plants was virtually unaffected by light of low to moderate intensities. He used a mass spectrometer, which is a device for separating gaseous particles of different mass (page 90), to follow the disappearance of marked

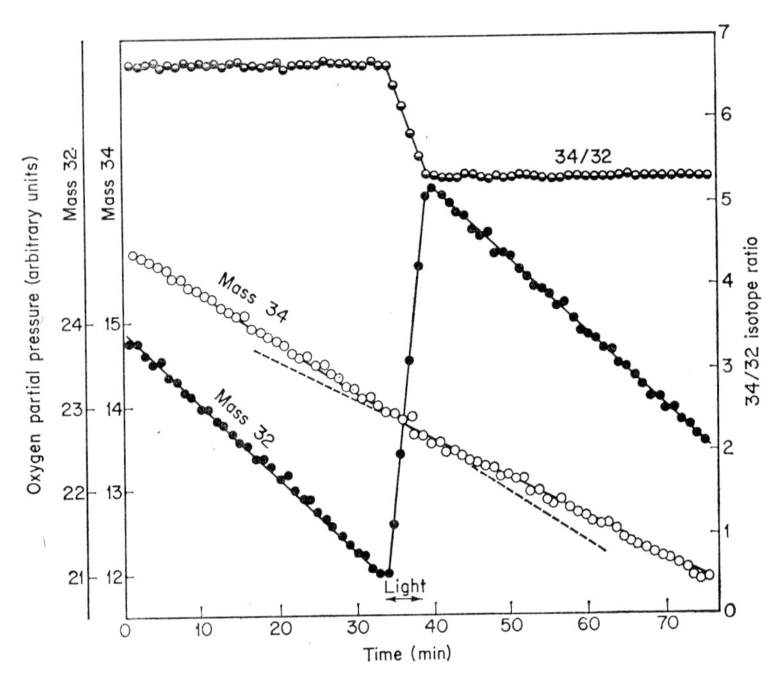

FIG. 3.1 Time course of metabolism of marked oxygen (mass 34) and ordinary oxygen (mass 32) by 22 cm² of tobacco leaf tissue, in 1·9 per cent enriched oxygen in helium, in dark and light. Mass spectrometer data. In darkness the two forms of oxygen were taken up at proportionate rates by respiration, so that the isotope ratio 34/32 was constant; in light the ordinary oxygen was greatly augmented by photosynthetic splitting of the unmarked water, whereas the respiratory uptake of marked oxygen continued at almost the same rate; however, the latter showed a slight change of slope due to the change in the relative partial pressures of the two isotopes in the 5 minute light period (tracer dilution). After Brown [42].

oxygen ($O^{18}O^{16}$, of mass 34) from a gas mixture containing a preponderance of ordinary oxygen ($O^{16}O^{16}$, of mass 32). The marked oxygen was taken up in respiration at the same rate both in darkness and in light, while the ordinary oxygen showed a proportionate depletion by respiratory uptake in darkness but was greatly augmented in light by photosynthetic splitting of the unmarked water (Fig. 3.1).

Brown's 1953 results strongly suggested that the usual correction, using the dark respiration rate, was valid but the fact that many of the chemical intermediates involved in photosynthesis and respiration are common to both processes makes it seem improbable that light should have no effect at all on respiration. We should expect a certain amount of rebuilding to occur, from intermediate compounds formed in the stepwise degradations that constitute respiration, before this process had gone all the way to carbon dioxide and water. To show experimentally that photosynthesis and respiration are truly independent would involve showing that both oxygen uptake *and* carbon dioxide evolution are concurrently unaffected by light. Brown and Weis [44] later (1959) carried out this test for the unicellular green alga, *Ankistrodesmus braunii*, using $O^{18}O^{16}$ and $C^{13}O_2$ with various light intensities. They found that as compared with darkness light caused a reduction of about 50 per cent in respiratory carbon dioxide evolution, an effect which was nearly independent of light intensity. Oxygen consumption was unaffected by light of low intensity but much enhanced at high intensity.

The results obtained by Brown and Weis clearly indicate that investigations of light effects upon respiration in terms of a single gas are likely to lead to very different conclusions according to whether oxygen or carbon dioxide is measured, and the data will be even more difficult to interpret if isotopic markers are not used to separate photosynthetic and respiratory effects upon the concentrations measured. Some of the evidence to date is discussed in more detail in Chapter 5.

In the following sections (B to H) of this chapter most of the methods of measuring photosynthesis, apart from biochemical methods of measuring its products, are mentioned at least in principle. Many of them have been modified in detail and combined in various ways, so that it would be impossible to review here the great proliferation of procedures that has resulted. Others have fallen into disuse but may still be valuable for special purposes.

B. METHODS BASED ON UPTAKE OF CARBON DIOXIDE

Measurement of carbon dioxide has the disadvantage that this gas is the substrate for photosynthesis; changes in its concentration are therefore much more critical than changes of oxygen in affecting photosynthetic rate, but changes there must be if the uptake is to be measured. This problem can be acute in *open circuit experiments* with leaves (Fig. 3.2*a*). In these, the concentration in an air stream of measured volume rate of flow is estimated by an appropriate method both before and after passage over the illuminated leaf surface in a suitable vessel (leaf chamber). The flow rate multiplied by the concentration change gives the rate of uptake. If the change is small the mean concentration over the leaf can be specified precisely but the errors in estimating the change will tend to be relatively large; if the change is large it is easy to estimate accurately but the mean concentration over the leaf surface is in doubt. An exceedingly sensitive and accurate method of measuring concentration combined with a high rate of flow, to give a small concentration change, are therefore desirable. It is usual to assume that the carbon dioxide diffusing into the leaf at any point is a constant proportion of the concentration over that point, i.e. that the concentration falls logarithmically and the mean ($\bar{C}$) is given by

$$\bar{C} = \frac{C_i - C_f}{\ln (C_i/C_f)} \tag{3.2}$$

where C_i and C_f are the concentrations before and after passing over the leaf. This assumption begs the question of the relation between apparent assimilation and carbon dioxide concentration by assuming direct proportionality. The matter is further complicated by such facts as that the stomatal aperture is likely to increase along the leaf in the direction of flow owing to the falling carbon dioxide concentration, and that in spite of this leaf temperature is likely to increase with the increasing humidity and consequent reduction in transpiration. These last-mentioned complications, at least, are absent if the experiment is carried out with carbon dioxide solution flowing over the leaf of a water plant, though the doubt as to the mean concentration persists. If air is bubbled through a suspension of unicellular algae, even assuming efficient equilibration between air and water, the same doubt will exist.

Instead of measuring concentration, in open circuit experiments it is possible to use an integrating method to collect during some considerable period the carbon dioxide remaining in the air stream after its passage over the leaf and measure its total quantity; this may be compared with the quantity similarly collected and measured for a parallel air stream that has not passed over the leaf, or with an amount calculated from the known initial concentration (for example, in a

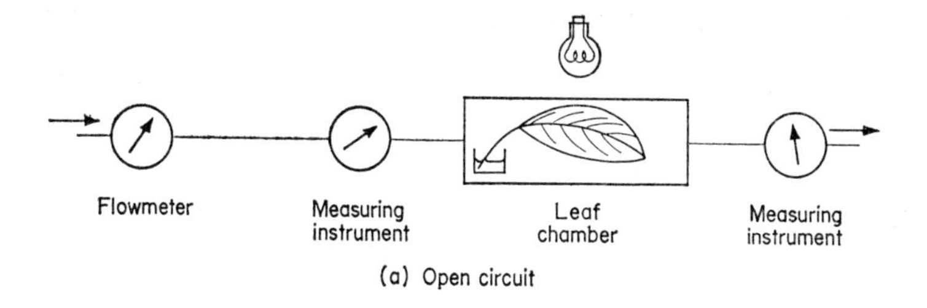

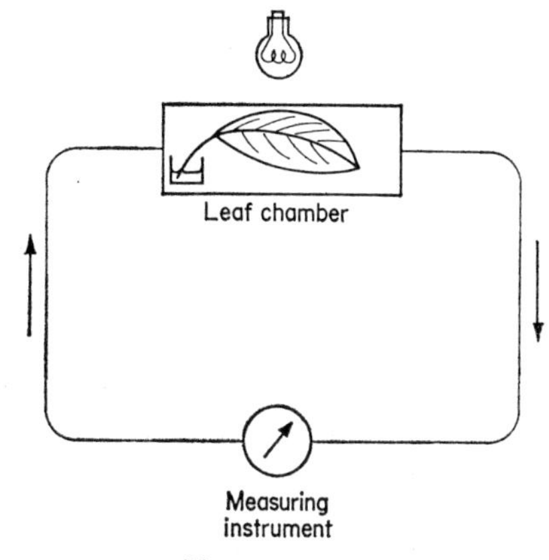

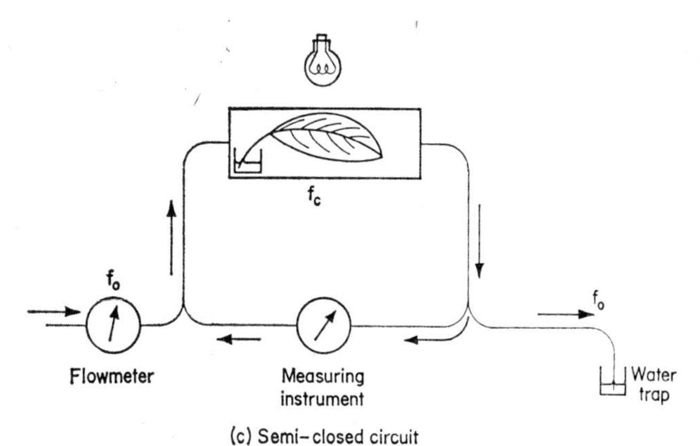

FIG. 3.2 Diagrams of apparatus for measuring carbon dioxide uptake or output by leaves. *a*, open circuit; *b*, closed circuit; *c*, semi-closed circuit. For further explanation see text.

cylinder of compressed air), the measured volume rate of flow and the time.

With land leaves in *closed circuit experiments* (Fig. 3.2*b*) there are even more complex sources of error. In these the same air is circulated continuously through a device for measuring concentration and the leaf chamber in series. Hence photosynthesis takes place with a continuously falling carbon dioxide concentration; its rate at any time has to be estimated from the known volume of the system and the slope of the curve for concentration *v.* time. This slope is not only a function of the rates of carbon dioxide uptake and output by the chloroplasts and mitochondria respectively and of the external concentration at the instant concerned, but also of the preceding changes in that concentration with time and of the various internal resistances to carbon dioxide movement. If a highly accurate method of concentration measurement is available, the volume of the closed circuit may be made large enough relative to the leaf area used for the fall in concentration to be very slow (and yet measurable). Otherwise, apparent photosynthesis should be measured in open circuit, or better still in semi-closed circuit apparatus (see below), in both of which a steady state during photosynthesis can be maintained by constant renewal of the carbon dioxide supply. It may be noted here that much the same considerations apply, in both open and closed circuit experiments, when oxygen concentration change is measured, for the output of oxygen is usually the equivalent of the carbon dioxide taken up. However, it is then possible, at least in principle and sometimes in practice, to 'buffer' the carbon dioxide concentration in a closed circuit to a nearly constant value without affecting the changes in oxygen concentration due to photosynthesis (see below).

In *semi-closed circuit experiments* (Fig. 3·2*c*) [248] the air is circulated as in the closed circuit type, with a high volume rate of flow (f_c) which need only be known very approximately; a carefully measured slow flow (f_0), say about one hundredth of f_c, containing a high known concentration of carbon dioxide (C_0), is introduced into the system beyond the measuring instrument but far enough from the leaf chamber for mixing to be complete before this is reached; an equal volume flow (f_0) of circuit air is allowed to escape through a water trap between the leaf chamber and the measuring instrument—this will always have the same concentration (C_d) as is currently shown by the instrument. When a steady rate of photosynthesis has been achieved the following relations will hold:

Suppose the air entering the leaf chamber has concentration C_c. Then assimilation (A) in unit time will be given by

$$A = f_c(C_c - C_d) \qquad (3.3)$$

but as f_0, C_0 and C_d are constant during steady-state photosynthesis, the carbon dioxide input (f_0C_0) must equal the sum of the assimilation and the output (f_0C_d)—otherwise the reading (C_d) of the instrument would change, that is $f_0C_0 = A + f_0C_d$.

$$A = f_0(C_0 - C_d) \qquad (3.4)$$

As the difference in concentration ($C_0 - C_d$) is large it can be accurately assessed; it is also easy to measure with precision the slow flow rate f_0. Hence a precise estimate of assimilation is obtained, without knowing the rate of circulation f_c. Owing to this last being very rapid, the concentration change over the leaf surface ($C_c - C_d$) is small and the choice of method for calculating the mean concentration is not important:

$$\frac{(C_0 - C_d)}{(C_c - C_d)} = \frac{f_c}{f_0} \quad \text{(from 3.3 and 3.4)} \qquad (3.5)$$

If we suppose for example that f_c is 100 l h^{-1}, f_0 is 1 l h^{-1}, C_0 is 1,000 p.p.m. (0·1 per cent) by volume and C_d is 200 p.p.m. (0·02 per cent), then

$$C_0 - C_d = 800 \text{ p.p.m.} \quad \text{and} \quad \frac{f_c}{f_0} = 100$$

Therefore

$$(C_c - C_d) = 8 \text{ p.p.m.}$$

The arithmetic mean of 208 and 200 is 204; the value of $\bar{C}$ from equation (3.2) is 203·98. If f_c had been overestimated by a factor as high as 2 × the true value of the mean concentration would be 208 p.p.m. and the estimate 204 would be only 2 per cent too low. Thus f_c need only be known approximately.

The mean carbon dioxide concentration over the leaf can be changed by altering f_0 and (or) C_0, but not changed by predetermined amounts (except by trial and error) as its value when the steady state has been reached markedly depends upon the rate of photosynthesis. The method is therefore inconvenient for factorial experiments (page 117). However, a modification of the semi-closed circuit method [218] consists in passing an extremely slow measured flow (f_0) of pure carbon dioxide into the circuit. With steady-state photosynthesis this is completely assimilated and f_0 is the rate of assimilation. If this is adjusted to predetermined values factorial experiments can be carried out with assimilation rates as the independent variate and the measured steady-state concentrations as the dependent variate [248]. The difficulty of obtaining predetermined carbon dioxide concentrations also applies, of course, in open circuit experiments with the added disadvantage that the effective concentrations used are then unknown.

In all three types of assimilation experiment just considered the leaf or other plant material used is enclosed in a vessel, but a recent development has been the *estimation in the field of carbon dioxide flux* into a crop, both downward from the atmosphere and upward from the soil [233, 231]. Samples of air taken at various heights above and within a field crop throughout the day are analysed for carbon dioxide content with an infra-red gas analyser (page 80) to yield estimates of the potential differences between the points concerned. Corresponding 'resistances' are estimated from concomitant measurements of wind speed and humidity. Flux is calculated as (potential difference)/resistance.

So far this method appears only to have been used in observational studies to obtain carbon dioxide uptake data for correlation with meteorological factors and dry matter accumulation, but there is no reason why it should not be applied to experimental investigations in which the field crop concerned is subjected to a number of different fertilizer, irrigation or other treatments.

i) *Chemical methods used as integrating methods*

Absorption in saturated baryta of the carbon dioxide from two parallel air streams, which had and had not passed over the plant in an open circuit experiment, and titration of the alkali remaining was a

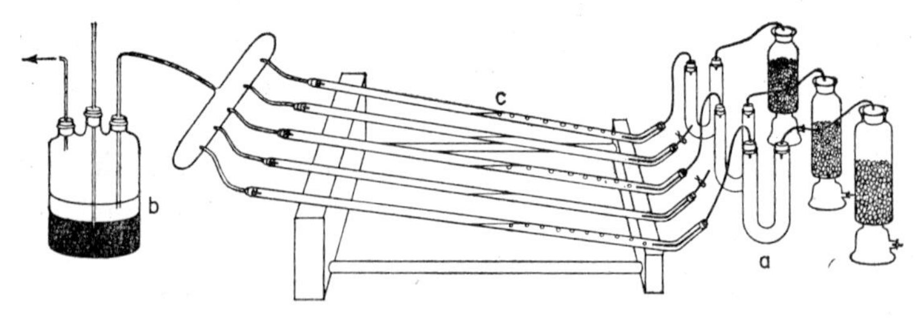

FIG. 3.3 Pettenkoffer's respiration apparatus. Carbon dioxide-free air is drawn through the plant chambers (*a*) and then bubbles through titrated baryta water in Pettenkoffer tubes *c*, about 1·5 cm diameter and 1 m long. These can be changed at suitable intervals and the solution removed for titration. *b* is a pressure regulator.

method extensively used, from the early researches of Blackman in 1895 [17, 18] to those of Maskell [213] in 1928. The classical apparatus for absorption used by most of these investigators was the Pettenkoffer tube (Fig. 3.3) but this is only efficient for rates of air flow of less than about 2 l h^{-1}. By using a tower filled with glass beads (Fig. 3.4) to break up the air stream as it passed through the absorbent (N/1 caustic soda solution), Porter *et al.* [260] found they could obtain complete absorption of the carbon dioxide from a 50 l h^{-1} stream of ordinary air. The proportion of the absorbent run out of the apparatus for titration was determined by weight, which avoided the necessity for washing out, and the carbonate was precipitated with barium chloride before the excess of soda was titrated with standard acid.

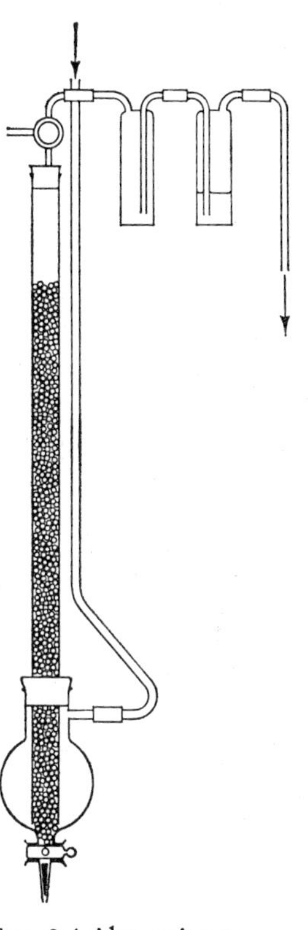

FIG. 3.4 Absorption tower, filled with glass beads to break up air stream passing through 1 N caustic soda solution. After Porter *et al.* [260].

ii) *Physico-chemical methods (integrating)*

Instead of titrating the alkali in which the carbon dioxide has been absorbed, its change in electrical conductivity may be determined—this is directly proportional to the carbon dioxide absorbed over a considerable range. The conductivity is obtained by measuring the resistance between two electrodes, immersed in the alkali solution, with an alternating current Wheatstone bridge. Newton [242] devised a cell in which the alkali was made to circulate continuously past the electrodes by the bubbles of the air stream rising in a glass spiral; the method could be made sensitive to 0·5 μl (1 μg) of carbon dioxide (the quantity in less than 2 ml of ordinary air), in an air flow of 2 l h^{-1}, with a very small standard deviation [138]. A self-recording alternating current bridge and conductivity cells of another design which were refilled automatically every 16 minutes were used for photosynthesis measurements on field crops [293].

iii) *Physical and physico-chemical methods (concentration measurement)*

(*a*) *pH change*. In unbuffered solution the carbon dioxide concentration may be estimated from the pH. This provides the basis for a method of estimating photosynthesis of water plants, or of land leaves if air is circulated over the leaf and bubbled through water in a closed circuit. The method has been used both with colorimetric estimation of pH change [31] and with a glass electrode. Blinks and Skow [29] used a glass electrode in direct contact with a photosynthesizing water-lily leaf (*Castalia*) or with algal cells settled from a suspension. They were able to record rapid changes of pH, for instance, with light flashes down to 0·02 s for the marine alga *Stephanoptera*. The smallest perceptible response was about 0·001 pH unit.

(*b*) *Electrical conductivity*. This also provides a measure of the concentration of carbon dioxide dissolved in pure water. The water may be made to circulate, by the bubbles of air, past a pair of electrodes and through a mixed bed deionizing resin column. The ions from the carbonic acid, formed by the dissolved carbon dioxide, increase the conductivity between the electrodes; they are removed as the water circulates through the resin column.

(*c*) *Katharometers* (*diaferometers*). The rate of cooling of a hot wire depends upon the heat conductivity of the surrounding gas mixture. In this instrument changes in the rate of cooling of an electrically heated wire, corresponding to changes in the composition of a gas mixture, are measured in terms of its electrical resistance. Changes in carbon dioxide concentration affect the heat conductivity of the air much more than the equivalent changes in oxygen. Hence the method has been developed for measuring carbon dioxide in photosynthesis [284]. It is naturally extremely sensitive to ambient temperature. In the more recent differential instruments [281] this source of error is minimized by having four resistance wires arranged as a Wheatstone bridge; these are in separate chambers in a single large block of copper, to maintain an even temperature. The galvanometer responds to changes in resistance of the two wires in the unknown gas mixture relative to those in the control gas. The response varies as the cube of the current which must therefore be accurately controlled. The method may be made sensitive to less than 1 p.p.m. by volume of carbon dioxide [302].

(*d*) *Infra-red gas analysers*. Several instruments have been developed to measure carbon dioxide by means of its infra-red absorption. In the earliest of these [205] air that had passed over assimilating wheat seedlings was circulated through the tube of a recording infra-red spectrophotometer, set so as to isolate the wave length band 4·2 to 4·3 μ where carbon dioxide absorbs strongly. Another type of instrument [61, 272] used the much simpler if less familiar principle of measuring energy of *all* the wave lengths radiated from a 'selective' source, namely a Meker gas burner which had a high output between 1 and 5 μ agreeing fairly well with the absorption spectrum of carbon dioxide in that region. The radiation from the burner passed through two tubes (one for the 'control' gas mixture and one for the 'unknown') and fell on paired thermopiles connected to a meter in opposition. Thus only differences in absorption were recorded. Water vapour was removed from the gas streams before they passed into the tubes because of its strong infra-red absorption.

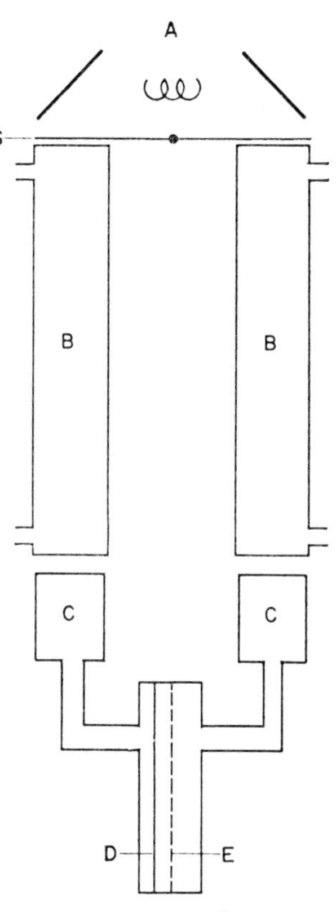

FIG. 3.5 Diagram of an infra-red gas analyser. For explanation see text. After Martin *et al.* [211].

Some more recent instruments [211] have a differential selective detector [Fig. 3.5]. Infra-red radiation from an electrically heated coil A passes in balanced amounts into two tubes B, for the control and unknown gas mixtures respectively. Thence it passes into the two compartments of the detector C, which are separated by a thin diaphragm D and filled with the gas to be measured: in this case pure carbon dioxide. This becomes heated by absorption of the infra-red radiation: in equal amounts if both tubes contain the same gas mixture. If, however, one tube contains more carbon dioxide than the other, less infra-red radiation will pass through and penetrate into the detector on that side; therefore the gas will remain cooler and the diaphragm will bulge into it. This alters the distance between two condenser plates, one attached to the diaphragm D and the other fixed (E), and so alters the electrical capacity. An alternating signal is obtained by fitting a rotating sector shutter S between the tubes and the source A; the capacity therefore fluctuates with a frequency determined by the shutter and an amplitude determined by the difference of carbon dioxide concentration between the two gas streams.

In such an instrument (as also with a differential katharometer), air can be passed through the control tube, over the leaf and then through the unknown tube. The difference in carbon dioxide concentration due to photosynthesis or respiration can then be read directly from the meter, after suitable calibration. Alternatively, the control tube may be filled with carbon dioxide-free air, or a known mixture, and the actual concentration in the unknown be determined. By high amplification great sensitivity may be attained, so that the meter may be read to less than 1 p.p.m. of carbon dioxide, but this involves some loss of stability and such readings are only reasonably accurate if averaged over a period. Owing to the overlap in the absorption spectra of carbon dioxide and water vapour the latter must be removed from the gas, or equalized in the two tubes; alternatively, filter tubes full of water vapour, or interference filters, can be used on both sides to cut out infra-red in the water vapour absorption bands.

C. METHODS BASED ON OXYGEN OUTPUT

i) *Biological methods*

In an elegant semi-quantitative method used by Englemann [86] a filamentous alga is mounted under a cover-slip in a suspension of motile aerobic bacteria such as *Proteus vulgaris*. In darkness the oxygen in solution is soon exhausted and the bacteria become motionless. If the filament is then illuminated the oxygen produced by the alga photosynthesizing causes the bacteria to move about. The distance from the cell at which this occurs gives some measure of oxygen production. Englemann found in this way that if the filament was illuminated with a spectrum, most oxygen was produced in the red light with a second but smaller maximum in blue.

Beijerinck [13] devised a rather similar method using luminescent bacteria such as *Micrococcus phosphorens*. The preparation is placed in darkness to remove oxygen by respiration, illuminated and then observed in renewed darkness. It has been stated that the bioluminescence becomes visible with only 1×10^{-8} mol l^{-1} of oxygen in solution.

ii) *Chemical methods (used for concentration measurement)*

These have the disadvantage that the medium (generally water) must be sampled—they are not therefore very convenient for following short period changes in rates of photosynthesis.

(*a*) *Winkler's method* for dissolved oxygen is described in standard chemistry texts [295]; it was made sensitive to 0·5 μl of oxygen by James [180], who used it for measuring photosynthesis of the water moss *Fontinalis* in solutions of carbon dioxide and of sodium bicarbonate. These solutions were sampled for analysis before and after passing over the plants in a continuous flow open circuit apparatus. For the bicarbonate solutions, rate of flow (including absence of flow) was found to be virtually without effect on rate of oxygen output, indicating that the carbon dioxide concentration at the plant surface was effectively 'buffered' to constant values by the bicarbonate. The concentrations used (0·02–0·60 per cent) were calculated to give 0·06 to 1·92 per cent free carbon dioxide and the pH of about 8·3 appeared to have no harmful effects.

(*b*) *Linossier's phenosafranin method* for dissolved oxygen, as modified by Miller [227], is convenient and very rapid; it may be used in the field and a test can be completed in about 2 minutes.

iii) *Physical and physico-chemical methods*

Many of these methods lend themselves to the collection of continuous records and are therefore particularly suitable for kinetic experiments.

(*a*) *The bubble-counting method* for water plants, suggested by Sachs in 1864 and since much used and modified, is described in Chapter 2 (page 42). Although now completely outdated as a research tool, it is nevertheless of interest as illustrating the many sources of error that reside in apparently simple experiments—it should not be too readily assumed that more modern methods are freer from these simply because the equipment is more expensive. The principal improvements in the method were made by Wilmott [330] in 1921, who used a fine glass nozzle over the cut end of the shoot, which resulted in bubbles of much more uniform size, and by Audus [8, 9] in 1940, who collected the gas in a capillary tube and measured the volumes produced in noted times instead of counting bubbles. With Audus's modification, excellent experiments on the interactions (see Chapter 4) of light intensity, carbon dioxide concentration and temperature may be carried out and curves resembling Figs. 4.4 and 4.5 obtained; even better results should be obtained with Linossier's method in which there would appear to be fewer systematic errors.

The bubbling method is subject to two main sources of error: one is that some of the oxygen goes into solution instead of appearing as bubbles; the other is that nitrogen diffuses from the water, where it is in equilibrium with the 80 per cent nitrogen content of the air, into the intercellular spaces which are being continually supplied with pure oxygen. The second error is illustrated in data derived from analysis of the gas bubbles escaping from an illuminated *Elodea* shoot (Kniep [189]):

Mean number of bubbles per second	O_2%	CO_2%	N_2%	O_2 *per second*	N_2 *per second*
3·9	54	0·6	46	2·11	1·78
2·9	49	0·9	50	1·42	1·45
1·3	36	1·0	63	0·47	0·82

When the amount of oxygen produced decreased by a factor of 4·5 × (5th column), the rate of bubbling decreased only threefold (1st column) owing to the higher percentage of nitrogen in the bubbles.

Of particular interest with this method are the effects of carbon dioxide tension, with and without stirring of the water. When the illuminated shoot has been photosynthesizing in still water long enough for a steady rate of bubbling to be established there are three diffusion processes going on, each with a constant gradient, namely diffusion of carbon dioxide and of nitrogen towards the plant and of oxygen away from it. Such gradients take time to establish; if they are levelled out by a brief stirring the subsequent changes in rate of bubbling will depend on the carbon dioxide content of the water. If this is low (for example 300 p.p.m. by volume) the rate of bubbling will *fall* in time as the carbon dioxide content of the water near to the plant becomes depleted. This effect will, however, be opposed by the oxygen accumulating in solution, so that less dissolves and more emerges as bubbles; if the carbon dioxide concentration is high enough for this gas to be always present in excess (for the light and temperature prevailing) the oxygen effect will result in a *rising* rate of bubbling. By choosing a suitable carbon dioxide concentration, the carbon dioxide and oxygen effects may be made to cancel out so that the rate is steady from the beginning. (All three types of result were obtained by early users of the method, with consequent controversy.) Superimposed on these two effects is a smaller nitrogen effect, for as the water round the plant is depleted of nitrogen, the rate of bubbling will tend to fall. Renewed stirring will cause the converse changes, namely, an increase in rate of bubbling with low carbon dioxide concentration, or with high concentration a fall in rate owing to the removal of accumulations of dissolved oxygen. We should note that the carbon dioxide effects are real effects on rate of photosynthesis but the oxygen and nitrogen effects are artifacts which affect the rate of bubbling only. These errors may be reduced by using water shaken with pure oxygen for making up the carbon dioxide solution (Wilmott [330]).

(*b*) *Pressure change methods. The Warburg apparatus* is the best known and the most widely used—originally developed for measuring respiration, it was adapted by Warburg [306] in 1919 to the measurement of photosynthesis, for which it is in several ways unsuitable. A typical Warburg unit, as used for respiration, is shown in Fig. 3.6*a*. Six or twelve of these are fitted with their flasks immersed in a water bath, held constant to $\pm \frac{1}{50}$ °C or less, and are rocked back and forth by a mechanical device (except sometimes during readings) to hasten equilibration between the liquid and gas phases. As the plant material respires, more oxygen goes into solution while carbon dioxide emerges into the gas phase and is absorbed in the alkali in the centre well. The pressure therefore falls and the manometer is read periodically after adjustment to constant volume (Fig. 3.6*b*) from the reservoir at the base. Pressure changes are converted into volume changes by means of a flask constant for the particular vessel and gas concerned (see reference [300] for calculations) and hence the volume of oxygen taken up ($-\Delta O_2$) in respiration is estimated. The method is very sensitive and can be used for measuring changes over relatively short periods if the shaking is efficient.

When the Warburg apparatus is used for photosynthesis the carbon dioxide is required for the process and must not therefore be removed by an absorbent. On the other hand, absorbing the oxygen is undesirable as anaerobic conditions tend to inhibit photosynthesis and further, in the absence of oxygen dark respiration could not be measured for the estimation of 'true' photosynthesis. It is therefore necessary to keep both gases present and with a photosynthetic quotient

$$(Q_p = \Delta O_2 / -\Delta CO_2)$$

of unity this would not normally be expected to result in pressure changes from which oxygen uptake could be estimated. Further, as the vessel constitutes a closed system, photosynthesis would normally be taking place with a falling carbon dioxide concentration. Three main methods have been used in attempts to circumvent these disadvantages: (1) Owing to the much smaller solubility of oxygen than of carbon dioxide in water, with a Q_p of 1·0 there will in fact be some increase of

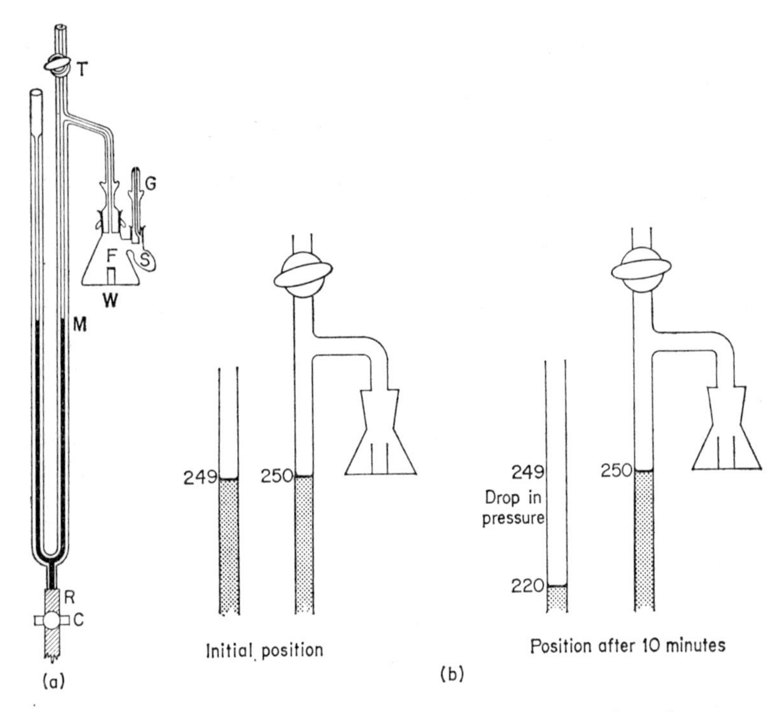

FIG. 3.6 Warburg constant volume respirometer. *a*, A typical Warburg unit. F, flask; S, side arm; G, side arm stopper with gas vent; W, centre well; M, manometer; R fluid reservoir; C, screw clamp to alter fluid level in manometer; T, three-way tap. *b*, Method of measuring pressure change at constant volume. After Umbreit *et al.* [300].

gas pressure with photosynthesis by the plant material in the liquid phase. This effect can be increased by having a large volume of liquid relative to the volume of gas. Even so, it is necessary to start with unphysiologically high concentrations of carbon dioxide (1–5 per cent), buffered to an acid pH of about 5, in order to avoid too rapid depletion of the supply. It is also necessary to know, or assume, the value of Q_p, which has usually been taken to be 1·0, but this is not invariably correct. (2) If *two* flasks are used, with different ratios of liquid to gas, a pair of simultaneous equations enables both oxygen output and Q_p to be estimated. Elaborate precautions are needed to ensure that light absorption and other conditions are as similar as possible in the two vessels, for instance the volume of cell suspension should be the same with different gas volumes (Fig. 3.7). Even so, when used for short light periods the method is very sensitive to the time lags in the responses of the manometers of the two vessels and to other sources of

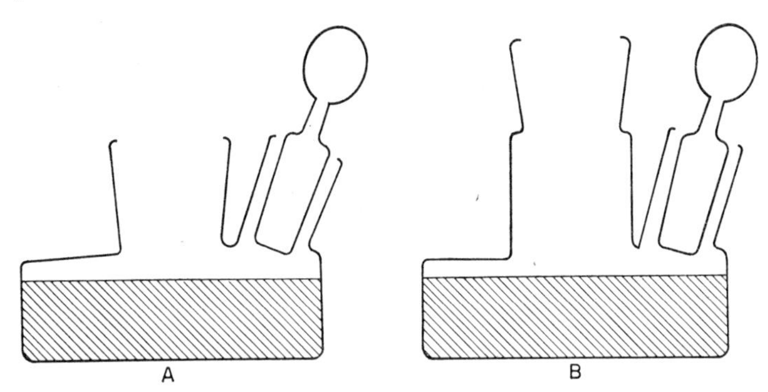

FIG. 3.7 Pair of flasks for the two vessel method of estimating oxygen and carbon dioxide exchange with Warburg apparatus. Equal volumes of cell suspension are used with different gas volumes; light absorption and mixing characteristics are similar for the two flasks. After Nishimura *et al.* [244].

error [244]. (3) If a 'carbon dioxide buffer mixture' is used to maintain a (nearly) constant carbon dioxide concentration, pressure changes can be assumed due to oxygen alone and only a single vessel is needed. The buffer generally used is a mixture of carbonate and bicarbonate giving a very alkaline pH of about 9; this damages some algae though *Chlorella pyrenoidosa* seems to tolerate it for many hours. An alternative method employed by Burk *et al.* [50] uses Pardee solution in a centre well or side receptacle of the flask to maintain 'constant' carbon dioxide concentration; this is a carbon dioxide buffer, based on diethanolamine, which can yield or absorb very large quantities of carbon dioxide, but the question of rate of equilibration is important [199].

Such manometric methods have the great advantage that change of pressure is measured independently of the total pressure—the sensitivity is therefore constant and there is no need to work at low oxygen pressures (see below). On the other hand there is a risk that the pressure changes may be due to some unforeseen reaction, for the methods yield no direct information as to the gases concerned. Their use in nearly all the attempted determinations of the maximum quantum yield of photosynthesis (page 218) contributed greatly to the duration of the controversy, for with so many possible sources of error to be invoked almost any result obtained by the opposing group could be questioned.

(*c*) *Haemaglobin method.* R. Hill [159] devised a biochemical method of measuring the oxygen produced when a suspension of isolated chloroplasts split water in the now well-known 'Hill reaction' which laid the foundation of so much modern work on the biochemistry of the light reactions. The chloroplast suspension was mixed with haemaglobin in an evacuated (Thunberg) tube and a spectrometer was used to estimate the formation of oxy-haemaglobin. Appreciable quantities of oxygen were only found if a hydrogen acceptor was included to take up the hydrogen released from the water. This haemaglobin method was later adapted to the measurement of photosynthesis by a suspension of *Chlorella* cells in the presence of carbon dioxide [162, 321]. The method was very sensitive to oxygen changes between the partial pressures of 2 and 30 mm Hg.

(*d*) *Hersch cell.* When manometric methods, such as those described above, were used for the study of effects of intermittent light (Chapter 9) it was only possible to measure the net oxygen production from a great number of light flashes and dark periods. With the Hersch galvanic cell [155] the increase in oxygen pressure resulting from a single 'long' flash of light (lasting 35 ms!) may be measured and hence the effects of different combinations of flash lengths and durations of dark periods can be explored. Thus Whittingham and Brown [324] found in 1958 that a short flash (less than 5 ms), which by itself gave no measurable oxygen production, could double the yield of a subsequent long flash. The method could detect a concentration of oxygen of 0·1 p.p.m. by volume (7×10^{-5} mm Hg partial pressure) in an air stream bubbled through the suspension of algae. The useful range of oxygen covered by the method is between the very low partial pressures of 7×10^{-4} and $3{\cdot}5 \times 10^{-1}$ mm Hg.

(*e*) *Polarographs and the platinum electrode.* The method of polarographic analysis can be used for measuring the concentrations of dissolved oxygen in aqueous solutions between 5×10^{-5} mol l^{-1} and saturation (that is $2{\cdot}4 \times 10^{-4}$ mol l^{-1} at 25°C). A polarograph consists of an electrolysis vessel with a dropping mercury cathode in which the surface is kept clean by constant renewal. This cathode can be polarized with a steadily increasing voltage and attracts positive ions, in this case hydrogen ions, which block it and prevent any appreciable increase of current until a critical voltage is reached; they are then discharged and reduce oxygen to form hydrogen peroxide, or at a higher potential, water. The current then rises almost vertically until the discharge of the hydrogen ions is limited by the rate of diffusion of oxygen to the cathode, when the latter becomes once more blocked. Little further increase of current with voltage can then occur. The voltage at which oxygen is first reduced is a characteristic for this substance and is called its decomposition voltage. The increase in current is almost a linear function of the oxygen concentration and an absolute calibration can be obtained by bubbling known mixtures of oxygen and nitrogen through the solution. In practice only two voltages need be applied—0·3 volts, rather below the decomposition voltage for oxygen, and 0·7 volts in the region of the limiting current.

Petering, Duggar and Daniels [256] and also Blinks and Skow [30] used a polarograph for measuring photosynthesis; the last-mentioned authors later replaced the mercury cathode with a bright platinum electrode. The tissue was placed directly in contact with this so that the diffusion distance and hence the response time were greatly reduced. This made it possible to follow short period changes in photosynthetic rate during the first few seconds of illumination (induction effects), as well as responses to single flashes of light. Haxo and Blinks [137] used a modification in which flat marine algae, one or two layers of cells thick, were held against the platinum surface by a band of water-permeable cellophane and a large volume of sea water could be passed across to give constant external oxygen and carbon dioxide concentrations. The steady-state current in darkness was taken as a base line, the oxygen diffusing across the tissue being partly used up in dark respiration. On illumination the oxygen produced diffused mostly towards the platinum electrode, where the concentration was maintained very low by combination with hydrogen ions. The increase of current over the dark value was taken as a (relative) measure of photosynthetic oxygen production and was found to be linear with light intensity.

Other modifications, for use with unicellular algae and isolated chloroplasts respectively, were described by Myers and Graham [241] and Fork [90]. Jones and Myers [183] noted apparent toxic effects of the platinum electrode with two blue-green algae (*Anacystis* and

Anabaena) and thought these might be due to hydrogen peroxide; they could be avoided by interposing a 0·1 mm thick layer of agar between the electrode and the algae. These effects were not shown by *Chlorella*.

(*f*) *Some other physical and physico-chemical methods* that have been used for measuring oxygen production in photosynthesis can only be mentioned briefly. In a method devised by F. F. Blackman [36] the leaf was exposed to light in an atmosphere of hydrogen and carbon dioxide; union of the oxygen produced with hydrogen to form water was catalysed by palladium black and the reduction in gas volume (three times that of the oxygen) measured with a gas burette.

A *phosphorescence quenching method* [1] gave sensitivity comparable to the Hersch cell, but required almost completely anaerobic conditions (10^{-6} to 10^{-4} mm Hg partial pressure of oxygen). The *mass spectrometer* method has already been mentioned (page 72) and is discussed further in section *D* below.

D. SIMULTANEOUS MEASUREMENT OF CHANGES IN CARBON DIOXIDE AND OXYGEN

Physical methods for concentration measurement

(*a*) *Katharometer* (*diaferometer*). The heat conductivity of pure oxygen differs from that of ordinary air by less than 5 per cent, the corresponding figure for pure carbon dioxide being 39 per cent [200]. The katharometer is not therefore very suitable for measuring small changes of oxygen concentration in air even if the carbon dioxide is removed first [302], although it has been so used [281]. However, by substituting helium for nitrogen in the gas mixture an increase in sensitivity to oxygen change by a factor of about 100× may be made and the sensitivity to carbon dioxide change increased by about 10× [303]. An arrangement of two Wheatstone bridges is used [281], one to determine net changes due to both oxygen and carbon dioxide, the other to determine changes in oxygen only in a portion of the air stream freed of carbon dioxide; hence the changes in carbon dioxide only can be calculated.

(*b*) *Mass spectrometer*. In this instrument [43] positively charged ions of the gases in the mixture are created by bombardment with electrons, accelerated in an electric field and focused into a beam which is then separated into its various mass components by deflection in a magnetic field. For a given magnetic field strength the deflection depends on the velocity, mass and charge of the ion; lighter particles travel faster and are deflected more. By regulation of the accelerating voltage the ions of any desired component may be brought to a focus on a collector; the ion current is proportional to the rate of formation of those particular ions and hence to the partial pressure of that gas.

Apart from the cost and elaborateness of the apparatus the mass spectrometer appears to provide in many ways the ideal method for measuring photosynthesis and respiration, in terms of both oxygen and carbon dioxide simultaneously, with or without isotopic markers. It may be used, by a scanning method, for the measurement and automatic recording at short intervals (fractions of a minute) of the partial pressures of a predetermined series of mass constituents of the gas mixture in the experimental system. Alternatively it can give a continuous record for a single component when very rapid changes are of interest. The gas required for analysis can be allowed to leak into the mass spectrometer through an exceedingly fine capillary and can be negligible in amount.

The main disadvantage of the ordinary mass spectrometer method for kinetic studies on *aquatic* plants is that it requires samples of gas. If these are taken from the gas phase over the water containing the plants, the diffusion barrier at the liquid–air interface [124] increases the response time and makes it difficult to study rapid changes in concentration. Moreover, owing to the different solubilities in water of oxygen and carbon dioxide, these are affected differently and Brown and Weis [44] found it necessary to make elaborate corrections. Hoch and Kok [166] have described a method in which the vessel containing the plants is completely filled with liquid and gas passes into the mass spectrometer, which is at very low pressure (about 10^{-5} mm Hg), through a semi-permeable membrane supported by a ceramic disc.

The sensitivity and accuracy of the determination of a change in partial pressure by the mass spectrometer method vary with the partial pressure itself (unlike the Warburg method). However, reasonable accuracy for rates of oxygen change may be attained even with as much as 10 per cent of oxygen in the mixture [43]. There is therefore no need, as there is with so many methods for oxygen, to approach anaerobic conditions, which tend to inhibit photosynthesis and which leaves of some species cannot long survive; for carbon dioxide, low levels such as prevail under natural conditions are actually an advantage.

E. ISOTOPIC MARKERS FOR CARBON DIOXIDE AND OXYGEN

The great advantages of either stable isotopes (C^{13} and O^{18}) or radioactive ones (C^{11} and C^{14}) are: first, that they can be detected and measured with high sensitivity and accuracy by the mass spectrometer or radiation counters respectively, and secondly, that they enable simultaneous output and uptake of the same gas to be measured.

The use of radioactive carbon in $C^{14}O_2$ has made possible the remarkable elucidation of the biochemistry of photosynthesis that has taken place since 1940. This has involved the assumption that C^{14} and $C^{14}O_2$ behave chemically in the same way as the normal C^{12} and $C^{12}O_2$. Although satisfactory for qualitative chemistry, in a quantitative sense this assumption cannot be strictly true—the disrupting of a bond between C^{14} and an adjacent atom would need a higher activation energy than with C^{12} and therefore be less likely to occur. Such discrimination might be cumulative in a sequence of reactions. It is not suggested that this invalidates the main chemical conclusions, though this might be expected with tritium (H^3), which has three times the mass of the normal isotope, but it should not be forgotten. For kinetic experiments it is more important. The attempted measurement with the mass spectrometer of the amount of discrimination, in the rate of photosynthesis, against $C^{14}O_2$ or $C^{13}O_2$ as compared with $C^{12}O_2$, gave 0·85 : 0·96 : 1·00 but involved the assumption that light did not affect carbon dioxide production by respiration [245]; later work (see Chapter 5) has indicated that light has a large effect and the amount of discrimination is therefore doubtful. There must also be differences in the rates of diffusion of the forms of carbon dioxide, but as the rate is proportional to the square root of the mass, the ratio for $C^{14}O_2$ (mass 46) to $C^{12}O_2$ (mass 44) will be 6·8 : 6·6 or a difference of only 3 per cent.

Another difficulty which besets experiments with marked carbon is that 'dark uptake' of carbon dioxide can occur in many plants. Thus in darkness $C^{13}O_2$ or $C^{14}O_2$ added to the gas mixture may be taken up rapidly while total CO_2 of respiration continues to enrich the mixture at a constant rate, implying an increased production in exchange for the intake. Such exchange may be several times as rapid as the net output [245]; there appears to be no way of deciding whether the rate of exchange is affected by light and the safest course seems to be to conduct photosynthesis experiments with plant material that is found not to show exchange in darkness [44].

For marked oxygen (O^{18}) rather less discrimination would be expected than for C^{14} as the difference in mass from the normal isotope is relatively smaller; also, exchange does not appear to occur with oxygen.

F. CHEMICAL ANALYSES AND ACCUMULATION OF DRY MATTER

In early work changes in carbohydrate content were used as a measure of photosynthesis, or even starch content as in Sachs's iodine method, which was later developed on a quantitative basis by Maskell [212]. Owing to the rapid incorporation of carbon in other compounds, such as organic acids and proteins, however, such methods are only useful for comparative purposes.

Accumulation of dry matter, on the other hand, provides an excellent measure of net carbon fixation for use over relatively long periods. Increase in dry matter due to uptake of soil nutrients is usually treated as negligibly small compared with that due to products of photosynthesis, though it may amount to as much as 10 per cent of the total.

In Sachs's half-leaf method (1884) the areas and dry weights of a series of half leaves (cut off along the midribs) taken at dawn are compared with the corresponding figures for the other halves taken at sunset. Sources of error are shrinkage in area (up to 5 per cent) due to water loss during the day and, of course, losses due to translocation to the rest of the plant and respiration. Even if estimates of these last are obtained at night, it is almost certain that translocation will be more rapid by day [125] and respiration also if the temperature is higher. However, the method has the great advantage that it can be used in the field. It was recently much improved [261] by the measurement, by a tracing method, of the areas of the portions of leaf left attached to the plant, both morning and evening.

It is very much better to sample the whole plant instead of leaves only and so include translocated dry weight. This is the basis of net assimilation rate (NAR) or unit leaf rate (E) as used in the technique of growth analysis [311]. Net assimilation rate is defined as net increase in total dry weight per unit leaf area per unit time and is usually calculated as

$$\frac{W_2 - W_1}{t_2 - t_1} \times \frac{\ln a_2 - \ln a_1}{a_2 - a_1}$$

where W_1 and W_2 are total dry weights at times t_1 and t_2, a_1 and a_2 are the corresponding leaf areas and the second part of the expression is the reciprocal of the mean leaf area for the period t_1 to t_2 assuming that the area increases exponentially with time. This measure is most suitable for periods of one to two weeks and has the advantage for agricultural, horticultural or ecological experiments that it enables net assimilation to be estimated without disturbing the environment, except by the removal of sample plants, and hence under the conditions in which the plants actually grow. It should, however, be emphasized that net assimilation rate includes the effects of respiratory losses by the whole plant; if the ratio of total leaf weight to total dry weight ('leaf weight ratio') falls, as it commonly does as the plant ages, this will itself cause a fall in net assimilation rate even though the rate of photosynthesis per unit leaf weight and the rate of respiration per unit dry weight are unchanged.

For plants growing in constant controlled environments, Hughes and Freeman [175] have developed a regression method of obtaining net assimilation rates, making use of the relation:

Relative growth rate = Net assimilation rate × leaf area ratio or, at an instant:

$$\frac{1}{W} \cdot \frac{dW}{dt} = \frac{dW}{dt} \cdot \frac{1}{A} \times \frac{A}{W} \quad [318] \qquad (3.6)$$

where A is leaf area, W is total dry weight and dW/dt is the rate of change of total dry weight with respect to time (or absolute growth rate). Quadratic (or higher term) regressions on time are fitted to the logarithms of leaf areas and of total dry weights as obtained from frequent small samples of plants. From these fitted curves, leaf area ratios (A/W) are obtained by taking antilogs of the differences of the log values (that is, $\ln A - \ln W$) at corresponding times. Relative growth rates are obtained by differentiating the ln (total dry weight) curve, for

$$\frac{d(\ln W)}{dt} = \frac{1}{W} \cdot \frac{dW}{dt} \quad [25] \qquad (3.7)$$

Hence the time curve for net assimilation rate is obtained from

$$\frac{dW}{dt} \cdot \frac{1}{A} = \frac{1}{W} \cdot \frac{dW}{dt} \div \frac{A}{W} \tag{3.8}$$

This curve is subject to the combined errors of the relative growth rates and leaf area ratios; hence with the very small samples used [175] the shape of the curve was in considerable doubt and only large treatment effects could be detected, but with larger or more frequent samples the method should be very useful. All the computations are carried out by computer.

The calculation of amount of photosynthesis per unit leaf area is common practice for higher plants and has the merit that both carbon dioxide and light must enter the leaf through its surface. It is, however, not always agreed that leaf area provides the best basis. Net assimilation rate has been calculated per unit of protein nitrogen content [327], as a measure of amount of living matter, or per unit leaf dry weight as a matter of convenience, or even per unit area of land when it is called crop growth rate [312]. Photosynthesis measured by other methods has also been expressed per unit of chlorophyll and for algae the basis may also be fresh weight or dry weight or volume of centrifuged cells.

G. TOTAL ENERGY FIXED

This can be estimated directly by carrying out combustion of sample plants in a bomb calorimeter at the beginning and end of a period. If a correction is to be made for respiration this must be measured by another method.

H. LIGHT ENERGY ABSORBED

Measuring the light absorbed minus heat given out, estimates the energy fixed without the necessity for destroying the plants and thus avoids sampling errors. On the other hand the technical difficulties are considerable. The method depends on estimating the difference in heat produced in a transparent calorimeter, illuminated with a given quantity of light, when it contains plant material (for example, a *Chlorella* suspension) or an inert absorber such as indian ink [209].

Measurement of the heat output by the plant material in darkness gives a direct measure of the energy lost in respiration which may be used to correct the net fixation as measured in light [294]; another method used is the inhibition of photosynthesis with ultra-violet light (assumed not to affect respiration) [2].

J. ENVIRONMENTAL CONDITIONS

In addition to the estimation of the photosynthetic rate by one or more of the five measures mentioned at the beginning of this chapter and discussed in sections B–H, it is necessary to measure the intensity of the more important external factors and, except in purely observational work, to control some of these with greater or less precision according to the nature of the experiment. Detailed discussion of the methods for such control is beyond the scope of this book. Six important external factors are listed at the beginning of Chapter 4; of these, light measurement is dealt with below and measurement of carbon dioxide concentration has already been discussed in section B. Standard physical methods are available for measurement and control of wind speed. The same may be said of temperature, though here it should be stressed that for many experimental purposes air temperatures should be varied to give a constant leaf temperature. For discussion of control of water supply see refs 173, 317 and 331 (the first a volume that will be found useful on many aspects of technique); for nutrient factors a first approach may be made through reference 158.

i) *Light measurement*

Light can only affect a system on which it falls in so far as it is absorbed; light reflected, scattered or transmitted by the system is without effect on it. The energy of absorbed light may be converted into heat energy (the most common occurrence) or transferred to electrons as in the primary processes of photosynthesis; it may be changed back into light energy and re-radiated at a longer wave length as in fluorescence, or it may be fixed as chemical potential energy, which occurs in the later stages of photosynthesis.

It is perhaps more difficult to obtain an absolute measure of the light used in photosynthesis than of any other important external factor and for this reason most experimental or observational work is in terms of relative light intensities only. The most obvious absolute measure might on first consideration seem to be the light energy falling on unit surface in unit time: J m^{-2} s^{-1}. This would be almost satisfactory for monochromatic light falling on a known area of chloroplast surface, if the light lost by reflection, scattering and transmission were also measured and subtracted from the total to give the quantity absorbed. This statement indicates three classes of difficulty: difficulties associated with the wave lengths of light used, with estimation of the effective absorbing surface and with correction for losses. In addition to the sources of loss mentioned, however, in leaves or intact algal cells there may be non-photosynthetic pigments which absorb some of the light before it can reach the chloroplasts and such light will be converted to heat; indeed there could be such pigments in the chloroplasts themselves, though modern work is indicating that all chloroplast pigments play some part in photosynthesis.

(*a*) *The use of light of different wave lengths.* The energy of a quantum of light varies with wave length, being about 1·4 times as high at the blue end of the visible spectrum as at the red. The amount of a photochemical reaction that can be carried out by light depends upon the *number* of quanta absorbed by the reacting system; the wave length of the light and energy content of the quanta are only important in that the former determines how much of the light is absorbed by the particular pigment concerned and the latter has to be above a threshold value for the reaction in question to occur at all.

Where photosynthesis is compared in different narrow wave bands, if the light energy absorbed can be estimated in each, these values can easily be converted to numbers of quanta by the expression

$$\text{Number of quanta} = \frac{\text{energy in joules}}{h\nu}$$

where ν is the frequency (velocity divided by the wave length, that is s^{-1}) and h is Planck's constant ($6{\cdot}6 \times 10^{-34}$ Js). Such a conversion should always be made when action spectra are determined, whether in terms of incident or absorbed light.

Where a continuous spectrum is used, either in a wide band of coloured light or in so-called white light, it is essential to specify the spectral distribution and desirable to estimate the total quanta absorbed throughout the range of wave lengths effective in photosynthesis. This last is seldom possible. Difficulties of estimating absorption are discussed below and even to obtain the total incident quanta within a given range involves a complex procedure. The total incident energy as measured with a blackened thermopile includes all the infra-red and ultra-violet radiation, which are inactive in photosynthesis, and gives no information as to the distribution of energy in the different visible wave lengths. To measure the energy separately in each narrow wave band of the visible spectrum involves considerable technical difficulties; spectroradiometers are becoming commercially available for this purpose but the cost is high.

A practicable method, in experiments with artificial light, is first to filter off the infra-red with a solution of cupric chloride, or with special heat-absorbing glass which will also remove the ultra-violet if present. The glass filter will probably have to be water-cooled and the water will help to cut out infra-red of wave length longer than 1,400 nm. Other colour filters may be added in an attempt to limit the range approximately to that effective in photosynthesis (say 400 to 720 nm) or to alter the wave length distribution. If a light source of known spectral distribution is used, such as a tungsten filament lamp at a given colour temperature [94], it is possible to convert the thermopile reading, obtained with the filters in position, into total incident quanta as described below. The curve relating relative energy output to wave length for the unfiltered source (e_λ in Fig. 3.8) is redrawn, allowing for the percentage transmission (τ) of the combined filters, as $e_\lambda\tau$ in Fig. 3.8. The area under this second curve is then equated to the reading given by the thermopile in J m^{-2} s^{-1} and is divided into strips corresponding to narrow wave bands. The area of each strip is divided by $h\nu$ and the quantum numbers so obtained are totalled. Obviously this procedure need only be carried out once for a given source and filters, after which quantum numbers are taken as proportional to thermopile readings. Apart from double interference filters for narrow wave bands (which transmit very little light), filters are not available to give a sharp cut off at a given wave length, such as 720 nm. However, if it is assumed that light of greater wave length is really ineffective, the area under the curve ($e_\lambda\tau$) may be taken up to this point only. This same procedure could be followed with unfiltered light (e_λ), but with a tungsten source as in Fig. 3.8 the accuracy would be extremely low owing to the very large proportion of infra-red included in the thermopile reading. With sunlight the proportion of visible light is very much higher, though even so some unwanted ultra-violet and considerable amounts of infra-red, which would moreover vary with atmospheric humidity, would be included in the readings. Monteith [232] suggests that it is safe to assume that 45 per cent of total solar radiation lies in the range 400 to 700 nm throughout the year in temperate latitudes, and this is probably true at the level of accuracy of estimates of photosynthesis of a field crop.

It should be emphasized that in the procedure just outlined the same filters must be used in the experiment as over the thermopile.

Selenium barrier layer type photocells are more convenient than thermopiles in that they can be obtained insensitive to infra-red, but they have the disadvantage that they are usually made with a sensitivity

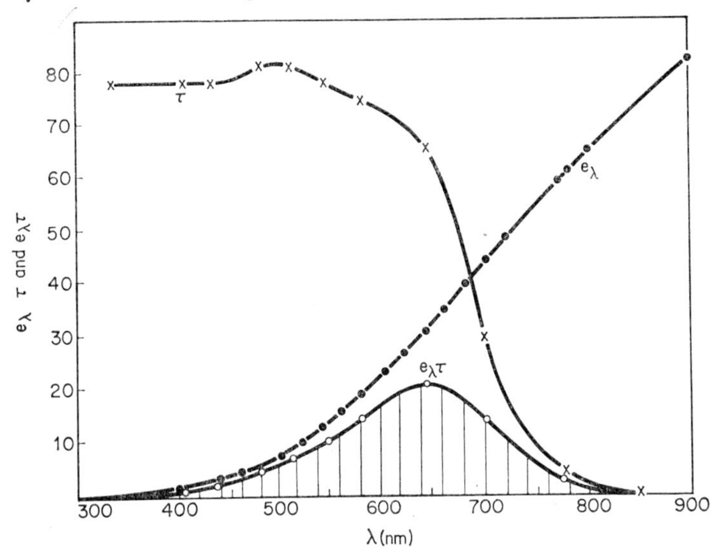

FIG. 3.8 Curve e_λ shows relative energy output for a black body radiator at 2,400°K (e.g. a tungsten filament lamp, approximately) as a function of wave length λ; curve τ shows the percentage transmission of 4 mm thickness of Schott KG3 glass filter; $e_\lambda\tau$, the product curve, shows the relative spectral distribution of the filtered light—the area under this curve may be equated to a reading given by a thermopile, with the same source and filter, and can be divided into narrow wave bands for conversion to numbers of quanta.

curve similar to that of the eye of a 'standard observer' (the so-called standard photopic luminosity factor V_λ—Fig. 3.9); such cells are very insensitive to red light which is strongly absorbed by chlorophylls. The meters are usually calibrated in luminosity units: lumens per square metre or lux (metre candles), or lumens per square foot (foot candles); these specify brightness as apparent to the human eye and have no relevance to the plant. They may be used as a scale of relative amounts of light where comparisons are confined to the same source, such as tungsten filament light, or even daylight if diurnal and seasonal changes in quality are ignored. The photocell and meter will only give correct readings in luminosity units if they have been calibrated for the source used. It is then possible to convert the readings to energy values and hence to numbers of quanta as described below.

It is necessary to take readings with the photocell of *unfiltered* light from the source. The energy distribution of the unfiltered source must be known (e_λ in Fig. 3.8) and the relative energy values e_λ at successive wave lengths multiplied by the corresponding values of the standard

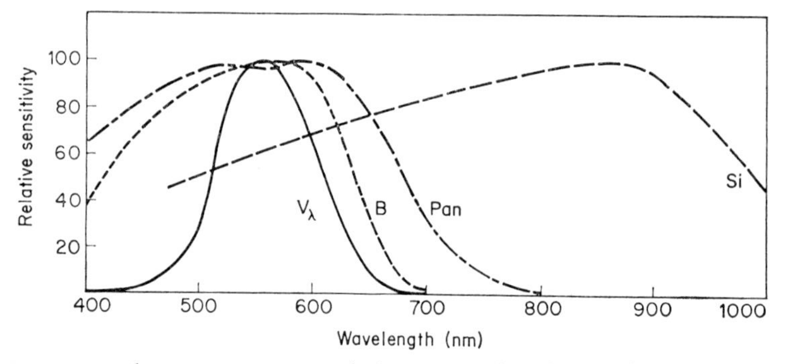

FIG. 3.9 Relative sensitivities to different wave lengths of radiation: V_λ, for the eye of a 'standard observer' (the standard photopic luminosity factor); B, for a normal type barrier layer selenium photocell; Pan, for a 'pan' type barrier layer photocell; Si, for a silicon cell.

photopic luminosity factor V_λ [335] and so converted into relative luminosities. These values of $e_\lambda V_\lambda$ are plotted against wave length and the area (a_0) under the curve is then proportional to the luminous flux. If a system of filters is used the values of $e_\lambda V_\lambda$ must next be multiplied by the corresponding transmission values τ_λ and a new curve of $e_\lambda V_\lambda \tau_\lambda$ plotted against wave length. If the area under this curve is a the total illumination with the filters in position is then Pa/a_0 where P is the reading given by the photocell without the filters. This total can be divided up in proportion to the areas of narrow wave band strips under the last-mentioned curve and converted into absolute energy units. The relation used [335] is

$$dF = KJ_\lambda V_\lambda d\lambda$$

and hence

$$J_\lambda d\lambda = \frac{dF}{KV_\lambda} \qquad (3.9)$$

where $J_\lambda d\lambda$ is the total hemispherical energy flux per m² from a total radiator, between the wave lengths $\lambda \pm d\lambda/2$ and dF is the total luminous flux radiated per m² in this wave length interval. K is taken as 660 lumens per watt at a wave length of 555 nm (where $V_\lambda = 1{\cdot}0$) and at this wave length

$$\text{W m}^{-2}\, d\lambda = \text{lm m}^{-2}\, d\lambda . \frac{\text{W}}{\text{lm} \times 1{\cdot}00}$$

V_λ is a proportionality factor which allows for changes in visual sensitivity with wave length by adjusting K.

To convert lm m^{-2} (or lux) into J m^{-2} s^{-1} we have

$$x \text{ lm m}^{-2} = \frac{x}{660 V_\lambda} \text{W m}^{-2} = \frac{x}{660 V_\lambda} \text{J m}^{-2}\text{ s}^{-1}$$

The energy for each narrow wave band is then divided by $h\nu$ and the quantum numbers so obtained are summed.

It has been stressed that a photocell must only be used as above when the meter has been calibrated for the particular light source used and even then the procedure is very troublesome. If, however, it could be covered with a system of filters which give it a uniform sensitivity for all wave lengths in the visible spectrum, in terms of energy or better still in terms of quanta, a photocell could be used for comparing the energy or quantum numbers from light sources of different spectral composition, such as daylight at different times of day or which had passed through different numbers of leaves. If fitted with a time-integrating device such a filter-covered photocell could be used to estimate total light energy or quanta from daylight under foliage, and allow for changes in quality. Fig. 3.10 shows the relative sensitivities, in terms of energy and of quantum numbers, of a first approximation to such an instrument [262], together with the energy sensitivity of the barrier layer photocell used. The most serious source of error is the marked fall in sensitivity in the longer-wave red light, beyond about 640 nm. Other barrier layer photocells are obtainable with better red sensitivity (Fig. 3.9 Pan) and would therefore be more suitable. Silicon cells have the advantage of a much bigger output, but although they have even better red sensitivity they are very insensitive to blue (Fig. 3.9); moreover they are very sensitive to infra-red which would therefore have to be removed by a suitable filter. The integrating device used is shown in Fig. 3.11 and is in principle a voltameter, with detachable copper electrodes dipping into copper nitrate solution.

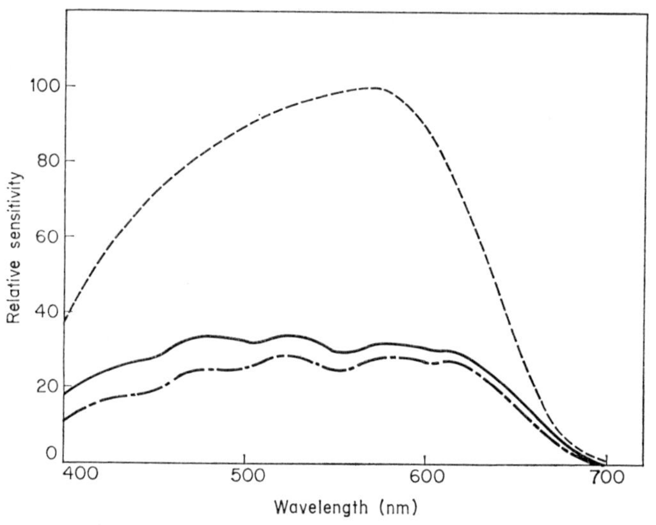

FIG. 3.10 Relative sensitivities of an integrating photometer to different wave lengths of light, in terms of incident energy (solid line) and incident quanta (broken line); also of the barrier layer selenium photocell used (short-dashed line). After Powell and Heath [262].

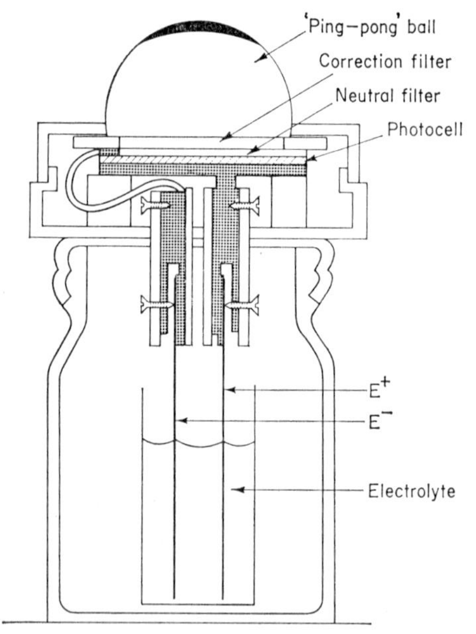

FIG. 3.11 Diagram of integrating photometer. E^+ and E^- are detachable copper electrodes; the electrolyte is copper nitrate solution. From Powell and Heath [262].

The current from the photocell transfers copper from the anode to the cathode and the amount is estimated by detaching the electrodes and weighing. This apparatus has been adapted for use in trees out of doors, with remote voltameter cells at ground level [179*a*].

(*b*) *Estimation of light absorbed and effective area.* The difficulties of obtaining a reasonably precise estimate of the absorbing area involved in photosynthesis are so great that this is seldom attempted. The same may be said of estimation of the amount of light actually absorbed and used in photosynthesis. The two problems are interconnected in so many ways that they will be discussed together at successive levels of decreasing complexity—first for a field crop, then for an isolated plant, for an isolated leaf, for unicellular algae and for isolated chloroplasts. Before this is done, however, we should consider how to measure the intensity of light (irradiance, or $J\ m^{-2}\ s^{-1}$) intercepted by a plane surface. Parallel light produces the greatest irradiance when it is normal to the surface; with increasing obliquity the irradiance falls off because unit area of the surface intercepts a decreasing cross-sectional area of the light beam. The irradiance in fact varies as the cosine of the angle of incidence [333]; with the light at 90° to the normal the irradiance is of course zero. If a thermopile with a matte black surface (radiometer) is used, this absorbs practically all light falling on it whatever the angle of incidence and the voltage produced obeys the cosine law just stated. With selenium barrier layer cells the output falls off, as the angle of incidence increases, much more rapidly than the cosine law demands (Fig. 3.13); such cells are said to have a cosine error and this can be corrected to a near approximation with special designs of transparent refracting or opalescent light-scattering cover, which increase appropriately the proportion of oblique rays reaching the cell (Figs. 3.12 and 3.13 [258*a*]).

A radiometer or a cosine-corrected photocell (suitably colour-corrected and calibrated: page 101) if placed parallel with a leaf can be used to measure the amount of light per unit surface area that it receives; if placed horizontally in the open they can give a measure of the irradiance on the earth's surface; the further the leaf is from the horizontal or the more it is shaded by others the more the two measures will differ and the less simple will be the relation between them. These considerations relate only to the intensity of light falling on the external surface of the leaf and the proportions then lost by reflection, scattering and transmission must be estimated separately.

For a field crop the ultimate limit in simplifying assumptions is to consider the crop as a homogeneous medium which not only completely covers the ground but everywhere absorbs and transmits light uniformly according to Lambert's Law [118]:

$$I_1 = I_0 e^{-kl} \quad \text{or} \quad I_1 = I_0 10^{-\varepsilon l}$$

where I_0 and I_1 are the intensities of light first absorbed and after passage through a depth of 1 cm respectively, k is the absorption coefficient

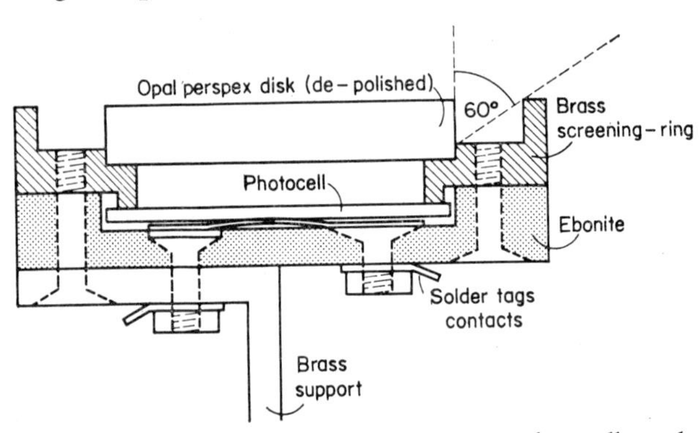

FIG. 3.12 Cosine-corrected barrier layer selenium photocell, with 4 mm thick Perspex disc and 60° angle of screening. Filters can be fitted between the disc and the cell as in Fig. 3.11. After Pleijel and Longmore [258*a*].

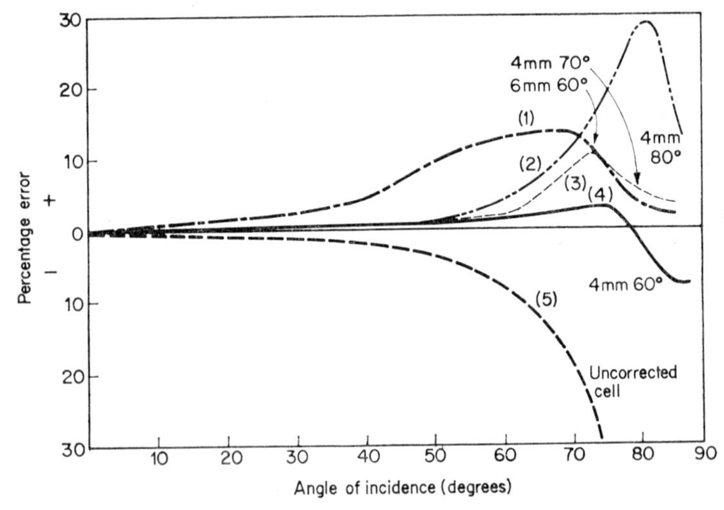

FIG. 3.13 Cosine error corrections given by different combinations of disc thickness and screening angle in the device shown in Fig. 3.12. After Pleijel and Longmore [258*a*].

and ε the extinction coefficient of the medium for the particular wave length of light concerned. Note that I_0 is exclusive of any losses due to reflection or scattering and is therefore less than the incident intensity. Instead of 1 cm the depth L is measured in units of leaf area index (LAI), this being the leaf area per unit of land area [311]. I is then the mean intensity of radiation on a horizontal plane L leaf area index units below the top of the crop.

Monteith [232] points out the inaccuracies that can result from using a uniform average light intensity for a given horizontal leaf layer within the crop in calculating rates of photosynthesis. These inaccuracies arise from the non-linear relation between photosynthesis and light intensity (Fig. 4.5); if the light were uniformly distributed at the mean intensity the total photosynthesis would be much greater than if it were mainly concentrated in sun flecks on unshaded leaves with other heavily shaded leaves at very low light intensity. Change in quality as the light is filtered through successive leaf layers will also introduce errors by changing k or ε. Monteith has devised a more elaborate model in which the crop is considered as made up of L layers, each contributing unity to the total leaf area index. In each layer a certain proportion s of the incident light passes through without interception and $(1 - s)$ is intercepted by leaves or stems. The distribution of light in the crop is given by expansion of the binomial expression.

$$I(L) = [s + (1 - s)\tau]^L \, I(0) \tag{3.10}$$

where $I(0)$ is the intensity above the crop, $I(L)$ that after L layers have been penetrated and τ is a transmission coefficient for the leaves concerned. The distribution of LAI within the crop was estimated by sampling it in layers and s calculated from radiation profiles measured with a thermopile. Values of s were 0·8 to 0·6 for barley, with relatively upright leaves, and 0·6 to 0·4 for kale in which the leaves are nearly horizontal. τ was obtained from laboratory measurements with light falling normally on individual leaves of the same crop plants. The following assumptions are necessary:

First, that in each layer of unit LAI, light is never intercepted more than once; secondly that s for a given crop is constant throughout the day; thirdly that the change in quality of light that has passed through a leaf is negligible for photosynthesis by another leaf; fourthly that after two interceptions light intensity is negligible for photosynthesis (because τ is about 0·1); fifthly that the proportion τ of incident light transmitted through any leaf in the crop is adequately represented by a laboratory value for normally incident light. With regard to this last assumption Monteith suggests that downward reflection from leaves not normal to the light helps to compensate for smaller transmission through their tissues. In spite of these various assumptions, calculations of dry matter production agreed reasonably with field values. For these calculations a relation between gross photosynthesis and incident light intensity derived from laboratory experiments was used, solar radiation was assumed to vary sinusoidally through the day and respiration was assumed proportional to leaf area.

Interesting predictions from Monteith's theory are that about the same total photosynthesis is to be expected in the long summer days of a cloudy temperate climate and the short days of a sunny equatorial climate; also that when the LAI is less than 3, photosynthesis increases as *s* decreases (the leaves become more prostrate and so intercept more light), but when LAI is greater than 5 photosynthesis increases with *s* (the leaves become more erect so that more of them receive some light without previous interception).

Whereas Monteith started with measurements of carbon dioxide concentration within the crop and tried to devise a model of simplified geometry that would approximately satisfy his observations, de Wit [332] started with the geometry and arrived at a very much more elaborate model intended for use with a computer; this would require too much space for description here but it should be referred to if it is desired to study the geometrical problems involved. It is claimed that the 'daily photosynthesis of a canopy with known characteristics can be computed for any time and place on earth from the relevant meteorological data'. It seems likely that further elaboration will not be worth while, at least until very much more physiological information is available on such matters as rates of photosynthesis and respiration of leaves of different ages as well as effects of light on respiration.

Radiometers or cosine-corrected photocells may be used in sets of three to estimate approximately the light absorption by a field crop. One detector is exposed horizontally to the sky to measure the total incoming light, a second is inverted above the canopy to measure light lost to the crop by reflection and scattering, while the third is placed at various levels within the crops to measure light transmitted. Hence light absorbed is obtained by subtraction. The values obtained are in terms of light per unit area of land. Even for a very uniform crop many sample positions are essential.

Isolated trees and other plants receive and absorb light coming from all parts of a hemisphere and the individual leaves or leaflets are often arranged normal to the prevailing light. The same is often true of plants in a glasshouse or even of many of the leaves in a field crop. To measure in absolute terms the total light intercepted by the plant a radiometer or cosine-corrected photocell would have to be placed parallel with each leaf, the measured irradiance multiplied by the leaf's area and the values summed for all leaves. Even so the problem of losses by scattering, reflection and transmission would remain. Relative losses might be estimated as described above for a field crop but with the detectors arranged parallel to 'tangents' to the crown of the tree.

The light measurements per unit area of land, as used for a field crop, are of little value in relation to photosynthesis by an isolated plant. Probably the best single measurement that can be made is the light falling on a three-dimensional surface, such as a hemisphere, with equal weight given to light coming from any direction above the horizontal. As a first approximation to this, the cell in Fig. 3.11 was covered with a dome made of just over half a ping-pong ball, which became in effect the luminous source. However, it was found that vertical light still had most effect on the reading, because the surface presented in plane projection to such light was twice that presented to a horizontal beam. An approximate correction for this effect was achieved by painting a black disc of suitable size in the top of the hemisphere.

The use of such a dome over the cell raises problems owing to the difficulty of specifying the absorbing area, so that the measurements can only be used as relative values, but this is really a problem inherent in the plant. It is obviously impossible to measure daylight so as to allow for its constantly varying angles of incidence and for the posture and area of every leaf on the plant. It seems better to treat isolated plants, and perhaps even many crop plants, as three-dimensional objects even though this means using only relative light intensities.

The meter, as shown in Fig. 3.11 or cosine-corrected, may be calibrated to yield time-integrated values of light energy (or quanta) received over periods of about a week. Other, more elaborate and costly devices can integrate light over much shorter periods [23], but low cost makes it possible to have large numbers of meters and thus sample more effectively the very variable light climate in a glasshouse or among a crop. Such total light values will be less well correlated with total photosynthesis the greater the variation in light intensity and hence, under natural conditions, the longer the period; the reasons are similar to those advanced by Monteith in criticism of the use of spatially averaged light intensity (see above). Nevertheless, it is better for plant physiological purposes to measure light in units that have a definite physical meaning than to attempt to design a meter that responds to light intensity in the same way as the plant. Apart from the fact that the relation of photosynthesis to light intensity depends greatly on the levels of many other factors (Chapter 4), such an attempt would involve assuming the relations that we wish to study. The human physiologist studying the physiology of the eye measures light in quanta and not in foot candles, though the latter are useful to the illumination engineer for such purposes as specifying factory illumination [207, 144, 208].

In experiments with single leaves the area of the leaf itself is almost always taken as the effective area for photosynthesis. A consideration of leaf structure shows, however, that there are complexities here on a micro-scale even greater than those on a macro-scale in a field crop (see also Chapter 8). The best that can be done is to estimate light absorbed by the leaf and even this is far from easy. Some of the methods available, which can also be used for films of algae, are shown in Fig. 3.14 (i) and (ii). They involve placing the leaf inside a matte white Ulbricht integrating sphere in which all the light introduced, except that absorbed by the leaf or black paper, is scattered until it ultimately reaches the photocell; the latter is shielded from direct light from the leaf. These methods can be used with monochromatic light of various wave lengths to obtain absorption spectra (Fig. 1.17). The method of Shibata, described below for algae, has also been used for leaves [275, 276].

Algal films, one cell thick, probably provide the most favourable living material for measurement of light absorption. They can be used in an Ulbricht sphere or in the rather simpler apparatus of Haxo and Blinks [137] shown in Fig. 3.14 (iii). Another simple method which can be used for suspensions of algae is that of Shibata [275, 276] shown in Fig. 3.15. Scattered light is collected by an opal glass plate, or for ultra-violet light by a sheet of oiled paper between two quartz plates, placed on the side of the cell nearest to the detector as in Fig. 3.15*b*; a similar proportion α of the diffuse light from the opal glass reaches the detector both with and without algae in the cell. The opal glass plate does not collect light reflected back towards the source and the method shown in Fig. 3.15*c*, where it is used to illuminate the sample with diffuse light, seems preferable. In neither method is scattered light which emerges from the sides of the cell collected and the latter must therefore be thin. Shibata [275] gives a method of correction for multiple reflections between the opal plate and the sample, but this

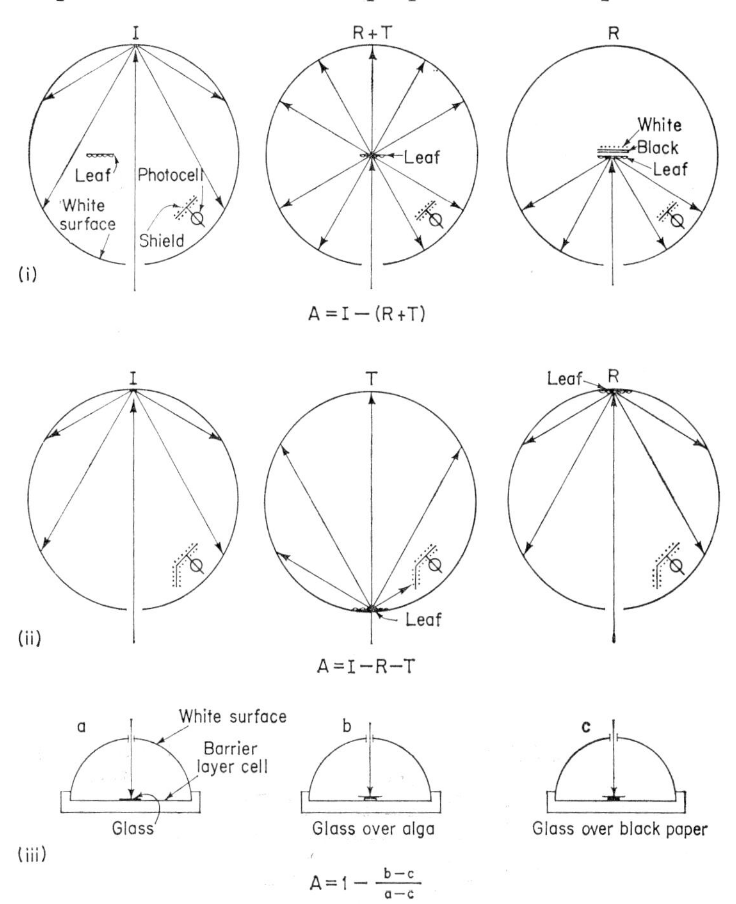

FIG. 3.14 Methods for measuring the reflection, R, and transmission, T, of light by a leaf or film of algae and hence deriving the absorption, A. (i) and (ii), two methods of using an Ulbricht integrating sphere; (iii), the half-sphere method of Haxo and Blinks [137]. After French and Young [106].

seems not to be necessary except for absolute measurements and for very highly reflecting samples; he also describes a method of obtaining reflection spectra, which tend to be mirror images of the

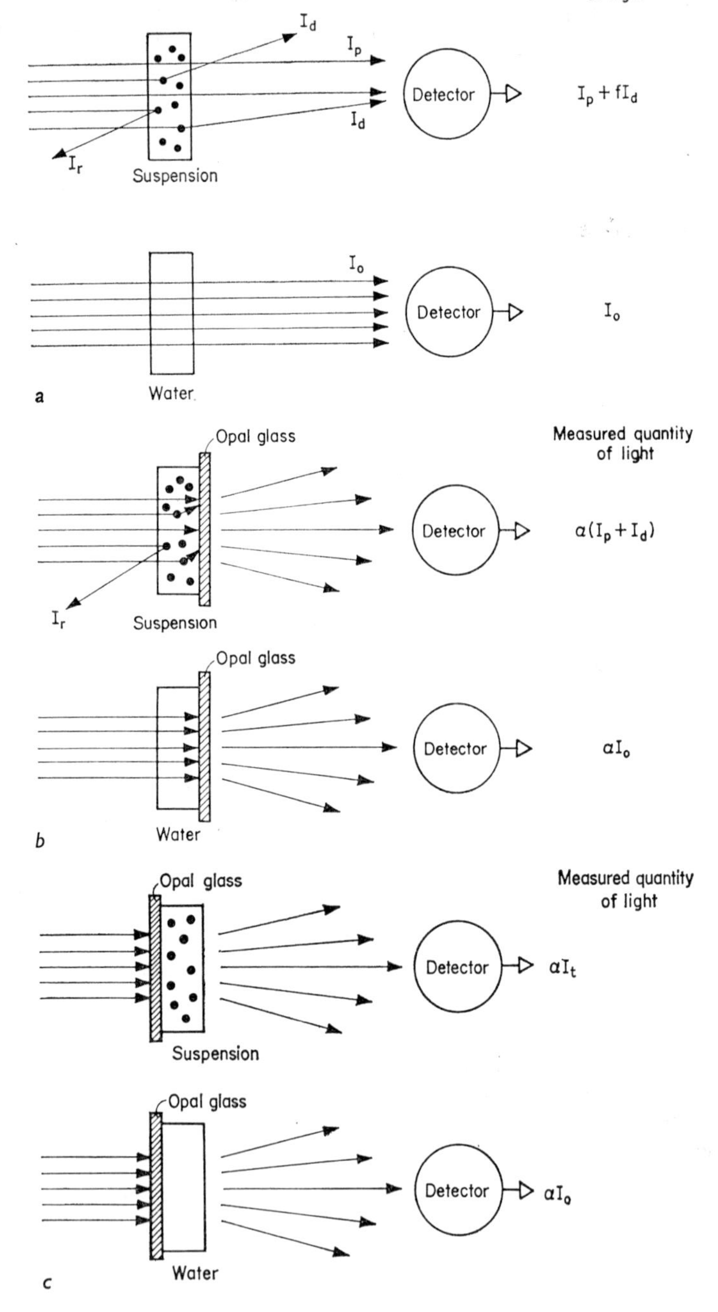

Measured quantity of light
I_d
I_p
Detector
$I_p + fI_d$
I_d
I_r
Suspension
I_o
Detector
I_o
a
Water
Opal glass
Measured quantity of light
Detector
$\alpha(I_p + I_d)$
I_r
Suspension
Opal glass
Detector
αI_o
b
Water
Opal glass
Measured quantity of light
Detector
αI_t
Suspension
Opal glass
Detector
αI_o
Water
c

absorption spectra, and this is useful for tissues which transmit very little light.

In much early work on quantum efficiency with suspensions of algae in Warburg vessels, such thick suspensions were used that all the incident light was absorbed; this was sometimes measured with a chemical actinometer in a similar vessel. This technique resulted in the cells moving between light and darkness as the culture was shaken, which is now known to have complex effects on photosynthetic gas exchange.

For isolated chloroplasts the same techniques as for unicellular algae would presumably be suitable, though chloroplasts are not normally used in layers one plastid thick. Absorption by a single chloroplast can be measured with a microspectrophotometer [334*a*] with a beam of light narrower than the chloroplast diameter.

Even within the chloroplast there must be selective absorption of the light as it passes through successive lamellae and both selective and non-selective scattering at the surfaces between the lamellae and the stroma. The wave length distribution and intensity available for photosynthesis will thus change from one side of the plastid to the other, and owing to the greater absorption by the grana the problem is not altogether dissimilar from that of a field crop. However, absorption of the light that actually enters a chloroplast is extremely efficient (page 31) and a single layer of chloroplasts should provide the most accurate system for the study of the absolute light use in photosynthesis because of the reduction in the errors due to multiple scatter-

---

FIG. 3.15 Shibata's opal glass method for estimating light absorbed by algae in suspension. *a*, Ordinary method, without opal glass. The amount of incident light I_0 is measured without the algae and as this is parallel light the reading does not depend on the distance of the detector; with the algal suspension the reading is due to parallel transmitted light I_p plus a fraction f of the diffuse transmitted light I_d which has hit cells in the suspension; f decreases with increased distance of the detector from the suspension. Of the transmitted light, $(1-f)I_d$ fails to enter the detector, as does also the reflected light I_r; absorption is therefore overestimated. *b*, Opal glass between sample and detector. A similar proportion α of I_0, without the algae, or of I_p and I_d with them, enters the detector. I_r is still lost. *c*, Opal glass between source and sample which is assumed thus illuminated with completely diffuse light; the transmitted light I_t is also assumed completely diffused and the same proportion α of both I_r and I_t enters the detector. Some of I_r will pass through the opal glass and be lost. After Shibata [275].

ing; isolated chloroplasts have the further great advantage already mentioned that they do not respire.

K. INTERNAL FACTORS

Measurement of internal factors demands a wide variety of techniques. The modern chemical methods used for investigating photosynthesis are best sought in the original papers; measurement of leaf water content is discussed in ref. 197 and measurement of stomatal aperture in refs. 143 and 147. It is necessary in Chapter 4 to refer to work in which stomatal movements were followed with a porometer. In most forms of porometer a small cup is sealed to the leaf and air is forced out of or drawn into the cup through the leaf by a known increase or decrease of pressure. The volume rate of flow of this air may be measured or the resistance to flow may be estimated (e.g. by a Wheatstone bridge with air flow instead of electric current [152]). Such measures, with appropriate corrections, may be used to estimate the relative resistances of the stomata to diffusion; other instruments yield direct measures of diffusive resistance [279, 220, 12].

4: Interaction of Factors

THE rate of photosynthesis is controlled by a large number of (i) external and (ii) internal factors, of which the more important are: (i) light (intensity and quality), temperature, carbon dioxide concentration, wind velocity, water supply, nutrient supply; (ii) age, chlorophyll content, enzyme factors, leaf water content, leaf structure, stomatal aperture.

The distinction is less clear than this suggests for the internal factors often change, either on a short-term or a long-term basis, in response to changes of the external factors. Examples of short-term changes are provided by the stomata, which alter their apertures in response to changes in the first five of the six external factors listed above. Long-term changes are, for instance, the alterations in leaf structure caused by sun or shade conditions or by variations in nutrient supply. Because of these long-term changes, there have been two main types of experimental approach—*either* the plants have all been grown under a single set of conditions and whole plants or leaves then subjected for short periods to the various experimental treatments while the rates of photosynthesis are measured, *or* they have been grown under these treatments from the beginning. The second method often gives a result of more practical or ecological value, but many more differences in leaf structure and in other internal factors are liable to be produced, and hence it is difficult to disentangle the direct effects of the external factors upon photosynthetic rate from indirect effects.

The classical method of biological experimentation, which derived from that in physics and chemistry, was to investigate the effects of varying one factor while all others were held as constant as possible. Up to 1900 physiologists (for instance Pfeffer [257]) attempted to interpret the effects of single external factors upon rate of photosynthesis in terms of the so-called Cardinal Points, namely the minimum, optimum and maximum. This approach was based on the supposition that for any one external factor there existed a minimum level or intensity below which the process concerned could not proceed, an optimum at which the rate was greatest and a maximum level above which the process was again inhibited. The supposition arose directly from the method of experimentation, for if attention is confined to a single factor it is indeed possible to evaluate the cardinal points. In fact, however, these are not unique but depend on the levels of other factors operating, so that the arbitrary selection of these levels decides the values obtained. Thus it should be possible, at least in theory, to specify an optimum *set* of conditions for the rate of a process, such as photosynthesis, but not to specify an optimum light intensity or carbon dioxide concentration in isolation. Of more scientific interest and practical value is a study of the mode of interaction of the factors concerned.

Interaction of factors may be defined as the modifications of the responses to changes in level of one factor produced by various levels of another factor or factors. If there is *no* interaction between two or more factors, then each produces its effect independently and their effects are therefore additive at all levels. A fictitious example of this is shown in Fig. 4.1 where Y (which might be rate of photosynthesis) is plotted against one factor A for different levels of another factor B. It is seen that adding the successive doses of A causes exactly the same increases in Y (that is the same responses) at each level of B. Thus the total response to the first 2 doses of A is 6 units of Y; the response to the first dose of B is 5 units—these effects can be added to give the response to A_2B_1 or 11 units above the initial A_0B_0 value of Y. Clearly, a parallel set of curves would also be obtained for different levels of A by plotting Y against B. Without specifying the forms of the curves, but only that they were parallel and thus the effects additive, we could write

$$Y = f(A) + f(B) + \text{constant} \tag{4.1}$$

where f denotes some unspecified function of the amount of the factor concerned. This could be generalized for any number of factors as

$$Y = f(A) + f(B) + f(C) + \ldots + k \tag{4.2}$$

Such a set of data would provide some experimental evidence that A and B were affecting Y through two quite independent systems.

This is a rather unusual state of affairs—more often the curves are not parallel and thus show an *interaction* between the factors, for the responses to one factor will now differ according to the level of the other. A typical example of this is to be seen in Figs. 4.4 and 4.5, which are discussed in more detail below. The type of interaction could in theory take any number of forms—there is no limit to the ways in

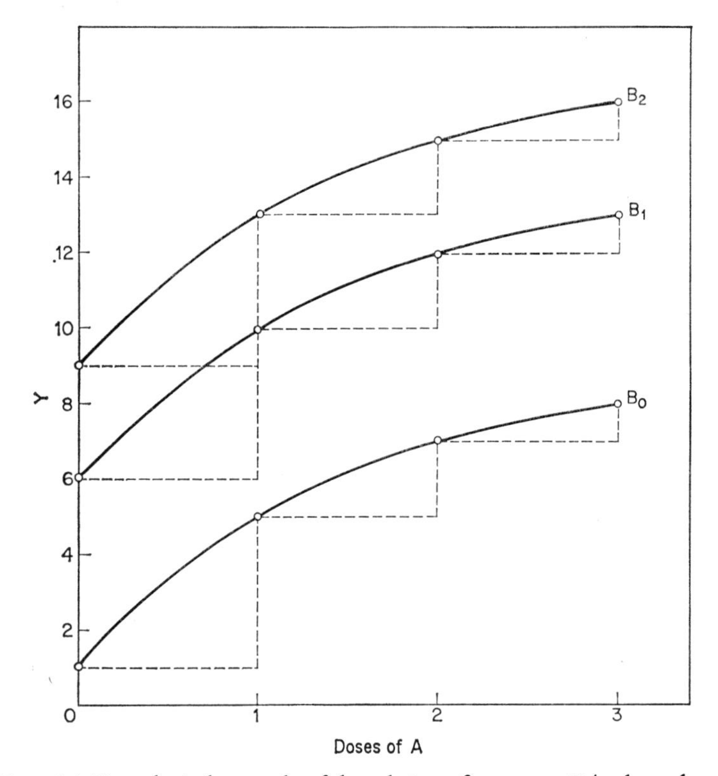

FIG. 4.1 Hypothetical example of the relation of a process Y (such as photosynthesis) to two factors A and B which affect it independently, that is their effects are additive and there is no interaction between them. Note that the curves are parallel.

which curves can be non-parallel—but there is one type of particular importance, which occurs very generally for rate of photosynthesis and other continuously variable plant responses. I shall now outline the history of the elucidation, so far as it has yet proceeded, of this type of interaction.

The German agricultural chemist, von Liebig, had postulated before 1840 that the yield of a crop was completely determined by the level of that particular nutrient element which was present in the lowest concentration relative to its optimum amount—a hypothesis which was called the Law of the minimum. This concept was not applied to the effects of factors on photosynthesis until more than sixty years later when F. F. Blackman [19] enunciated in 1905 his 'principle of limiting factors' in the following 'axiom': 'When a process is conditioned as to its rapidity by a number of separate factors, the rate of the process is

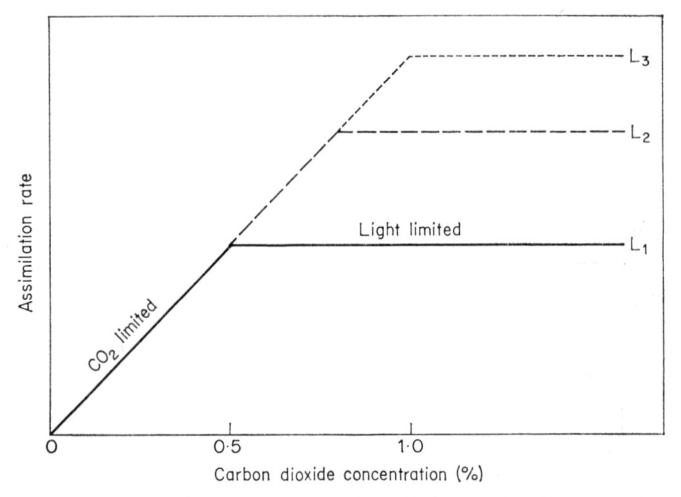

FIG. 4.2 Diagram to illustrate F. F. Blackman's 'principle of limiting factors'. For explanation see text.

limited by the pace of the "slowest" factor.' This slowest or limiting factor could be identified experimentally as follows: 'When the magnitude of a function is limited by one of a set of possible factors, increase of that factor, and of that one alone, will be found to bring about an increase of the magnitude of the function.' His meaning is best illustrated by a diagrammatic example (Fig. 4.2). At light intensity L_1 assimilation increases in direct proportion to carbon dioxide concentration up to say 0·5 per cent. At this concentration the light energy is supposed only just sufficient to allow of the assimilation of all the carbon dioxide supplied to the chloroplasts; thus any further increase in concentration cannot increase the rate of photosynthesis, which is now light limited. If, however, a higher light intensity L_2 is substituted for L_1, carbon dioxide supply once more becomes the limiting factor; as it is increased, assimilation can again rise proportionately until, at say 0·8 per cent, light once more becomes limiting. A further repetition of this is shown in Fig. 4.2. A similar diagram could be obtained by plotting assimilation against light intensity for different levels of carbon dioxide concentration. Blackman placed these two factors in the category of 'conditions of supply of material or of energy'; for such factors the rate was supposed proportional to the intensity of the factor (and hence there was a constant response or slope of the line) up to a level at which some other factor became limiting; the response to the first factor then suddenly changed from a constant value to zero.

Blackman seems to have been led to formulate his theory by experimental results for the interaction of light and temperature [215, 19] which are more conveniently considered elsewhere (Chapter 7). However, in 1911 Blackman and Smith [21] published data for carbon dioxide uptake by water plants supplied with known concentrations dissolved in circulating water; to these data they fitted the postulated pairs of straight lines as shown in Fig. 4.3 (*a* and *b*). Both graphs were made from a number of experiments carried out at considerable intervals and therefore with different plant material. The data for *Elodea* included the results of four experiments carried out at somewhat higher light intensity than the rest, but the single pair of lines was fitted on the assumption that light was non-limiting (or saturating) below the assimilation rate represented by the horizontal line. This was criticized [49] on the grounds that two different curves were appropriate for the two light intensities and it is clear that the data were too imprecise to provide much support for Blackman's hypothesis. Nevertheless this was until recently widely accepted as the *law* of limiting factors; as such it was used (and still is) to account for a great deal of facile thinking while his important warning about 'taking deliberate thought to the other factors, lest surreptitiously one of *them*, and not the factor under investigation, becomes the real limiting factor to an increase of functional activity' was commonly forgotten.

Continental workers later carried out similar experiments, some of them using bicarbonate solutions instead of dissolved carbon dioxide, and obtained much smoother curves in which the rate of increase of assimilation with carbon dioxide supply fell off gradually as the latter was increased. Of these, Harder [135] published in 1921 the results of a large factorial experiment, perhaps the first of its kind. He investigated the effects upon rate of photosynthesis, for *Fontinalis*, of all possible combinations of several light intensities with several carbon dioxide supplies, the latter given as potassium bicarbonate of different concentrations. Apparent photosynthesis was estimated as oxygen out-

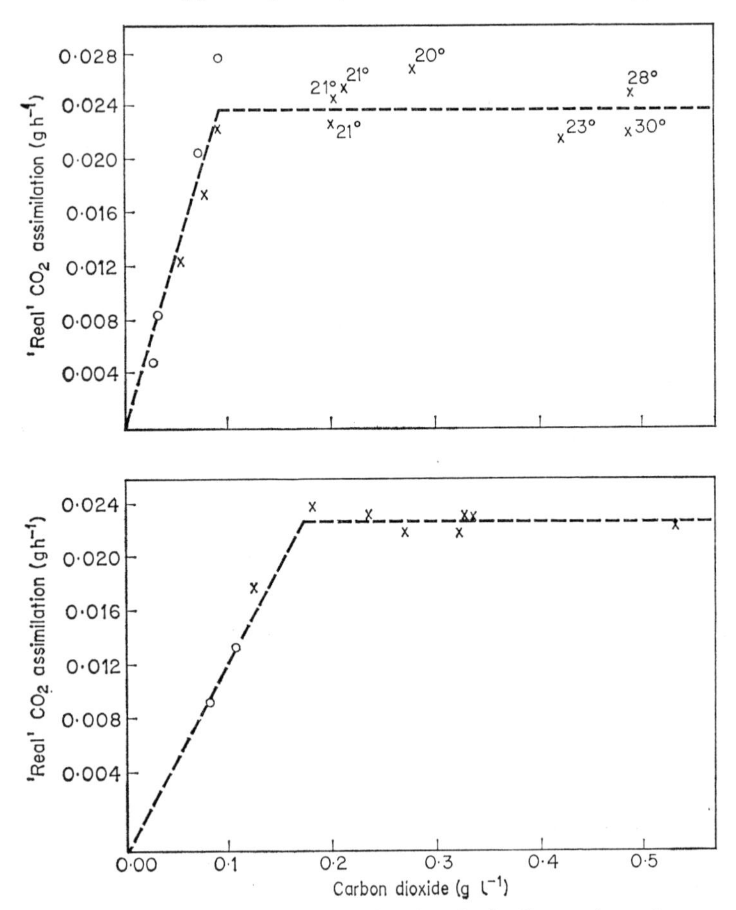

FIG. 4.3 Limiting factor relations fitted to data for 'real' assimilation by water plants plotted against carbon dioxide concentration. Circles, 8·1 units of light intensity; crosses 5·7 units. *a*, *Elodea*. Temperature 19°C except where otherwise noted. *b*, *Fontinalis*. Temperature 22 or 23°C. After Blackman and Smith [21].

put into a closed vessel, measured by Winkler's method (page 83) and corrected to 'true' photosynthesis. His data showed that if, at any combination of the two factors, he raised the level of *either* factor it brought about an increase in assimilation rate. Thus the theory of limiting factors, in its original and rigid form, was disproved.

Harder's data have been plotted in Figs. 4.4 and 4.5 for carbon dioxide and light respectively. The divergence of the curves indicates a considerable interaction—compare Fig. 4.1 where interaction is absent. The light and carbon dioxide curves may be combined in a single three-dimensional diagram as a solid model with assimilation plotted

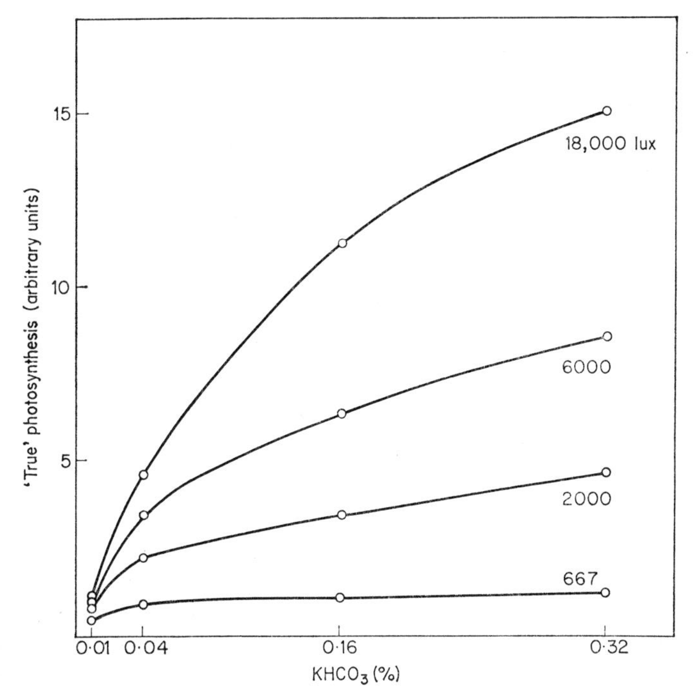

FIG. 4.4 Relation of rate of 'true' photosynthesis, for *Fontinalis*, to carbon dioxide at four light intensities. Smooth curves drawn through experimental points. Data of R. Harder [135].

vertically and the two factors plotted horizontally at right angles. This can be shown on paper as an isometric projection (Fig. 4.6) and is then a useful way of presenting the results of factorial experiments (for another, more advanced method see ref. 269). Factorial experiments provide the best and indeed almost the only way of investigating the interaction of factors. The curves have been drawn through the actual experimental points, without any smoothing, and this magnificent set of data provides a model for the way in which many quantitative plant responses are related to pairs of interacting factors. It is therefore worth taking a little trouble to become familiar with it. The easiest way of looking at the diagram is to consider the individual curves for assimilation plotted against either carbon dioxide supply or light intensity. We can compare Figs. 4.4 and 4.5 with Fig. 4.6. It is seen that with very low carbon dioxide supply (0·01 per cent $KHCO_3$) the first increase

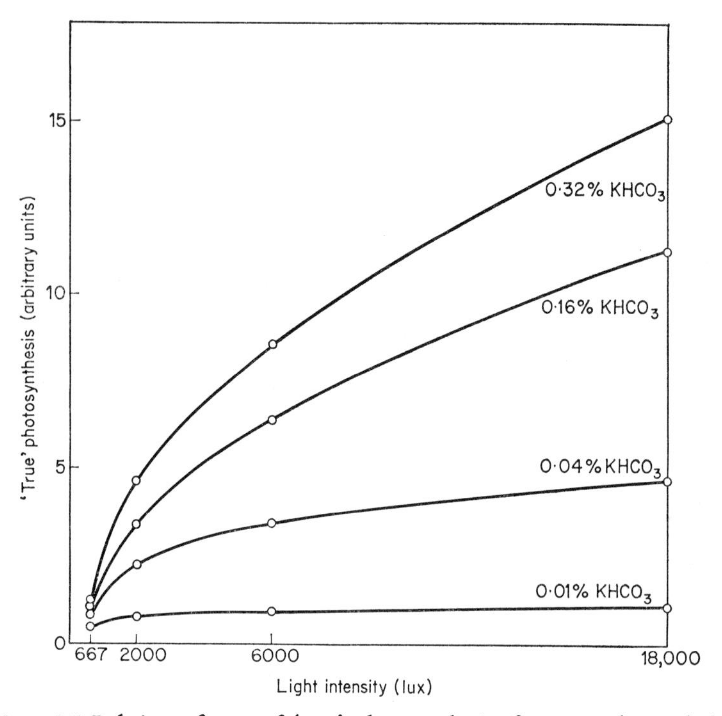

FIG. 4.5 Relation of rate of 'true' photosynthesis, for *Fontinalis*, to light intensity at four carbon dioxide levels. Smooth curves drawn through experimental points. Data of R. Harder [135].

in light (from zero to 667 lux) causes an appreciable amount of photosynthesis but further increases have small and decreasing effects, as the rate is severely carbon dioxide-limited. At each successively higher level of carbon dioxide supply this limitation is less and less marked, though even with the highest concentration the curve still bends over as light is increased, that is carbon dioxide limitation apparently has some effect on the rate of photosynthesis throughout the range explored. Similar considerations, but with light-limitation, apply if we examine the assimilation *v*. carbon dioxide supply curves in Fig. 4.6.

With practice it is not difficult, and is much better, to consider the shape of the surface made by all the curves. Near the front on the extreme right, in the most carbon dioxide-limited part of the surface,

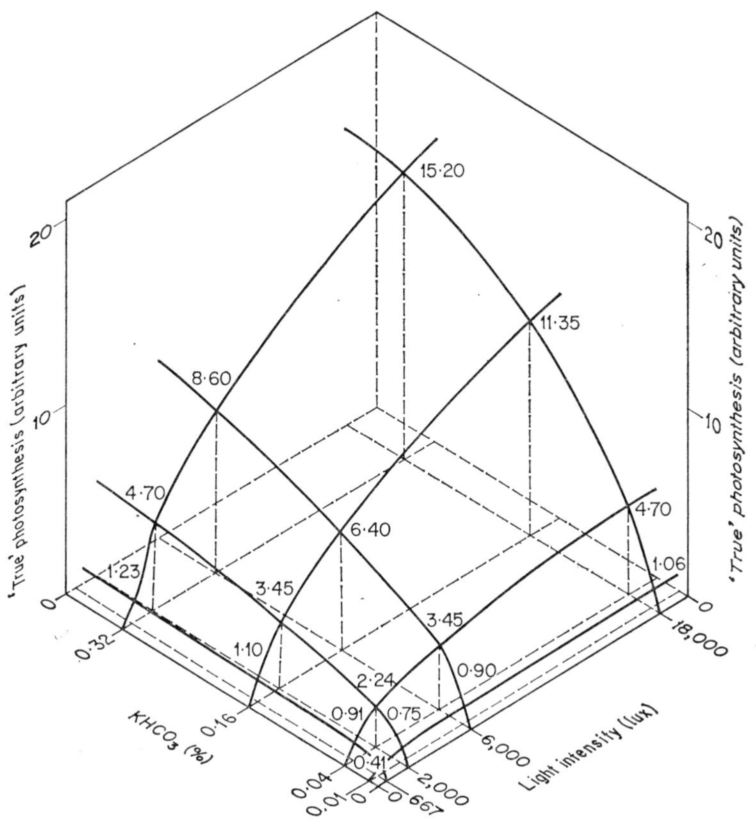

FIG. 4.6 Three-dimensional diagram in isometric projection, combining the curves of Figs. 4.4 and 4.5. Experimental data inserted at points of intersection. Data of R. Harder [135].

there is a large response to increase of carbon dioxide but little response to increasing light; near the front on the extreme left, in the most light-limited part, the converse holds. Up the middle, from front to back, both carbon dioxide and light increase in more or less balanced proportions and with both at high level the highest rate of assimilation is achieved. Note that even so the response falls off as light and carbon dioxide are increased, probably because some other factor, perhaps temperature, is to some extent limiting. If the levels of all the external factors were increased together in balanced proportions the rate of photosynthesis should be increased to the maximum possible for the internal mechanism, that is the rate would be entirely limited by some internal factor—perhaps the quantity of an enzyme. Any further increase in the level of external factors would either be without effect or would cause a fall in rate due to harmful effects. A further complication is that at high temperature the rate of photosynthesis would fall off with time (see Chapter 7).

Harder's data show a large interaction between the two factors light intensity and carbon dioxide supply, such that the higher the level of either factor the greater the response to a given increase in the level of the other. The curves in Figs. 4.4, 4.5 and in Fig. 4.6 whether for assimilation rate plotted against light or against carbon dioxide, thus diverge markedly. We might postulate that instead of being additive, as in Fig. 4.1 and equations (4.1) and (4.2), the effects of the factors have a product relationship:

$$Y = f(A) \times f(B) \times f(C) \ldots \times k \quad (4.3)$$

Then instead of a given increase in a factor having a constant absolute effect whatever the level of the other factor it will have a constant relative or percentage effect. This hypothesis (first put forward by Mitscherlich [229] for growth) could be tested by plotting log (assimilation rate), as the effects should be additive in terms of log values and the curves therefore parallel:

$$\log Y = \log[f(A)] + \log[f(B)] + \log[f(C)] + \ldots + \log k \quad (4.4)$$

Fig. 4.7 where this has been done shows that the curves are indeed much more nearly parallel, showing that to a first approximation a product relationship held; however, they still show a tendency to diverge, especially at the lower values, indicating that the relative or percentage effect produced by each factor increased with the level of the other factor. This is often more marked, in a common type of interaction better fitted by another relation now to be considered.

The discrepancies, between the findings of Blackman and his pupils on the one hand and the Continental workers such as Harder or Warburg [135, 306] on the other, were at last in 1928 resolved by the work of Maskell [213, 214] in Blackman's laboratory. Maskell used detached leaves of cherry laurel (*Prunus lauro-cerasus*) and estimated throughout the 24 hours the rates of uptake of carbon dioxide from air passed through an illuminated chamber containing the leaf. This was done by absorption in baryta and titration of the carbon dioxide from two parallel air streams, only one of which had passed over the leaf. He discovered that under constant light intensity there was a

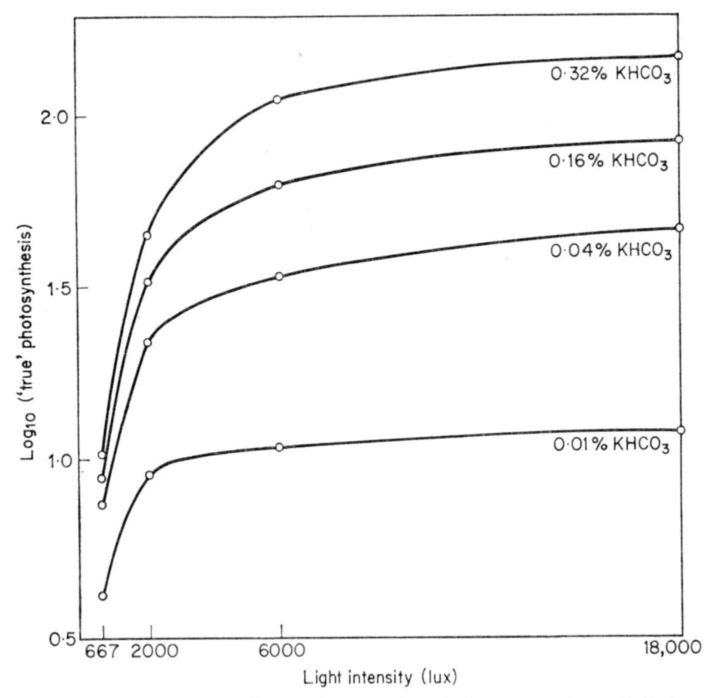

FIG. 4.7 Relation of $\log_{10}$ ('true' photosynthesis), for *Fontinalis*, to light intensity at four carbon dioxide levels. Smooth curves drawn through experimental points. Data of R. Harder [135].

diurnal rhythm of assimilation rate with a minimum at about midnight. The rhythm virtually disappeared when at low light intensity high concentrations of carbon dioxide were used; this suggested that it was due to a rhythmic change in some resistance in the diffusion path—with a high enough external concentration (2·3 per cent) enough carbon dioxide could pass even the highest resistance for light to be always the main factor limiting rate of photosynthesis. Maskell suspected that the varying resistance was due to changing stomatal aperture. He later showed with a porometer that the stomata had in fact a rhythmic movement which ran roughly parallel with the diurnal changes of assimilation, minimal apertures occurring at about midnight. This led him to an extensive experimental investigation of the forms of the curves, at two different light intensities, relating assimilation to the diffusive conductance of the stomata as estimated from readings of his porometer taken during the experiments. On the basis of this investigation he put forward what has come to be known as Maskell's resistance formula.

Maskell stated that: 'CO_2 may be said to be "limiting" . . . when a small change in the CO_2 concentration, other factors being constant, produces an approximately proportionate change in apparent assimilation. For land leaves this relation appears . . . to hold over a range of CO_2 concentrations very much exceeding that in the normal atmosphere, and apparently up to an assimilation rate approaching the temperature or light limit.' For such conditions he put forward a relation which forms the basis for much of Chapter 2, namely, in our symbols:

$$\frac{dq}{dt} = \frac{KPA}{L_{ext} + L_s + L_i + L_{aq} + L_{lt}} \qquad (4.5)$$

where dq/dt is apparent assimilation, K is the diffusion constant for carbon dioxide in air, A is the leaf area, P is the external concentration of carbon dioxide (in bar partial pressure) at a distance from the leaf. L_{ext} . . . L_{lt} are effective lengths in series:

L_{ext} for diffusion up to the leaf, L_s for the stomata, L_i for the intercellular spaces, L_{aq} for the aqueous diffusion path and L_{lt} for the photochemical and chemical phases in photosynthesis. This last is a function of light and temperature. When P is so high that apparent assimilation approaches the light or temperature limits it becomes necessary to make allowance for the changes in concentration at the chloroplast surface, which will also be affected by L_{lt}, but as long as approximate proportionality holds between dq/dt and P this is obviously unimportant.

Maskell went on to deal with the more complex problem, posed by high carbon dioxide concentrations, by combining such a diffusion equation based on Fick's Law (page 43) with an expression suggested by Michaelis and Menten for the relation between rate of reaction and concentration of substrate in a single enzyme system. This Michaelis–Menten equation had been found by Warburg [306] to fit approximately the relation between assimilation by *Chlorella* and carbon dioxide concentration, under high light intensity; it is now known, however, that many enzymes are involved in photosynthesis and the fact that an equation yields an approximate fit to data does not necessarily indicate that it represents the fundamental mechanism concerned. Maskell used it 'as a simple expression of a possible type of relationship between CO_2 and light at the chloroplast surface'; it took the form

$$\gamma = \frac{\gamma_{max} \times P_s}{P_s + k_1/k} \tag{4.6}$$

where γ was the rate of real assimilation per unit area of leaf, γ_{max} was the maximum rate for the light intensity used (the 'light limited' value) and hence a measure of that intensity at the chloroplast, P_s was the concentration of carbon dioxide at the chloroplast surface, k_1 was the velocity constant for the dissociation of a photochemical product which was formed at a rate proportional to the light intensity (γ_{max}) and k was the velocity constant for the combination of this product with carbon dioxide. k/k_1 then gave a measure of the efficiency of the photochemical mechanism at the chloroplast surface. Maskell assumed 'for simplicity' that any resistance in the diffusion path of respiratory carbon dioxide to the chloroplast surface was negligible, or small compared with that to the outside of the leaf (there is now evidence that this assumption is unjustified—page 43). He combined the Michaelis–Menten equation with the diffusion equation:

$$\gamma = \frac{K(P - P_s)}{L} + r \tag{4.7}$$

where L was the sum of the effective lengths up to the chloroplasts and r was the (dark) respiration rate. Substitution of an expression for P_s, found from equation (4.6), in equation (4.7) gave an equation for the curve relating real assimilation γ to increase of external carbon dioxide concentration P as follows:

$$\gamma = \frac{KP}{L} - \frac{K}{L\ k/k_1} \cdot \frac{\gamma}{(\gamma_{max} - \gamma)} + r \tag{4.8}$$

Thus the unmeasurable value of P_s was eliminated, though it was necessary to estimate γ_{max} experimentally with high carbon dioxide concentration. With L estimated, as discussed in Chapter 2, k_1/k would become the only unknown constant in the equation. The converse substitution, of an expression for P_s from (4.7) in (4.6), gave an equation for the relation of γ to increase of light intensity γ_{max}, namely:

$$\gamma = \gamma_{max} \times \frac{\left(\frac{KP}{L} + r - \gamma\right)}{\left(\frac{KP}{L} + r - \gamma + \frac{K}{L\ k/k_1}\right)} \qquad (4.9)$$

For the calculus of the effects of variation in L, γ_{max} and k/k_1 upon the shapes of the curves reference must be made to Maskell's paper, noting the difference in symbols used, especially that his $D =$ our K/L and his $K_L =$ our k_1/k.

On the basis of the theoretical curves derived from such combination of the two relations Maskell concluded: '. . . the divergences found in the form of the curves relating assimilation to CO_2, or to light, represent variations due largely to different relative values of the resistances in the diffusion and photochemical phases of the process. The *Elodea* result [Blackman and Smith] would correspond to a set of conditions with a high resistance to diffusion [L], or a high efficiency of light at the chloroplast surface [k/k_1] or a high light intensity [γ_{max}], while Harder's results would correspond to a set of conditions with a relatively low resistance to diffusion or a low efficiency of light at the chloroplast surface or both of these.'

These conclusions as to the effects of different relative resistances on the shapes of the curves may be illustrated from the simpler equation (4.5) by inserting a series of fictitious values and plotting the resulting light curves; for calculating curves relating assimilation to carbon dioxide the more complex equation would be needed. We may summate all the diffusive effective lengths, L_{ext} to L_{aq}, as L_d and assume that for the chemical resistance L_{lt} to be inversely proportional to light intensity I, that is, $L_{lt} = L_c/I$ assuming constant temperature. Then:

$$dq/dt = \frac{KPA}{L_d + L_c/I} \qquad (4.10)$$

Values of apparent assimilation calculated from this equation are given in Table 4.1. These show that when an internal or external factor (other than carbon dioxide concentration for this relation) is varied, the magnitude of the effect upon assimilation rate will depend upon the ratio of the resistance affected by that factor to the total resistance (compare also page 61 and Table 2.2). Thus the effect of a given change in light intensity will be greater at wide stomatal aperture when L_d is low than with nearly closed stomata when L_d is high and accounts for most of the total effective length.

TABLE 4.1

Apparent assimilation rates in $mm^3\ cm^{-2}\ h^{-1}$, *for ordinary air* (300 *p.p.m. carbon dioxide*) *at* 20°C, *calculated from equation* (4.10) *with varying light intensity* (I) *and diffusive resistance* (*effective length* L_d). $K = 0{\cdot}16\ cm^2\ s^{-1}$; $P = 3{\cdot}0 \times 10^{-1}$ *mbar*; $A = 1{\cdot}0\ cm^2$; 1 *hour* = 3,600 *seconds*; 1 $cm^3 = 10^3\ mm^3$; $L_d = 0{\cdot}67$ *cm chosen as a likely minimal value for a land leaf with wide open stomata* (*cf. page* 60 *and Table* 2.2); L_c *chosen as* 40 *cm to give* $L_c/100 = 0{\cdot}4$ *cm and hence* $dq/dt = 161\ mm^3\ cm^{-2}\ h^{-1}$ *at the highest light and lowest diffusive resistance* (*cf. Table* 2.3), *thus:*

$$\frac{dq}{dt} = \frac{0{\cdot}16 \times 3 \times 10^{-4} \times 1{\cdot}0 \times 3{\cdot}6 \times 10^3 \times 10^3}{0{\cdot}67 + 40/100} = 161$$

Light intensity I ($J\ m^{-2}\ s^{-1}$)	L_c/I (*cm*)	L_d (*cm*)			
		0·67	1·74	10·3	50·0
1	40	4·3	4·1	3·4	1·92
2	20	8·4	8·0	5·7	2·47
5	8	20·0	17·8	9·5	2·98
10	4	37·0	30·1	12·1	3·20
20	2	64·8	46·3	14·1	3·33
50	0·8	118·0	68·0	15·6	3·41
100	0·4	161·0	81·0	16·2	3·43
Light saturation	0	258	99	16·8	3·46

This can be seen by comparing the first and last columns in the table. Similarly, at high light, when L_c/I is small, a given change in stomatal aperture will have a larger effect on assimilation than at low light when L_c/I is a large part of the total (compare the last and first lines of the table).

If the assimilation rates given in Table 4.1 are plotted as tabulated, comparison of the *shapes* of the various curves is made more difficult by the great differences in the light-saturated values to which they tend (as is also true of Fig. 4.5). For Fig. 4.8 the rates have therefore been recalculated as fractions of these values so that all the curves tend towards the same asymptote. It is seen that the lower the diffusive effective length L_d, the more gradual is the change in slope of the curve; the higher L_d, the more the curve resembles the pair of straight lines postulated by Blackman. There is never a discontinuity, however, and the rigid limiting factor relation is a first approximation only, which is approached when one factor is at very low level relative to the others, that is, when the resistance that it affects constitutes a very large part of the total. We may therefore use the term 'limiting factor type of interaction' to cover the whole family of curves and this type of interaction is intended in further reference to limiting factors in this work, unless stated otherwise.

It may be noted that Harder (and also Warburg) used carbonate–bicarbonate buffer mixtures which liberate carbon dioxide at the plant surface as assimilation proceeds (page 83). Thus the diffusive resistance was probably low. Blackman and Smith, on the other hand, used

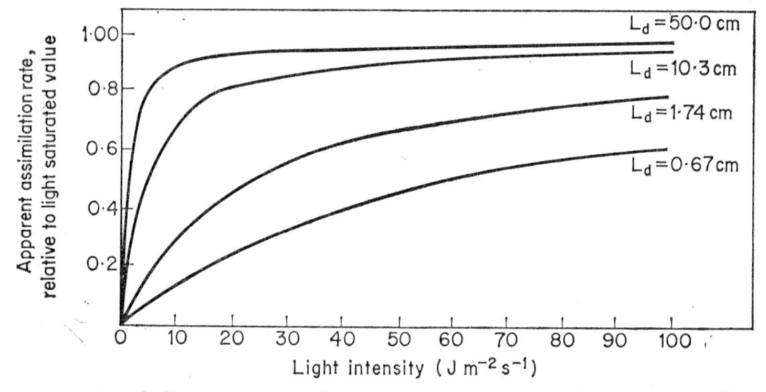

FIG. 4.8 Maskell curves for relation of apparent assimilation rate to light intensity for four different values of diffusive resistance (effective length L_d); recalculated as fractions of light-saturated values from Table 4.1.

carbon dioxide dissolved in the water and there may well have been a considerable diffusive resistance up to the plant.

It seems probable that no two external factors affect any activity of the plant, such as photosynthesis, entirely independently (as in Fig. 4.1); similarly, it is likely that there is interaction between all the internal factors and also between these and all the external factors; to complicate matters further, all the various activities of the plant, such as photosynthesis, respiration, translocation, transpiration, cell division, cell extension and so forth, are interdependent. The exigencies of language and of the human brain make it impossible to discuss everything simultaneously and although there are mathematical methods for dealing with the interactions between three or more factors I propose here to confine myself in the main to discussing not more than two at a time. Even so I shall not attempt to deal with all possible pairs but select a few instances of factors which are particularly closely associated.

Certain pairs or sets of factors can be considered as combining to be, in effect, a single factor. It is apparent that rate of carbon dioxide supply to the chloroplasts depends not only on the external concentration but on all the intervening resistances to diffusion, as well as on the chemical resistances which are controlled especially by light and temperature. For a land leaf, changes in either wind velocity or stomatal aperture alter the overall effective length of the diffusion path to the chloroplast (L_d of equation 2.6) and thus both are part of the diffusive resistance term. The external concentration of carbon dioxide only tells us the magnitude of the potential difference term if we can know or assume the internal concentration at some known point, for example at the chloroplast surface, and this will depend upon the light intensity, the efficiency with which the light is used and the temperature.

In view of the many variables it is not surprising that different workers, using a great variety of plant material and of methods, should have obtained very different shapes of curve relating photosynthesis to external carbon dioxide concentration, and very different interactions with light intensity. I shall not discuss further such differences. Most of them could be reconciled on the lines of the more complex scheme put forward by Maskell, although this does not necessarily mean that it is correct in all respects. With one or two exceptions the curves have in the past been obtained using concentrations of carbon dioxide greater than the normal 300 p.p.m. by volume and often very much greater. The shape of curve below this level has been little studied experimentally, because, until recently, sufficiently sensitive methods have not been available—it is generally assumed to be linear. Obviously, in this region the question of what correction, if any, to make for respiration is important, and the work of Brown and his co-workers and others discussed in the next chapter suggests that the answer is not a simple one. Unless concurrent measurements of respiration in light can be made, probably the safest course at present is to study the curves relating apparent photosynthesis to carbon dioxide concentration as was done by Maskell.

At the beginning of this chapter two types of experimental approach were mentioned. The consideration of the interaction of factors, both external and internal, should have made it clear that ideally the two types of experiment should be combined factorially. Plants should be grown under a number of experimentally controlled external conditions, producing modifications in their internal factors; photosynthesis should then be measured with the plants maintained under the same conditions or transferred to the others in all possible combinations. In this way the short-term responses (in terms of photosynthesis) of all the different types of plant produced, to all the various conditions, might be elucidated. Some work of this sort had been carried out by the methods of growth analysis applied to shading experiments in natural daylight [24] or to experiments with artificial light under controlled conditions [271].

5: Respiration in Light and Related Topics

A. THE MINIMUM INTERCELLULAR SPACE CARBON DIOXIDE CONCENTRATION (Γ)

BLACKMAN [18] in 1895 found that 'In bright light a fully green leaf assimilates all the CO_2 that is forming by respiration and none escapes from it.' Maskell [213] in 1928 failed to detect any carbon dioxide emitted by an illuminated leaf into a slow stream of carbon dioxide-free air, again implying that all respiratory carbon dioxide was reassimilated (unless respiration was suppressed by light). Maskell [214] therefore made his theoretical curves for *apparent* assimilation *v.* external carbon dioxide concentration pass through the origin. As early as 1851 Garreau had found that leafy shoots in light did not consume all the carbon dioxide in a closed system, but Blackman attributed this to carbon dioxide production by the non-photosynthetic plant parts. The same criticism could be applied to later similar findings with whole plants growing in soil [226, 292] but Gabrielsen [111] found that detached leaves of elder (*Sambucus nigra*) enclosed in a limited volume of air, either with or almost without carbon dioxide, and illuminated at 10,000 lux, brought the concentration to a steady-state value of 90 p.p.m. (0·009 per cent) (Fig. 5.1). Heath [138, 141] also found that if air was passed through the intercellular spaces of brightly illuminated leaves of *Pelargonium zonale* and *Begonia sanguineum* the carbon dioxide concentration was reduced to about 90 p.p.m. and 100 p.p.m. respectively, but no further. He interpreted this as the minimum concentration that could occur at the surfaces of the assimilating cells and put forward as the simplest hypothesis 'that the

0·01% CO_2 represents the concentration which must be maintained in the intercellular spaces to give a rate of assimilation high enough to balance the respiration rate; in fact that it is the "compensation point" for CO_2 concentration'. He was, however, impressed by the almost identical concentrations found for leaves with very different degrees of dispersal of the chloroplasts—the *P. zonale* leaves had chloroplasts in every epidermal and mesophyll cell while in *B. sanguineum*

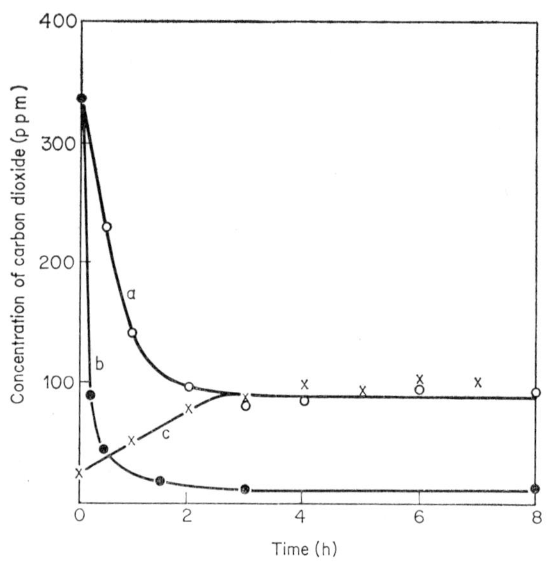

FIG. 5.1 Time course of carbon dioxide concentration changes in a closed vessel of air: *a*, with leaflets of *Sambucus nigra*; *b*, with filter paper moistened with NaOH solution; *c*, as *a* but with an initial concentration of 24 p.p.m. carbon dioxide. Light intensity 10,000 lux in all cases. After Gabrielsen [111].

there were extensive chlorophyll-free hypodermal layers [139] (Fig. 5.2). This seemed difficult to explain on the basis of a simple balance between respiration and assimilation and he suggested that the tension of free carbon dioxide in the cytoplasm at the cell surface might be in some way 'buffered' to a constant value of 100 p.p.m. Therefore, although the term carbon dioxide compensation point has been used by a number of workers [68, 265] for the measured steady-state concentration in a closed system, and effectively it is that, Heath and co-workers have preferred to use the non-committal symbol Γ (gamma) for this concentration in order to avoid any implications as to the precise relations of respiration and assimilation involved. Later work has indicated that to treat it as the resultant of the rate of photosynthesis and the (dark) rate of respiration may indeed be an over-simplification (see below).

Although the leaf of *B. sanguineum* could reduce the air that was passed through the intercellular space system to practically the same low carbon dioxide concentration as could that of *P. zonale*, it left about twice as much carbon dioxide in an air stream passed over the leaf surface. This may be attributed to the inefficient arrangement of stomata in localized patches and to the longer diffusion paths to the chloroplast-containing tissues (Figs. 5.2 and 5.3).

When a steady state has been reached in a closed circuit experiment the measured concentration Γ must also be the concentration in the intercellular space system (assuming that virtually all gas exchange takes place through the stomatal pores); Γ thus specifies the *minimum* possible concentration that could be reached in that system, for a similarly illuminated leaf in ordinary air at the same temperature, and tells us the maximum concentration difference that could occur across the stomata in such conditions. Clearly, these limits will not in fact be reached except with completely closed stomata (possibly on first illumination after darkness); with 300 p.p.m. (0·03 per cent) carbon dioxide outside the leaf the internal concentration will usually be much greater than Γ and the potential difference therefore much less than $(300 - \Gamma)$ p.p.m. Penman and Schofield [255] estimated that under field conditions 'a luxuriant crop grows with a concentration of gaseous carbon dioxide inside the leaf which is only 10–20 per cent less than that in the outside air. For normal crops the difference between outside and inside can only be a few per cent.' This was based on a comparison of transpiration for a field crop of sugar beet, as estimated for the whole season from meteorological data, with assimilation as estimated from the final dry weight of the crop (page 63). Hence they concluded that the external, stomatal and hydro-diffusion resistances were in the ratios 4 : 1 : 30 to 4 : 1 : 100. This almost certainly greatly overestimates the resistance in the *liquid* phase as the calculations were based on the assumption that photosynthesis was entirely limited by diffusion of carbon dioxide. During much of the growth period and especially for the inner leaves light-limitation of photosynthesis must have caused increased carbon dioxide concentrations at the chloroplasts and thus have been largely responsible for the high average concentrations in the intercellular spaces. To put it another way, a large part of the resistance attributed to diffusion in liquid must in fact have been in the chemical resistance term (page 60). More recent estimates made by Monteith [231] on

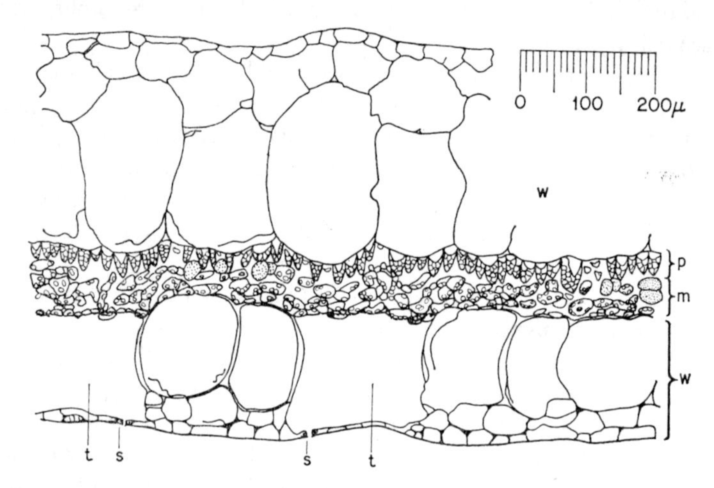

FIG. 5.2 Transverse section of leaf of *Begonia sanguineum*. W, water storage tissue; p, palisade parenchyma; m, spongy mesophyll; s, stoma; t, cylindrical tunnel abutting on stomatal group. From Heath [139].

FIG. 5.3 Lower epidermis of leaf of *Begonia sanguineum*, showing groups of stomata. From Heath [139].

single bright days for a bean crop show, at midday, the carbon dioxide falling from about 280 p.p.m. at the 'effective surface' of the crop to about 180 p.p.m. in the intercellular spaces. The estimate of intercellular space carbon dioxide depends upon a figure for stomatal resistance obtained from humidity measurements, transpiration being assumed to take place from leaves at the same effective crop surface as those concerned in photosynthesis. However, although the shaded leaves can contribute relatively little to photosynthesis, for the transmission coefficient of leaves for visible light (400–700 nm) is only about 0·1 [232], transpiration can also be caused by energy at longer wave lengths and in the range 400–2,000 nm the transmission coefficient of leaves is of the order of 0·25. Thus more leaf layers may be concerned in transpiration and hence more stomata, whose combined conductance must be greater than that of the stomata responsible for photosynthesis. It follows that intercellular space carbon dioxide may be somewhat overestimated by this method.

As another approach, we may calculate the concentrations in the sub-stomatal cavities corresponding to the assimilation rates in Table 2.3; this can be done by multiplying 300 p.p.m. by the ratios of the sum of the estimated internal effective lengths (0·845 cm—page 60) to the total effective lengths (Table 2.2). For 'still' air the values found are 98, 69 and 21 p.p.m. with stomatal apertures of 10 μ, 3 μ and 1 μ respectively; for a 3·5 km h^{-1} wind the corresponding concentrations are 215, 113 and 23. A similar calculation applied to Gaastra's [110] estimates of resistances suggests that with his lowest 'mesophyll resistance' (page 64) and at the highest light intensity, even though the stomata were very wide open, the intercellular space carbon dioxide would have had a concentration of 105 p.p.m. These estimates all ignore production of carbon dioxide by respiration within the leaf, which would of course make it impossible for the concentration to fall below Γ. They do suggest, however, that in bright light the concentrations within the leaf may approach rather near to Γ, especially in 'still' air.

It is interesting to note that stomata have been found to open progressively with reduction in the carbon dioxide concentration [142], (in the intercellular spaces rather than that outside the leaf, at least for *Pelargonium* [140]), but that such reduction below a value approximating to Γ seems to be without further effect. Thus stomata of wheat [153] at 26–30°C and of turnip [110] at 20°C showed no change of aperture between 100 p.p.m. and zero carbon dioxide concentration; a measured value of Γ for wheat leaves at 25°C was 78 p.p.m. [148]; Γ values for turnip are not available. Maize leaves can give Γ values approximating to zero (see below) and maize stomata have been found to show more opening in carbon dioxide-free air than at 30 p.p.m., at 25°C [217]. If of general application such results suggest that stomata respond to carbon dioxide only down to about the minimum intercellular space concentration that could occur in nature at the temperature concerned.

The disentangling of cause and effect in experiments on carbon dioxide concentration and wind velocity is made difficult by the mutual effects of photosynthesis, acting through the intercellular space carbon dioxide concentration, upon stomatal aperture and of stomatal aperture upon photosynthesis, again acting through that concentration. In so far as increase of wind velocity increases the carbon dioxide supply to the leaf (reduces the total effective length) it is also likely to cause partial stomatal closure (increase the total effective length). There are two experimental approaches to such problems: one is to estimate stomatal diffusive conductance concurrently with photosynthesis, and to use a curve-fitting method of analysis, as was done by Maskell. The other is to eliminate the stomatal factor, either by using plant material without motile stomata (algae, liverworts, mosses, submerged leaves of water plants) or, as was attempted by Heath [138, 141], by forcing air at a constant rate of flow through the stomata, and the intercellular space system of the leaf. This last method showed, as already noted, that the air emerging from the leaf had a nearly constant carbon dioxide content of about 100 p.p.m.—this held over a range of rates of carbon dioxide supply from 2 to 16 mg CO_2 dm^{-2} h^{-1} which 'implies remarkable efficiency of uptake once the CO_2 has entered the leaf' [141]. Perhaps the internal resistances for assimilation are not as high as suggested in Chapter 2.

The fact that a photosynthesizing leaf cannot reduce the carbon dioxide content below a certain value (Γ) has important bearings upon the technique of kinetic experiments, especially open circuit experiments with ordinary air of 300 p.p.m. carbon dioxide content. Fig. 5.4 shows the results of two open circuit experiments carried out by Heath [141] on the same 9 cm^2 area of *Pelargonium* leaf on successive days. Ordinary air was passed over the leaf surfaces in very shallow attached chambers at a constant flow of 2 l h^{-1}. In Fig. 5.4*a*, apparent assimilation, corrected by direct proportionality from the estimated mean carbon dioxide concentration (equation 3.3) to one of 300 p.p.m., is plotted against an estimate of stomatal diffusive conductance after the manner of Maskell [214]. The curves much resemble those of Maskell in that apparent assimilation rose rapidly with stomatal opening at first but that at the higher conductances further stomatal opening had little or no effect. Taken by themselves these curves would suggest that at the wider stomatal apertures some other resistance was severely limiting assimilation, but inspection of Fig. 5.4*b* shows that at about the point where assimilation ceased to rise the residual carbon dioxide in the air leaving the leaf chamber had fallen to 100 p.p.m.; in fact at this point *all* the available dioxide was being assimilated and any further reduction in resistance could have no effect because the potential

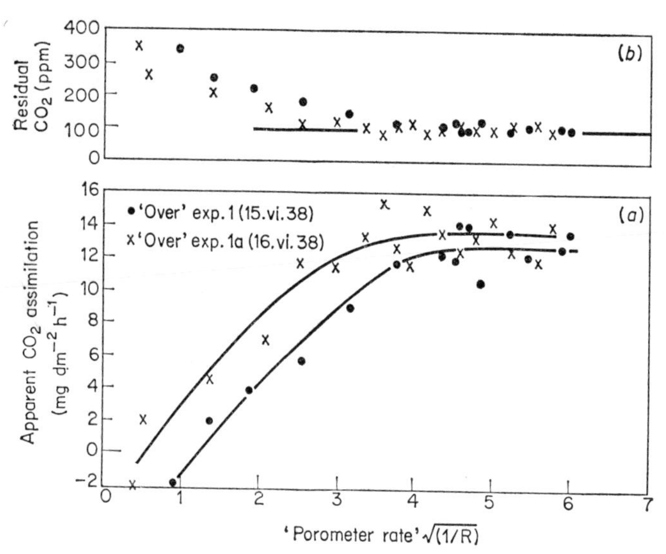

FIG. 5.4 *a*, Apparent assimilation, 'corrected' to a carbon dioxide concentration of 300 p.p.m., plotted against an estimate of stomatal diffusive conductance, for two open circuit experiments with the same leaf of *Pelargonium zonale*. *b*, Residual carbon dioxide concentration in the air that had passed over the leaf. From Heath [141].

difference had fallen to zero. This result emphasizes the importance of very high flow rates in such experiments. It is also important to have extreme turbulence in the leaf chamber, or else use a very shallow one, for with streamline flow in a deep chamber the air in contact with the leaf might be approaching Γ while that remote from the leaf had a much higher content—the mean residual concentration would then appear fictitiously high. The fact that only the carbon dioxide content above Γ is available for assimilation means that instead of equation (3.2) it would be an improvement to estimate the mean concentration over the leaf from

$$\overline{C} = \frac{C_i - C_f}{\ln\left(\dfrac{C_i - \Gamma}{C_f - \Gamma}\right)} \tag{5.1}$$

With one exception, Maskell's experiments were carried out with mean concentrations of 1,300 p.p.m. upwards and it seems unlikely that the concentration at the leaf surface ever approached Γ. There was, however, another source of error in that the porometer cup was left permanently attached to the leaf throughout each of his experiments. The stomata on which he based his estimates of diffusive conductance were thus enclosed in a small volume of still air and owing to the reduction in carbon dioxide content must have opened wider in the light than those on the rest of the leaf, for which assimilation was being measured. This would extend his curves to the right, somewhat as in Fig. 5.4*a*, and may well have affected the values he calculated in fitting the curves, though it is unlikely to have invalidated his general scheme. Such a criticism does not apply to the data in Fig. 5.4, for here the leaf chamber was itself the porometer cup attached to the leaf, so that the same stomata were concerned in both measurements.

While we are discussing questions of technique it may be mentioned that the criticisms of closed circuit experiments (page 76) do not apply when these are used to determine Γ after a steady state has been reached.

The nearly constant value of Γ at about 100 p.p.m., found in the earlier work, seems to have been largely fortuitous. It was soon found that Γ increased markedly with temperature at a given light intensity (see Table 5.1 and page 197 below), as might be expected for some kind of balance between photosynthesis and respiration. However, the published Γ values for thirteen species, in fairly bright light and between 20° and 30°C, lie within the range 48–122 p.p.m. Of the four species with Γ values outside this range (Table 5.1), *Chlorella* might be expected to reduce the ambient carbon dioxide concentration to a very low value because of the close proximity of the respiratory and assimilatory centres; no such obvious reasons suggest themselves for the other three, namely, maize and sugar cane which have extremely low Γ values, and Norway maple which has a very high one. Maize leaves are without chloroplasts in the ordinary epidermal cells, which presumably respire. Peculiarities found in maize leaves which may have a bearing on the very low Γ values achieved are: first, the starch-forming and almost grana-free chloroplasts in the endodermal sheaths of the vascular bundles (Plate 3B); secondly, the observation that the main early products of photosynthesis, found when $C^{14}O_2$ was fed for a short period (4 s), were the C4 dicarboxylic acids oxaloacetate, malate and aspartate, rather than the Calvin cycle products [135*a*]. Sugar cane shared this pattern of labelling as did a number of other tropical Gramineae of closely related tribes. Grasses of another group of tribes (including wheat) showed little or no label in C4 dicarboxylic acids and this was also true of plants from a wide range of other families. A species of *Cyperus*, however, was found to have the C4 dicarboxylic acids as very early photosynthetic products [135*a*]. The Γ figures for *Dactylis* and *Triticum* (Table 5.1) are fairly high, though Meidner (unpublished) has found somewhat lower values for another wheat species. The work of El-Sharkaway, Loomis and Williams [69*a*] with the tropical dicotyledonous plant *Amaranthus edulis* suggests that it should produce very low Γ values, for they were unable to detect carbon dioxide emission into carbon dioxide-free air in open circuit experiments with bright light; Γ measurements were not made.

Since Γ determines the maximum possible potential difference for the diffusion of carbon dioxide from ordinary air into the leaf, it has been suggested [148] that it gives a good (inverse) measure of the efficiency for net carbon dioxide absorption, under the given conditions of temperature and water deficit (see next chapter), with the stomatal factor eliminated and carbon dioxide limiting. (Light intensity has little effect on Γ except at low levels, though these rise with temperature: Chapter 7.) If this suggestion be accepted it is of interest to compare the generally rather small range of Γ in Table 5.1 with the similarly small range of mean net assimilation rates [145] for different species during the main growing season (that is in fairly bright light). Some values are presented in Table 5.2. It seems that, with a few notable exceptions, herbaceous species differ relatively little in their photosynthetic efficiency and the very great differences in relative growth rates of different species must therefore be mainly due to differences in leafiness [311] as measured by leaf area ratio (page 94). The value of NAR for maize at Ibadan (1·52) is one of the highest recorded and it is tempting to attribute this to a steep carbon dioxide gradient into the leaf, although it was somewhat overestimated as the measured leaf area did not include the leaf-sheaths. The very high NAR found for *Helianthus* in New South Wales (2·09) was associated

TABLE 5.1

Minimum intercellular space carbon dioxide concentration (Γ) for isolated leaves, thalli etc. in bright light and air of normal oxygen concentration at temperatures between 20° and 30°C

Species	*Place*	Γ (*p.p.m.*)	*Light* (*lux*)	*Temperature* (°C)	*Source*
Sambucus nigra	Copenhagen	90	10,000	24–28	Gabrielsen 1948*a* [111]
Begonia sanguineum	London	100	26,460	25–30	Heath 1951 [141]
Pelargonium zonale	London	90	26,460	25–30	Heath 1951 [141]
,, ,,	Frankfurt	⊙ca 60	13,000	20	Egle & Schenk 1953 [68]
,, ,,	,,	⊙ca 80	,,	25	,, ,, ,,
,, ,,	,,	⊙ca 120	,,	30	,, ,, ,,
,, ,,	London	49	9,700	20	Heath & Orchard 1957 [150]
,, ,,	,,	62	,,	25	,, ,, ,,
,, ,,	,,	86	,,	30	,, ,, ,,
Pelargonium sp.	New Haven Conn.	65	75,600	23	Moss 1962 [235]
Allium cepa	London	65	9,700	20	Heath & Orchard 1957 [150]
,, ,,	,,	90	,,	25	,, ,, ,,
,, ,,	,,	122	,,	30	,, ,, ,,
Coffea arabica	London	58	9,700	20	Heath & Orchard 1957 [150]
,, ,,	,,	82	,,	25	,, ,, ,,
,, ,,	,,	121	,,	30	,, ,, ,,

Pheonix reclinata	Pietermaritzburg	48	15,100	25	Meidner 1961 [216]
,, ,,	,,	78	,,	30	,, ,,
Nicotiana tabacum	New Haven Conn.	60	75,600	23	Moss 1962 [235]
Lycopersicum esculentum	New Haven Conn.	75	75,600	23	Moss 1962 [235]
Acer platanoides	New Haven Conn.	145	75,600	23	Moss 1962 [235]
Dactylis glomerata	New Haven Conn.	60	75,600	23	Moss 1962 [253]
Triticum sativum	London	78	9,700	25	Heath & Meidner 1961 [148]
Saccharum officinarum	New Haven Conn.	7	75,600	23	Moss 1962 [235]
Zea mays	New Haven Conn.	9	75,600	23	Moss 1962 [235]
,, ,, (Glasshouse)	Pietermaritzburg	0·5	1,600	25	Meidner 1962 [217]
,, ,, (Glasshouse)	,,	2·8	,,	30	,, ,,
,, ,, (Glasshouse)	,,	0·0	6,500	25	,, ,,
,, ,, (Glasshouse)	,,	0·0	,,	30	,, ,,
,, ,, (Out-doors)	,,	1·9	,,	20	,, ,,
,, ,, (Out-doors)	,,	2·7	,,	25	,, ,,
,, ,, (Out-doors)	,,	3·9	,,	30	,, ,,
Pellia epiphylla	Frankfurt	⊙ca 75	3,800	20	Egle & Schenk 1953 [68]
Fegatella conica	Frankfurt	⊙ca 70	3,800	20	Egle & Schenk 1953 [68]
,, ,,	,,	⊙ca 85	,,	25	,, ,, ,,
,, ,,	,,	⊙ca 105	,,	30	,, ,, ,,
Chlorella vulgaris	Frankfurt	⊙ca 5	2,940	20	Egle & Schenk 1952 [67]

TABLE 5.2

A. Mean net assimilation rates during the vegetative phase in the main growing season in temperate climates: g dm^{-2} week^{-1}

Species	*Conditions*	*Place*	*Year*	*Period*	*NAR*	*Source*
Hordeum vulgare	Sand culture	Rothamsted	1921–4	May–June	0·55	Gregory 1926 [131]
Sinapis alba		Bonn, Germany	1885	19/5–23/6	0·51	Hornberger (Boysen Jensen 1932 [33])
Mangold	Field crop	Rothamsted	1934	July–Oct.	0·60	Watson & Baptiste 1938 [313]
Sugar beet	,, ,,	,,	1934	July–Oct.	0·68	,, ,, ,,
Cucumber	Glasshouse	Cheshunt	1917	11/6–10/7	0·41	Gregory 1918 [130]
Lycopersicum esculentum	,,	,,	1932	August	0·61	Bolas *et al.* 1938 [32]
Dactylis glomerata	Dense sward	Rothamsted	1934	17/6–29/7	0·24	Richards (Heath & Gregory 1938 [145])
Poa trivialis	,, ,,	,,	1934	1/7–29/7	0·40	,, ,, ,, ,,
Lolium multiflorum	,, ,,	,,	1934	17/6–29/7	0·32	,, ,, ,, ,,

B. Maximal net assimilation rates for small plants without mutual shading in temperate climates: g dm^{-2} week^{-1}

Species	*Conditions*	*Place*	*Year*	*Period*	*NAR*	*Source*
Lolium multiflorum		Oxford			0·82	Blackman & Black 1959 [22]
Helianthus annuus		,,			0·98	,, ,, ,,
,, ,,		Invergowrie			1·19	,, ,, ,,

C. Maximal net assimilation rates under tropical conditions or with high insolation and temperature: g dm^{-2} week^{-1}

Species	*Conditions*	*Place*	*Year*	*Period*	*NAR*	*Source*
Zea mays		Ibadan, Nigeria			1·52	Blackman & Black 1959 [22]
Ipomœa cœrulea		,, ,,			0·82	,, ,, ,,
Lycopersicum esculentum		,, ,,			0·91	,, ,, ,,
Helianthus annuus		Kampala, Uganda			1·37	Huxley 1963 [176]
,, ,,		Adelaide			1·36	Blackman & Black 1959 [22]
,, ,,		Deniliquin, N.S.W.			2·09	Warren-Wilson 1966 [310]

with remarkably intense insolation. Data for NAR of sugar cane do not seem to be available but it may be noted this crop holds the record for dry matter production [22].

If Γ represents a simple balance between photosynthesis and respiration, carbon dioxide may be pictured as diffusing from the respiring mitochondria to the photosynthesizing chloroplasts and also out into the intercellular spaces, in proportions depending on the relative

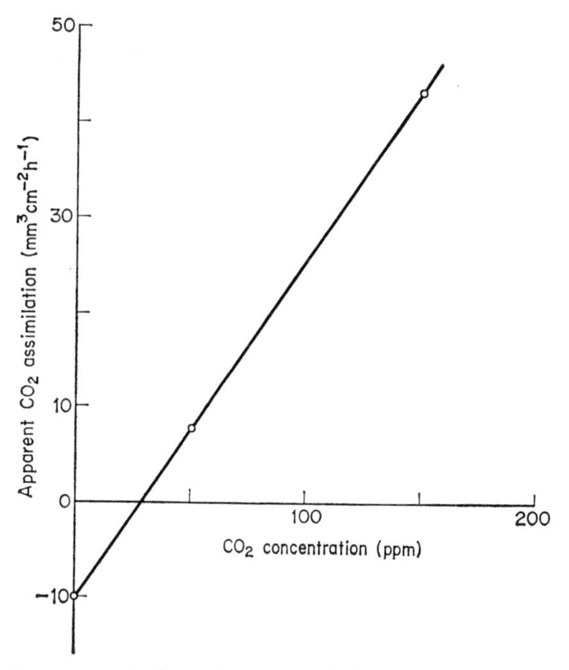

FIG. 5.5 Apparent assimilation by a turnip leaf at low carbon dioxide concentrations. Light intensity 40·5 $J\,m^{-2}\,s^{-1}$ (about 11,600 lux); inlet air temperature 20°C. After Gaastra [110].

gradients (potential difference ÷ effective length) for the two paths as suggested at the end of Chapter 2. For leaves of *Sambucus*, *Pelargonium* and wheat, output of carbon dioxide into virtually carbon dioxide-free air of normal oxygen concentration was found to proceed at approximately the same rate in bright light as in darkness [111, 247, 151]. *If* it could be assumed that light had no effect on respiration rate (as assessed by carbon dioxide production), this would imply that virtually no reassimilation of respiratory carbon dioxide occurred and hence that the resistances in the paths from the mitochondria to the chloroplasts were very high indeed compared with those to the exterior. This could explain the relatively high Γ values found for these species. If these assumptions held, the relation between apparent assimilation and low values of carbon dioxide concentration would be virtually linear (at high light intensity) because of severe carbon dioxide limitation; the line would cut the carbon dioxide axis at Γ and the assimilation axis at a negative value equal to the (dark) respiration rate. This type of relation was obtained by Gaastra for turnip, though as seen in Fig. 5.5 he had no observations at concentrations below Γ except the respiration rate measured in darkness. If carbon dioxide concentration tended to be 'buffered' at Γ (page 132) the curve would be steeper there than elsewhere and hence might be expected to be more or less S-shaped. This could be brought about by, for instance, an effect of carbon dioxide concentration on respiration rate in light. The matter is complicated by the possibility of carbon dioxide exchange by dark reactions (page 92), or of photoxidation (non-specific oxidations catalysed by chlorophyll in light) which is said to occur under such conditions of extreme carbon dioxide deficiency [96] (but see page 194). The technical difficulties of estimating carbon dioxide output at known ambient concentrations below Γ are considerable but not insurmountable and a first attempt is discussed below; a thorough exploration of this region of the curve, with the measurement of both carbon dioxide and oxygen, and preferably with the use of isotopes, might well be rewarding.

Addendum

Recently Tregunna and Downton [296*a*] have found Γ values approximating to zero for *Amaranthus edulis* (page 139) and nine other species of the same genus as well as for a number of other dicotyledons in the Amaranthaceae, Portulacaceae and Chenopodiaceae. *Atriplex rosea* gave a Γ of zero but *A. hastata* and *A. oblongifolia* gave Γ values of 35 and 40 p.p.m. respectively, showing that the division into low and high Γ plants can cut across taxonomic grouping. They have also [61*a*] correlated very low Γ, both in dicotyledons and monocotyledons, with the presence of bundle sheaths containing numerous starch-forming chloroplasts and (generally) absence of starch in the mesophyll; also with the data of Hatch *et al.* [135*a*] for C4 acid formation in monocotyledons (page 139). These observations suggest highly efficient removal of photosynthates from the mesophyll to the vascular bundles in low Γ plants, with starch formation when necessary to maintain a sugar gradient. Downton and Tregunna [61*a*] suggest that zero Γ may result from the normal substrate for carbon dioxide production being used in the C4 acid pathway of photosynthesis instead of being decarboxylated.

B. *INHIBITION OF RESPIRATION BY LIGHT—THE 'KOK EFFECT'*

Kok [190, 191, 192] first announced in 1948 that for *Chlorella*: 'Upon illumination, dark respiration is completely suppressed and substituted by a mechanism providing the cell with a form of energy, obtained with a high yield from absorbed light quanta.' He used a Warburg

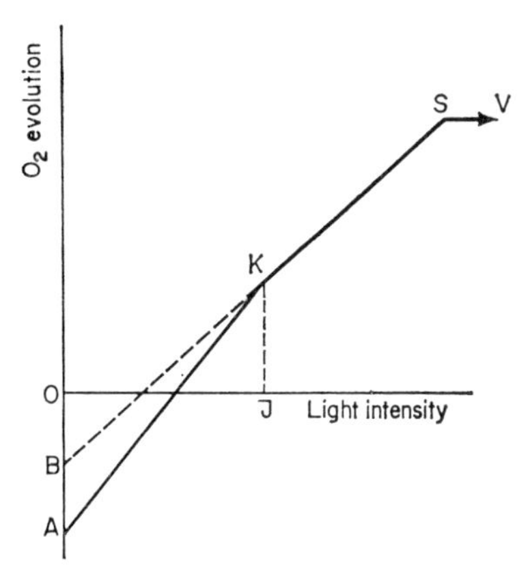

FIG. 5.6 Diagram of relation of net oxygen exchange to light intensity, for *Chlorella*. AK, steep portion; KS, less steep portion; SV, horizontal portion (light-saturated); OA, oxygen uptake rate measured in darkness; OB, oxygen uptake rate (estimated) for light. J, lowest light intensity used in the experiments. After Kok [190].

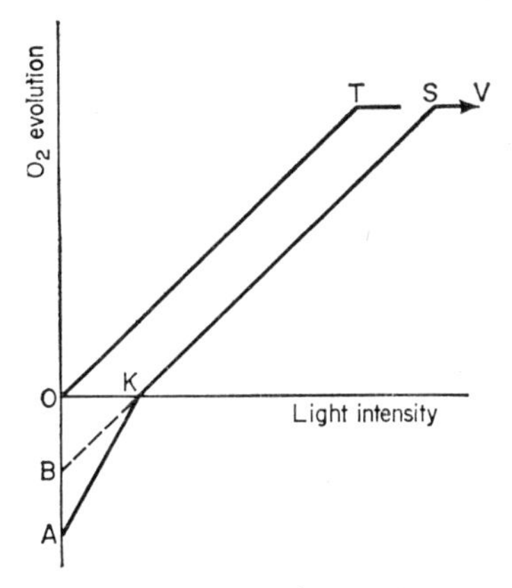

FIG. 5.7 As Fig. 5.6 but with point K moved to coincide with the light compensation point. OT, relation assumed for 'true' photosynthesis. The slope AK is twice that of KS and hence up to the compensation point the quantum requirement is halved. After Kok [190].

technique (page 85) and presented diagrams (Figs. 5.6 and 5.7) of net oxygen uptake or evolution plotted against light intensity. These not only showed a sharp break (S) where light became saturating, as postulated by F. F. Blackman (page 116), but another (K) at the lowest light intensity used (J in Fig. 5.6). He considered that K might be assumed to occur at the light compensation point (page 190) as in Fig. 5.7. An example from his 1951 paper, including observations below the light compensation point, is shown in Fig. 5.8. The ratio of the slopes of the two lines varied with the conditions of growth of the

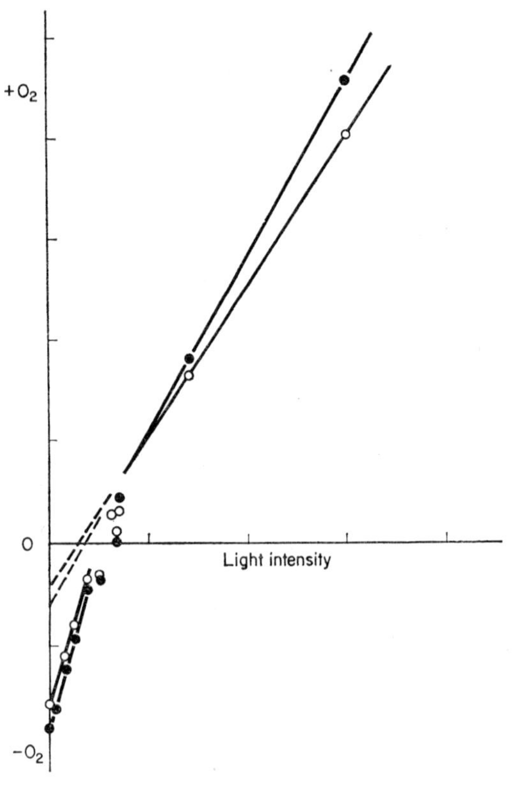

FIG. 5.8 Two series of measurements (open and closed circles) of net oxygen uptake or output in relation to light intensity, for *Chlorella*; in each the dark value was measured first and the light intensity then increased stepwise. After Kok [192].

algae up to a maximum of 2 : 1, when an extrapolation of the upper line to zero light intensity gave an estimate of respiration (oxygen uptake) in light as half that in darkness (OB as compared with OA in Fig. 5.7). Kok [190] concluded that up to the compensation point the quantum requirement per molecule of oxygen released was half the normal value of 8 (page 220) and that there was a coupling between photosynthesis and respiration leading to the situation set forth in the quotation at the beginning of this section thought that the 'form of energy' might well be energy—rich phosphate bonds to replace those formed in the dark by normal respiration.

Van der Veen [301] presented a similar diagram, showing this 'Kok effect' for carbon dioxide uptake by a fragment of tobacco leaf, at an unspecified temperature and in 3·6 per cent carbon dioxide; changes of

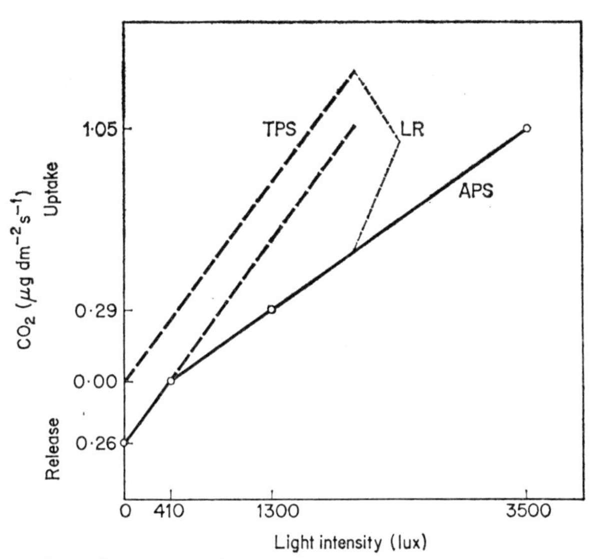

FIG. 5.9 The Kok effect in tobacco leaves. APS, observed relation to light of apparent photosynthesis, in terms of carbon dioxide, at 25°C and 300 p.p.m. carbon dioxide concentration; TPS, hypothetical true photosynthesis, implying that light respiration (LR) had the dark value up to the light compensation point and then increased linearly with light intensity. After Decker [58].

concentration were measured by the diaferometer method (page 80). Gabrielsen and Vejlby [114] were unable to repeat this result, with potato and beet leaves, unless they combined high carbon dioxide concentration with low temperature, and they suggested that the beginning of narcotization by carbon dioxide might be involved; it may be noted that Kok's experiments also were carried out with high carbon dioxide concentrations. However, Decker [58], working with attached tobacco leaves and 300 p.p.m. mean carbon dioxide concentration in a closed circuit, obtained results (Fig. 5.9) much like those of van der Veen.

Decker pointed out that the explanation adopted by Kok and accepted by van der Veen tacitly assumed that the relation of true photosynthesis to light was linear and that the line passed through the origin and was parallel to the upper limb of the curve for apparent photosynthesis (as shown in Fig. 5.7). However, a second possible assumption was that the respiration (carbon dioxide production or oxygen uptake) was equal in light and darkness, in which case the curve for true photosynthesis must change in slope, parallel to that for apparent photosynthesis. A third possible assumption, which Decker preferred, was that the relation of true photosynthesis to light was linear and the line parallel to the lower limb of the apparent photosynthesis curve; this implied constant respiration up to the light compensation point and then a regular increase with light intensity (Fig. 5.9). A fourth possibility was, of course, that the line for true photosynthesis was parallel to neither limb of the apparent photosynthesis curve and later work discussed below shows this to be true at least in some cases (Figs. 5.14 and 5.15).

Emerson and Chalmers [78], like Gabrielsen and Vejlby, were unable to confirm the Kok effect: under all the combinations of cultural and experimental conditions that they tried, *Chlorella pyrenoidosa* gave photosynthesis as a linear function of light intensity from zero up to several times compensation, with no change of slope at or near the compensation point. Emerson [74] concluded: 'The small amount of positive evidence for the "Kok effect", together with negative results in several instances of search for confirmatory evidence, leads to the conclusion that a substantial difference in quantum yield above and below the compensation point is not a general phenomenon. If it is something associated with special conditions, these conditions have yet to be specified. It seems equally possible that the apparent positive evidence arises from changes in rate of respiration between the dark and light intervals chosen for calculation of rate of photosynthesis.'

C. MASS SPECTROMETER STUDIES WITH MARKED OXYGEN AND CARBON DIOXIDE

The early experiments of A. H. Brown [42, 245], in which uptake of marked oxygen in light was measured with a mass spectrometer, indicated that this went on at the same rate as in darkness, both in *Chlorella* and higher plants (page 71). Later [44] it was considered that corrections were necessary both for tracer dilution (Fig. 3.1) and for preferential reutilization of the oxygen and carbon dioxide produced within the plant and thus nearest to the sites of respiration and photosynthesis. This preferential reutilization arose from the resistances to diffusion, of which the most important was considered to be at the liquid–gas interface in the reaction vessel. Brown and Weis [44] derived equations to correct respiratory and photosynthetic rates as measured in the gaseous phase to those that would have been measured in the liquid phase. They measured simultaneously uptake and output of both oxygen and carbon dioxide, and found that evolution of carbon dioxide by the green alga *Ankistrodesmus braunii* was inhibited by light to an extent (about 50 per cent) which was almost independent of light intensity up to about six times light compensation; on the other hand oxygen consumption was unaffected by light at low intensity but much enhanced at high intensity (Figs. 5.11 and 5.12).

Brown and Weis found their results, for *A. braunii* and also for an algal flagellate (*Ochromonas malhamensis*) [316], consistent with the stoichiometric equation

$$P_{O_2} = U_{CO_2} + \Delta P_{CO_2} + \Delta U_{O_2} \quad (5.2)$$

where P_{O_2} was photosynthetic production of oxygen, U_{CO_2} was photosynthetic uptake of carbon dioxide, ΔP_{CO_2} was the deficit in respiratory production of carbon dioxide in light as compared with the dark rate and ΔU_{O_2} was the light-induced increase in rate of respiratory uptake of oxygen (Fig. 5.13). Photosynthesis was considered as an oxidation–reduction system producing oxidant and reductant at equivalent rates. The oxidant had no effect on the respiratory reactions and all appeared as molecular oxygen (P_{O_2}); the reductant might inhibit decarboxylations which evolved respiratory carbon dioxide (ΔP_{CO_2}) or increase consumption of oxygen (ΔU_{O_2}), for example by forming water, or both, according to the stages at which it impinged upon the chemical pathways of respiration. The amount of reductant used in such interactions with the respiratory intermediates was not available for normal photosynthetic reduction of carbon dioxide and there was thus an equivalent diminution in the quantity of this gas assimilated (U_{CO_2}). The consistency of the data with the equation implies that the production of reductant by light was in fact equivalent to P_{O_2} and that it all appeared as changes in the terms on the right-hand side, i.e. it all appeared as measurable by isotopic gas analysis; we may therefore conclude that P_{O_2} provides a good measure

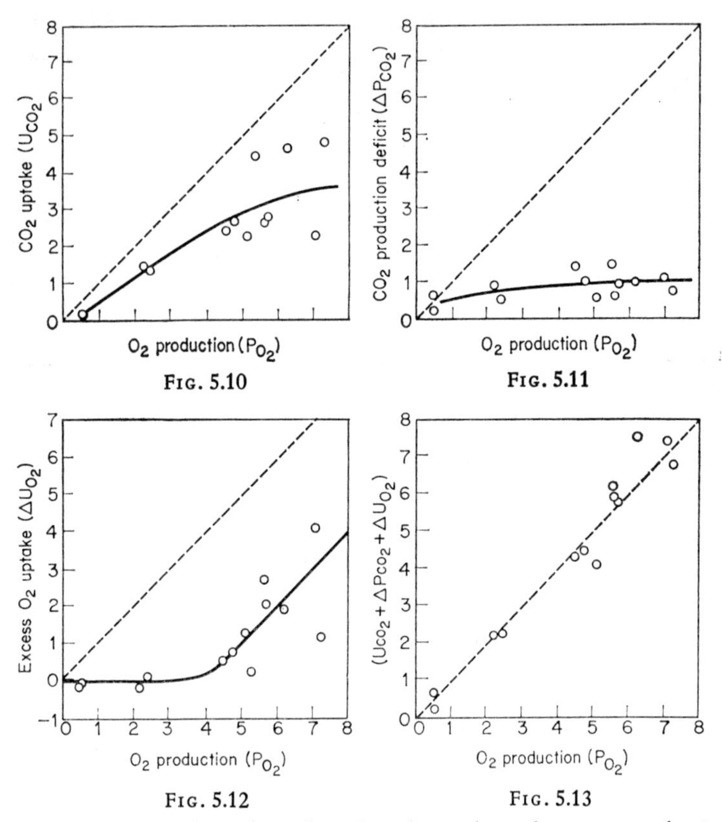

FIG. 5.10 Relation between carbon dioxide uptake and oxygen production by *Ankistrodesmus braunii* at various light intensities. The diagonal line shows the relation expected for a photosynthetic quotient of unity. Mass spectrometer data. After Brown and Weis [44].

FIG. 5.11 Relation between the deficit due to light in respiratory production of carbon dioxide and the rate of production of oxygen by *Ankistrodesmus braunii* at various light intensities. After Brown and Weis [44].

FIG. 5.12 Relation between the increase due to light in respiratory consumption of oxygen and the rate of production of oxygen by *Ankistrodesmus braunii* at various light intensities. After Brown and Weis [44].

FIG. 5.13 Relation to the rate of production of oxygen by *Ankistrodesmus braunii* at various light intensities of the sum of carbon dioxide uptake, carbon dioxide production deficit and oxygen consumption excess. The diagonal line shows the relation expected if equation (5.2) is correct. After Brown and Weis [44].

of production of reducing power (high-energy electrons) in photosynthesis, if we assume that this scheme applies to green plants in general and neglect possible cyclic phosphorylation in which reducing power production is not accompanied by oxygen production. Cyclic phosphorylation would appear to have been absent or negligible in Brown and Weis' experiments and may not occur at all *in vivo*.

If we can accept equation (5.2) as a basis for measuring photosynthesis the question that arises is how to obtain a good approximation to P_{O_2} without the use of isotopes. As long as ΔU_{O_2} is zero, which is suggested for low to medium light intensities, measuring either carbon dioxide *or* oxygen and correcting for the dark respiration rate will give an accurate measure of P_{O_2}, for ΔP_{CO_2} reduces both the photosynthetic uptake and respiratory production of carbon dioxide in light equally. When ΔU_{O_2} is not zero, measuring either carbon dioxide or oxygen and correcting for dark respiration rate will underestimate P_{O_2} by ΔU_{O_2}.

Similar experiments were carried out by Ozbun, Volk and Jackson [249] with detached leaves of *Phaseolus vulgaris*, using light intensities of 3,200 and 16,000 lux from fluorescent tubes. Oxygen uptake was much increased and sometimes more than doubled in light; carbon dioxide evolution (corrected for dark fixation) was reduced and sometimes halved; neither was affected by increasing the light intensity from 3,200 to 16,000 lux and the respiratory quotient thus remained constant (at 0·229 in one experiment). The photosynthetic quotient (page 85), so constant at 1·0 for apparent photosynthesis in classical work, fell from 1·84 at 3,200 lux to 1·41 at 16,000 lux (compare Fig. 5.10).

These results for bean leaves much resemble those of Brown and Weis for *Ankistrodesmus* although no corrections were made for diffusive resistances of the stomata nor those at the gas–water interfaces in the intercellular spaces.

Hoch, Owens and Kok [168] also used a mass spectrometer to distinguish between oxygen output and uptake in light, but did not measure carbon dioxide. Their instrument was modified to eliminate the resistance of the gas–water interface (page 91). Some of their data for the blue-green alga *Anacystis* are shown in Fig. 5.14. The cells were grown with 3 per cent carbon dioxide in air bubbled through the medium, but unfortunately no information is given as to the carbon dioxide concentration used in the actual experiments. The samples used were bubbled with nitrogen, which removed most of the oxygen

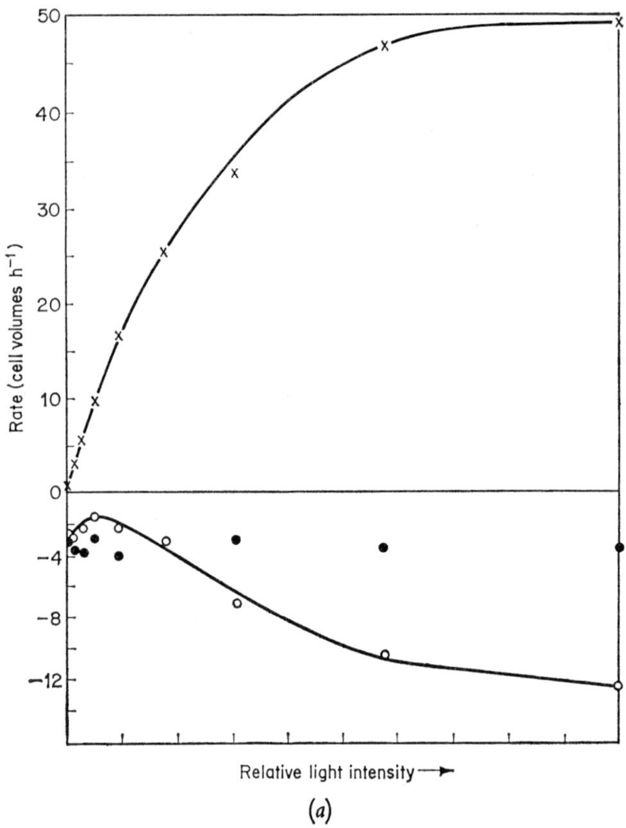

(*a*)

Fig. 5.14*a* and *b*. Relation to light intensity of oxygen output (positive) and uptake (negative) for two cultures of *Anacystis nidulans*. Closed circles, oxygen uptake in darkness immediately before illumination; open circles, oxygen uptake in light; crosses, oxygen output. Note difference in scales for uptake and output. A separate aliquot of a single culture was used at each light intensity. Averages over 2 minute light periods in *a* and 10 minute periods in *b*. After Hoch *et al.* [168].

(and presumably carbon dioxide also) and then O^{18} enriched oxygen was added. This suggests that the carbon dioxide concentration was low and not controlled. At low light intensity the oxygen uptake was reduced as compared with the dark values but at somewhat higher intensities it was increased. Fig. 5.15 shows how the reduction in uptake at low light brings about the Kok effect, though the 'break' is clearly

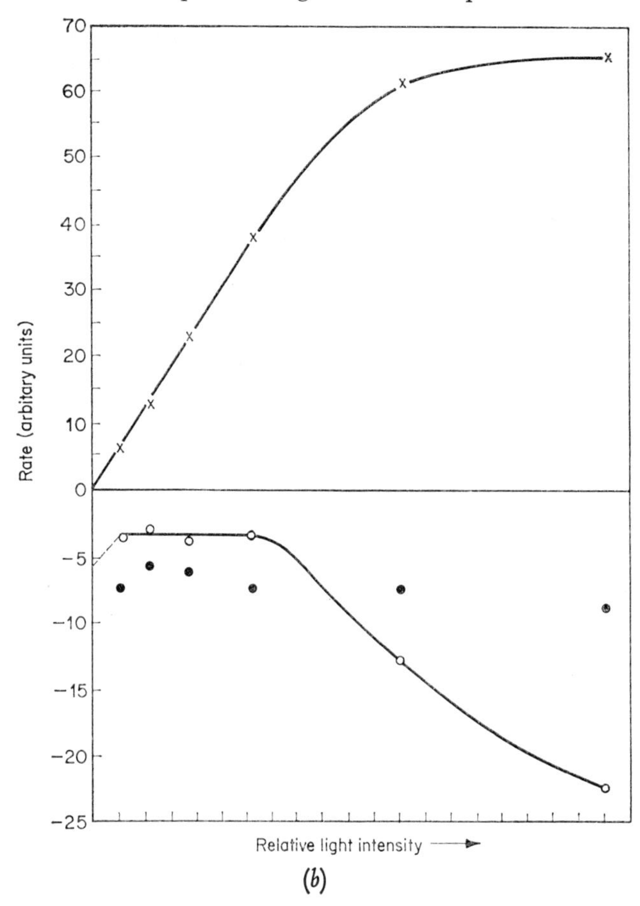

(*b*)

part of a smooth curve, as would be expected. For the green alga *Scenadesmus*, on the other hand, Hoch *et al.* found no change in oxygen uptake up to about the compensation point; there was then a rather sudden increase in uptake with increasing illumination which would also give rise to a change of slope, or Kok effect, in the net output curve. The increased uptake found for both species occurred at much lower light intensities, possibly because of higher oxygen concentrations, than in the experiments of Brown and Weis with *Ankistrodesmus*—there, increased uptake only occurred above light saturation of the net output. Hoch *et al.* suggested that the function of the light-stimulated respiration might be to increase the ratio of ATP to $NADPH_2$ produced in light to that needed for carbohydrate formation or to the even higher level needed for growth; this might be partly accomplished by re-oxidation of some of the $NADPH_2$ by oxygen and the phosphorylation of ADP to ATP with the energy so liberated. If this supposition was correct, oxygen uptake might be expected to

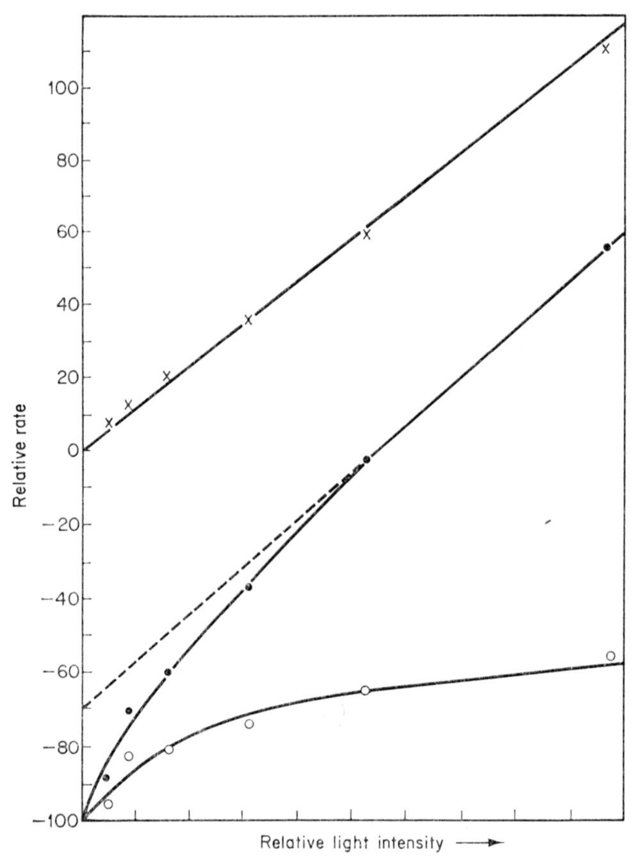

FIG. 5.15 Relation to light intensity (at low levels) of oxygen output (positive) and uptake (negative) by *Anacystis nidulans*. Rates expressed per cent of the dark oxygen uptake. Open circles, oxygen uptake; crosses, oxygen production; closed circles, net oxygen exchange. Averages from two experiments, with 2½ and 10 minute light periods respectively The broken line demonstrates the Kok effect due to light-induced reduction of oxygen uptake; at higher light intensities increase of uptake as in Fig. 5.14 would cause further bending of the net exchange curve which would not then be parallel to the oxygen production curve (page 157). After Hoch *et al.* [168].

increase with rate of photosynthesis. Hoch *et al.* also found that the suppression and stimulation of oxygen uptake were effected by different wave lengths of light; this important discovery is discussed in Chapter 9 (page 236).

The light suppression of respiration, in terms of oxygen uptake, that gives rise to the Kok effect where it occurs, and the interactions between photosynthesis and respiration found by Brown and Weis and by Ozbun *et al.*, must involve the movement of metabolites between the centres of photosynthesis (chloroplast lamellae) and those of respiration (mitochondria). Such movement must depend in part on the average separation of these centres in terms of effective lengths for diffusion of the molecules concerned; it would be expected to be most effective in blue-green algae where the lamellae are arranged throughout the cell and are not enclosed in a chloroplast membrane, less effective for green algae where they are so enclosed and least effective for leaves of higher plants where the distances are much greater and there are some cells devoid of chloroplasts. It should be noted that Myers [238] found a large Kok effect in *Anacystis* (page 235) and stated that it was even more dramatic in another blue-green alga, *Anabaena variabilis*; in *Chlorella pyrenoidosa* it appeared only as a minor non-linearity. However, the stimulation of oxygen output and inhibition of carbon dioxide uptake in light observed in bean leaves by Ozbun *et al.* make this argument much less convincing.

It must be noted that all the investigations discussed in this section except those of Hoch *et al.* were carried out with sub-normal oxygen concentrations.

D. STIMULATION OF RESPIRATION BY LIGHT

Of recent years a good deal of evidence, mainly from *net* carbon dioxide uptake in light or output in darkness and therefore not unequivocal, has suggested an increase of respiratory production of carbon dioxide in light. This is the opposite of the effect found with marked carbon dioxide in mass spectrometer experiments (see previous section).

i) *The relation of apparent assimilation to carbon dioxide at very low concentrations*

It has been suggested (page 144) that the shape of the curve relating

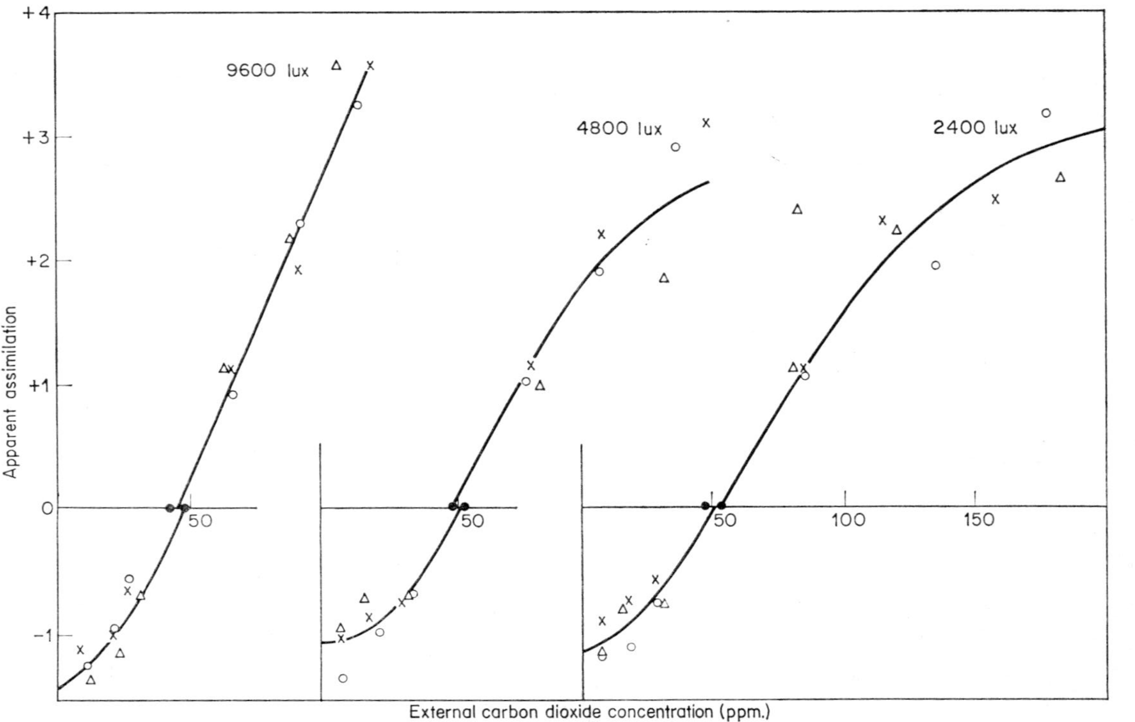

FIG. 5.16 Relation to external carbon dioxide concentration of apparent assimilation of carbon dioxide by wheat leaves at 25°C and three light intensities; uptake positive, output negative, in units of dark respiration rate. Different symbols refer to replicates; only extreme values of Γ are shown (solid circles); curves are parabolae or straight lines of best fit. After Heath and Orchard [151].

assimilation to carbon dioxide concentration below Γ is of interest, and in the absence of a more comprehensive investigation results of an experiment with wheat [151], carried out in semi-closed circuit (page 76) with the apparatus described by Orchard and Heath [248], are presented in Fig. 5.16. The curves are quadratic regressions fitted separately above and below Γ (of which there were two independent estimates in each replicate). The mean slopes at Γ for both the upper and lower regression curves are significantly steeper than that of a straight line passing through Γ and the intercept on the assimilation axis at the dark respiration rate (-1); the regressions below Γ have significant (positive) quadratic coefficients and are therefore curved. These tests disprove the usual hypothesis of a simple balance between a rate of photosynthesis proportional to external carbon dioxide concentration and a constant respiration rate (Fig. 5.17).

The regression curves below Γ do not cut the assimilation axis significantly below -1. However, the carbon dioxide values are the external concentrations; if these were corrected for the diffusive resistance of the stomata, to give the concentrations in the intercellular spaces, the curves would be steepened and probably give an estimate of carbon dioxide output into carbon dioxide-free air significantly greater than the dark respiration rate. This was found to be the case in a similar experiment with *Pelargonium* in which, however, a significant curvature was not found except above Γ at low light intensity (4,900 lux) where light limitation might have been involved.

As already stressed, the technical difficulties of such experiments are considerable and the possibilities of artifacts must be borne in mind.

The question of a curvature below Γ is important. It is usually stated that dark respiration is unaffected by carbon dioxide except at high concentrations, although Decker and Wien [60] found an increase of 26 per cent for *Eucalyptus* leaves in air at 16 p.p.m. of carbon dioxide as compared with normal air. Goldsworthy [123] found that tobacco leaf segments which had taken up $^{14}CO_2$ in light, subsequently released more $^{14}CO_2$ into carbon dioxide-free air in the light than into air of 300 p.p.m. carbon dioxide content. In most studies of effects of light on respiration of leaves it is assumed, often tacitly, that respiration in light is unaffected by carbon dioxide concentration; a linear extrapolation of the apparent assimilation curve from above Γ is made to cut the assimilation axis and a parallel straight line for 'true' photosynthesis drawn through the origin; light respiration is then given as the vertical distance between the two lines (Fig. 5.17). Decker [58], who used this method, but mentioned the above assumption, pointed out that to make 'true' photosynthesis zero at zero external carbon dioxide implied no reutilization of endogenous carbon dioxide; he therefore considered that it gave a minimal estimate of light respiration, which for tobacco leaves was 3·5 times the dark rate. Goldsworthy's [123] results showed approximately a 50 per cent increase in release of $^{14}CO_2$ when a darkened leaf was illuminated.

If the curvature below Γ in Fig 5.16 is real it implies, between zero

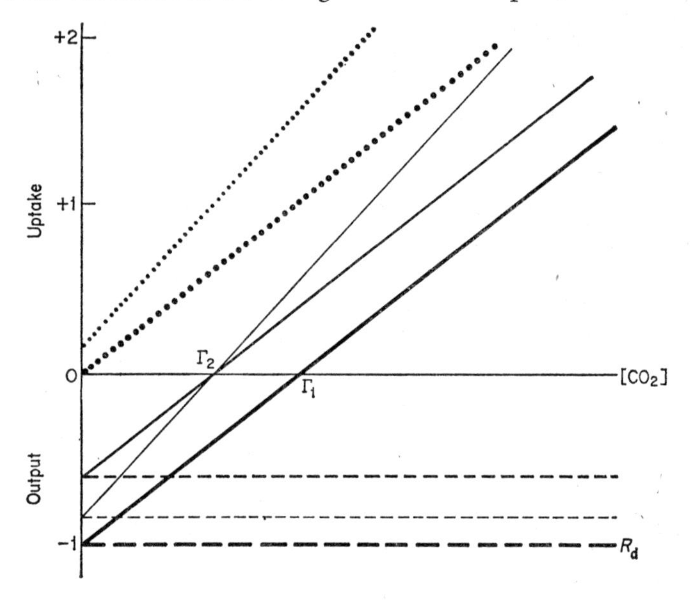

FIG. 5.17 'Straight balance' hypotheses for relations of photosynthesis and respiration (in terms of carbon dioxide uptake and output) to carbon dioxide concentration, resulting in the relation for apparent assimilation. Abscissa: carbon dioxide concentration. Ordinate: uptake (positive) or output (negative) of carbon dioxide in units of dark respiration rate (R_d). Solid line, apparent assimilation; dotted line, photosynthesis; dashed line, respiratory output. Heavy lines, no reassimilation; medium lines, reassimilation proportional to carbon dioxide production, with concentration measured outside leaf; fine lines, as medium lines but concentration measured within intercellular spaces. After Heath and Orchard [151].

ambient carbon dioxide and Γ, either an increasing true assimilation rate per unit of carbon dioxide supplied, or a carbon dioxide production which increases rapidly at first and then more and more slowly as Γ is approached, or a carbon dioxide production which is constant up to say $0{\cdot}8\Gamma$ and then decreases (Fig. 5.18); the observed changes of respiration with carbon dioxide concentration noted above are consistent with the last suggestion. Although the curvature above Γ looks like some sort of 'buffering' of carbon dioxide concentration at Γ (page 132) it cannot be distinguished from partial limitation of photosynthesis by light or temperature. If, on the other hand, the curvature

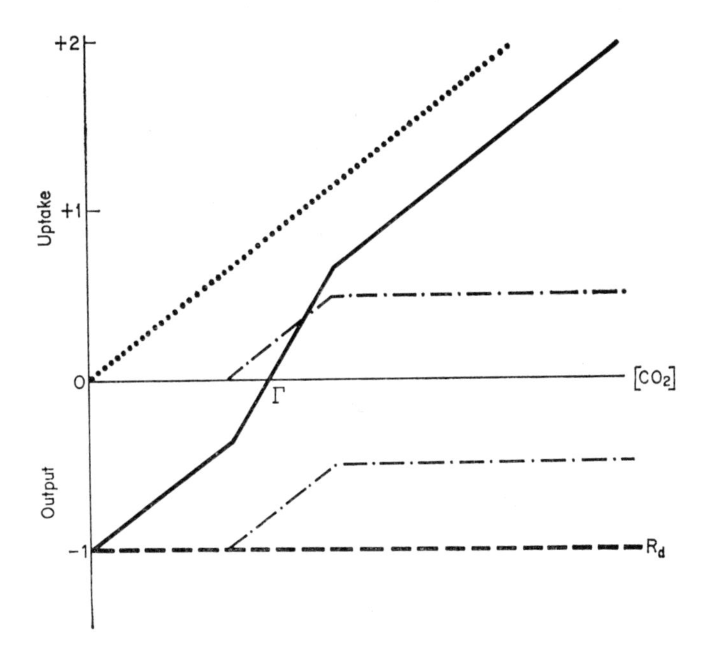

FIG. 5.18 Hypothesis to account for curvature of relation of apparent assimilation to carbon dioxide concentration. Abiscissa: carbon dioxide concentration. Ordinate: uptake (positive) or output (negative) of carbon dioxide in units of dark respiration (R_d). Solid line, apparent assimilation; dotted line, photosynthesis; dashed line, respiration; chain line, 'third process'—either an increased photosynthetic uptake or a reduced respiratory output in the neighbourhood of Γ. Straight lines are used for simplicity but smooth curves would be expected where a change of slope is shown. After Heath and Orchard [151].

below Γ is not real, the data are consistent with carbon dioxide uptake proportional to external concentration and a constant rate of carbon dioxide production ('respiration'). This in no way implies equality of such respiration in light and darkness.

ii) *The carbon dioxide burst in darkness*

Infra-red gas analysers (page 80) have made it possible to observe, for a number of species [247, 57, 60, 297, 298, 92, 236, 151] that when an illuminated leaf is darkened a carbon dioxide burst occurs—that is

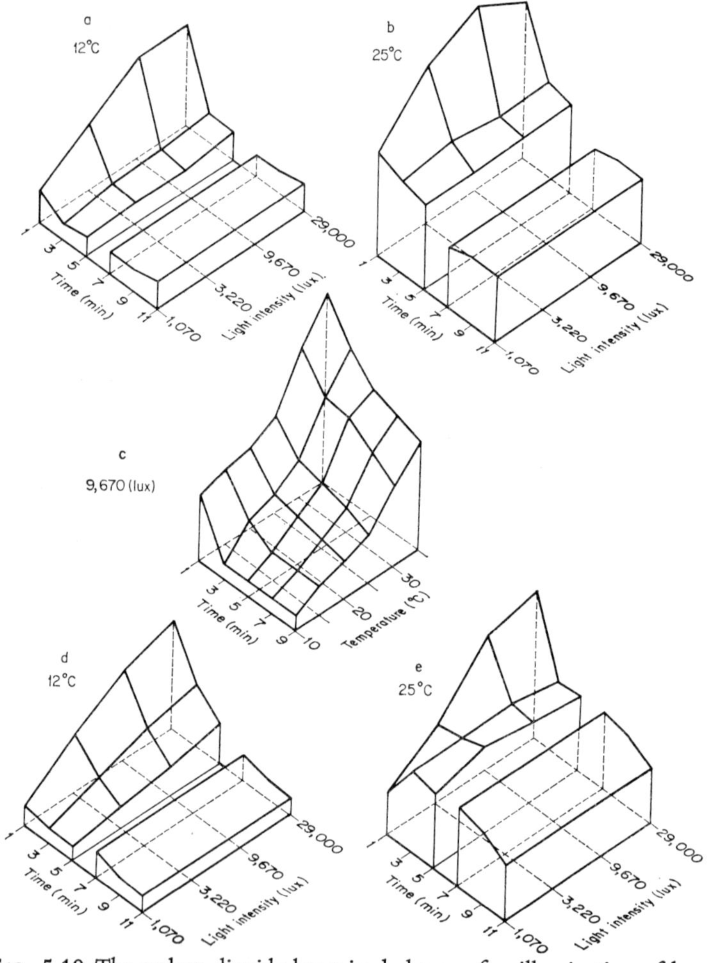

FIG. 5.19 The carbon dioxide burst in darkness, after illumination of leave in a closed circuit with the carbon dioxide concentration at Γ. Relative rate of carbon dioxide production, plotted vertically, as affected by time in darkness preceding light intensity and temperature. *a*, *b* and *c*, leaves of *Pelargonium zonale*; *d* & *e*, leaves of wheat; *a*, *b*, *d* & *e*, means for two replicates; *c*, means for three replicates. After Heath & Orchard [151].

an output of carbon dioxide at a rate that for a short period greatly exceeds the steady rate of dark respiration. Examples for *Pelargonium* and wheat are shown in Fig. 5.19 [151]. It is apparent that the burst was much increased following higher light intensities, but temperature had a smaller effect than might be expected (page 181)—the temperature coefficients (Q_{10}) for respiration rates in the first two minutes of darkness, after light intensities of 9,670 or 29,000 lux, were 1·3 to 1·4 for wheat and (with one exception) 1·2 to 1·6 for *Pelargonium*, whereas for steady dark respiration the mean values were 2·7 and 2·3 respectively. The Q_{10} values for the burst were in general less than the relative increase of Γ for 10°C rise of temperature. Heath and Orchard [151] argued that Γ should increase less with temperature than the carbon dioxide production process, owing to photosynthesis having a Q_{10} above unity in the severely carbon dioxide-limited conditions. Also, at a given temperature, Γ should be nearly proportional to the rate of carbon dioxide production, but the relations of Γ to light intensity (Fig. 7.6) and to oxygen were very different from those of the burst. They considered it unlikely, therefore, that the magnitude of the burst reflected that of light respiration.

Decker [59] also found that the carbon dioxide burst, for tobacco, was larger at higher temperatures and following higher light intensities; this last was confirmed by Tregunna, Krotkov and Nelson [297], who concluded that the initial burst was followed by a second which was independent of light intensity—this, however, was much less definite. Later [298] they found that neither the first nor the second burst was shown by yellow leaves from nitrogen-deficient soybean plants (Fig. 5.20) and in albino maize the rate of carbon dioxide evolution was constant in light and darkness. Normal maize shoots showed the second burst only (Fig. 5.21). They confirmed Meidner's [217] finding that normal maize had a Γ approximating to zero (Table 5.1) whereas green soybean leaves gave a Γ of 44 p.p.m. under their conditions. Forrester, Krotkov and Nelson [92] showed that the magnitude of the initial burst was increased in pure oxygen and in 1 per cent oxygen it had disappeared completely (Fig. 5.22); even pure oxygen did not cause the initial burst to appear with maize leaves [93]. The conclusions reached by this group, partly from other evidence considered below (iii and iv), were as follows: light, acting through the photosynthetic apparatus, completely suppresses ordinary dark respiration, but stimulates carbon dioxide output by another process (photorespiration); it is suggested that the initial carbon dioxide burst

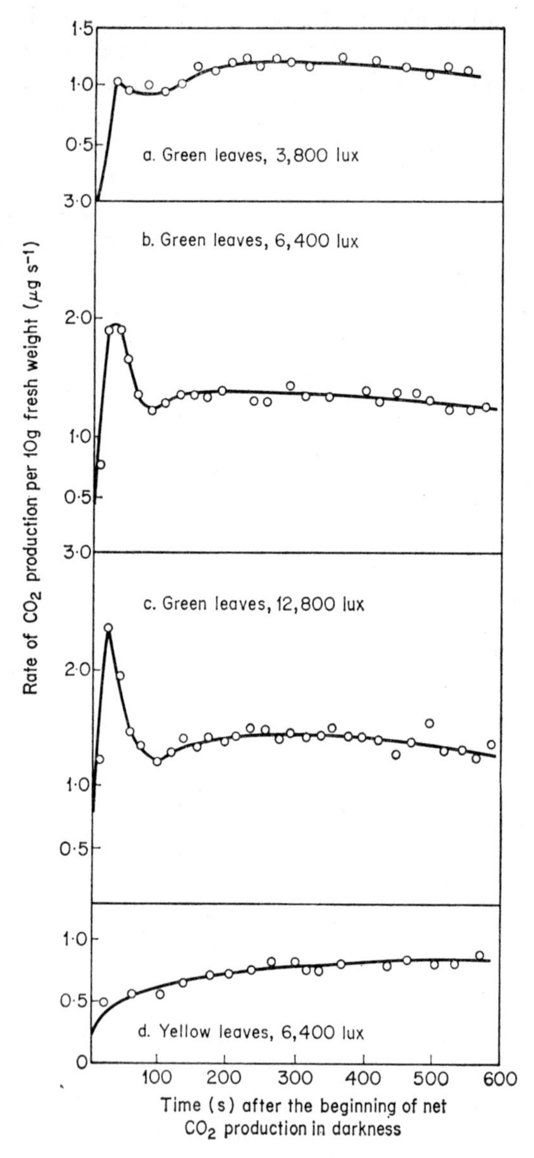

FIG. 5.20 Rates of carbon dioxide production by soybean leaves in darkness after illumination at different light intensities. Net production first recorded after 8 s of darkness. After Tregunna *et al.* [298].

represents the last remnants of this photorespiration; maize lacks photorespiration and thus has no initial burst and gives a zero value for Γ. Note, however, that the epidermal cells, which are without chloroplasts, would be expected to continue to respire in light as does the

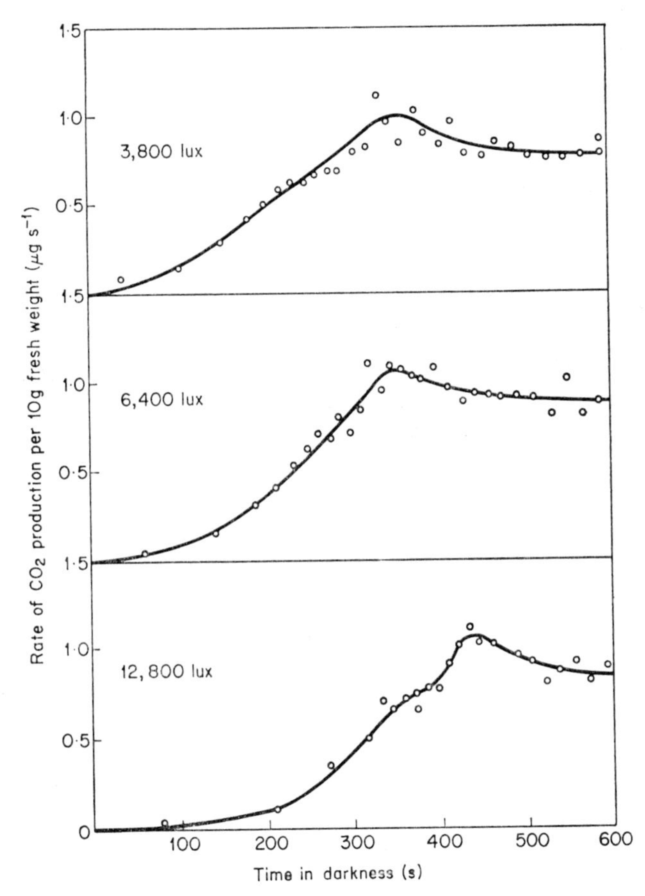

FIG. 5.21 Rates of carbon dioxide production by shoots of maize in darkness after illumination at different light intensities. After Tregunna *et al.* [298].

albino maize—it is not clear why some of their carbon dioxide of respiration does not escape from the leaf to give a Γ above zero. *Amaranthus edulis* apparently resembles maize in having Γ values at or close to zero (page 144) and on the above hypothesis should lack

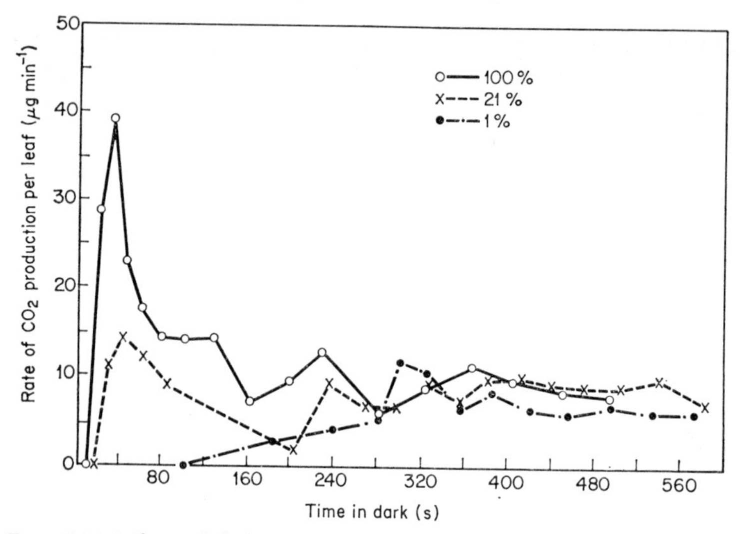

FIG. 5.22 Effects of different concentrations of oxygen on rates of carbon dioxide production by soybean leaves in darkness after illumination at 10,700 lux. After Forrester *et al.* [92].

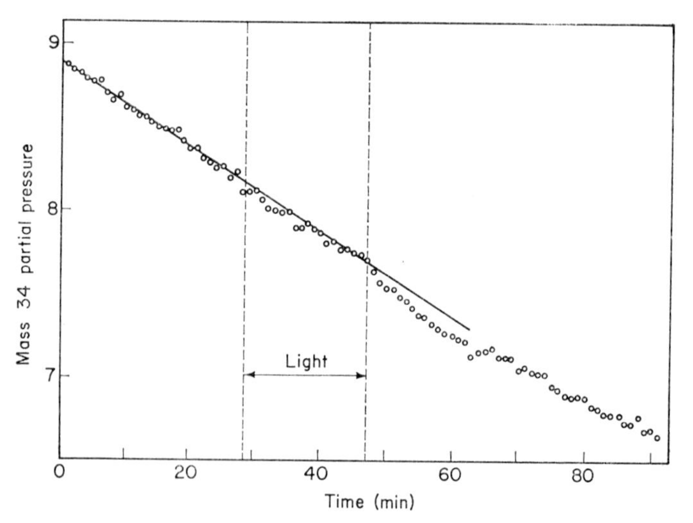

FIG. 5.23 Time course of partial pressure of tracer oxygen over a *Chlorella* suspension. Note the temporary increase in uptake immediately after darkening. Gas phase 2 per cent oxygen in nitrogen. Mass spectrometer data. After Brown [42].

photorespiration and give no initial burst; it shows, however, a well-marked burst on darkening, both in low and normal oxygen concentration [15*a*].

An alternative interpretation of the burst would be that when the light is first extinguished high concentrations of intermediates in the photosynthetic pathways lead to back reactions, though in that case high ambient carbon dioxide concentration would be expected to lead to a larger burst and Decker [59] found very little effect of carbon

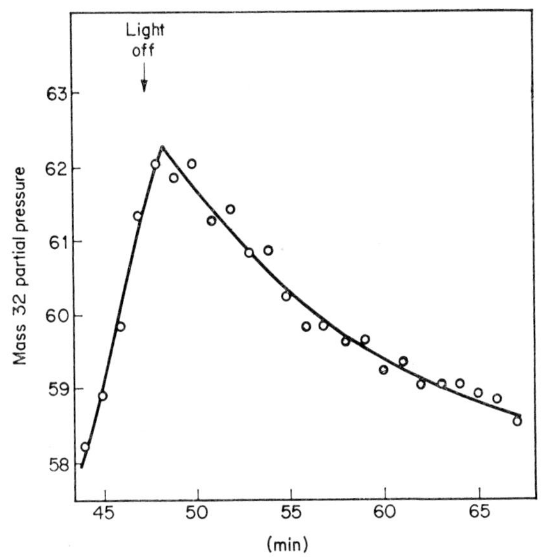

FIG. 5.24 Time course of metabolism of ordinary oxygen by *Chlorella*, in the same experiment as in Fig. 5.23; the abscissa values in the two figures correspond. Note the most rapid uptake immediately after darkening. After Brown [42].

dioxide. Brown [42] in mass spectrometer experiments with *Chlorella* found a temporary increase in the rate of uptake of marked oxygen which did not occur until the light was extinguished and could not therefore be photorespiration (Fig. 5.23). The rate of uptake of ordinary oxygen was also most rapid immediately after darkening the culture (Fig. 5.24). It is not apparent how these two effects could both arise as artifacts due to establishment of new diffusion gradients across the resistances in the system. It should again be noted, however, that Brown's experiments were carried out at low oxygen tensions such as prevent the carbon dioxide burst in leaves (Fig. 5.22). Tregunna *et al.* [298] fed $C^{14}O_2$ to tobacco leaves for some minutes just before darkening, and found a sudden output of C^{14} when the light was extinguished; if the $C^{14}O_2$ was fed for the last 40 seconds only of the light period there was no release of C^{14} until after 3 minutes of darkness. They concluded that some recently formed photosynthate, but not unreduced carbon or unfixed $C^{14}O_2$, was responsible for the initial burst.

Moss [236] made use of the magnitude of the burst in leaves supplied

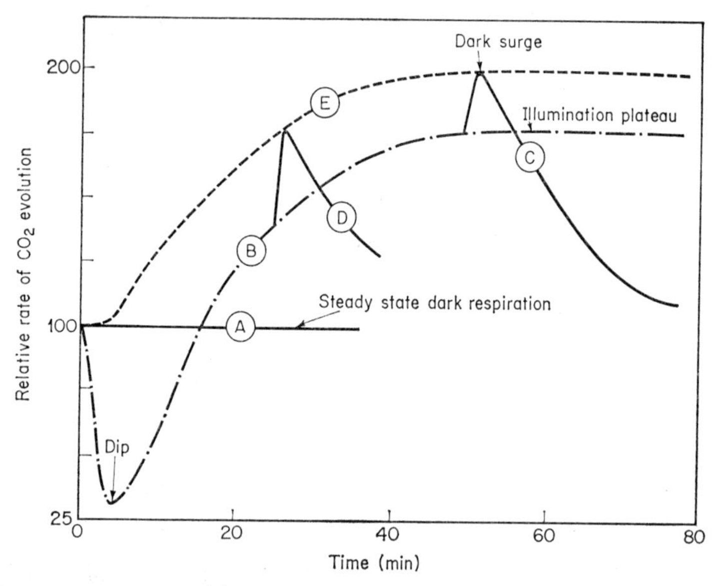

FIG. 5.25 Diagram of the time course of carbon dioxide output from a leafy shoot of *Pelargonium* into dry almost carbon dioxide-free air following illumination at 27,000 lux, calculated per cent of the dark respiration rate (3·2 mg dm^{-2} h^{-1}): curve B, in light from time 0 to 80 min; curve C, from 0 to 50 min and then darkened; curve D, from 0 to 25 min and then darkened. Curve E is drawn through the maxima of the 'dark surges' at these and other times. Leaf temperature 25°C. After Moss [236].

with carbon dioxide-free air to estimate light respiration. His diagram to illustrate the method is shown in Fig. 5.25. The relative fall below the dark rate of output ('dip') which occurred in the first few minutes of illumination was taken as a minimal estimate of 'true' assimilation. The 'illumination plateau' of net output was only reached after long periods of the order of an hour and varied widely in different experiments from 95 to 254 per cent of the dark rate for *Pelargonium*, 100 to 137 for sugar beet and 110 to 154 for tobacco, sunflower and *Liriodendron*. In successive runs with the same shoots or leaves the light was extinguished after different times and the peak of the burst or 'dark surge' taken as a minimal estimate of the relative excess of light respiration over that in darkness. These peaks were joined to give a curve for light respiration, which rose for *Pelargonium* in different experiments to 138 to 277 per cent of the dark rate, for sugar beet to 120 to 177 and for tobacco to 150 to 173. There was no obvious relation to the temperature differences between experiments. The slow rise to the plateau was attributed to the formation in light of a substrate for light respiration, which continued to be rapidly metabolized when the light was extinguished, giving the dark surge. This substrate must have been formed from an endogenous carbon source since the air entering the chamber was free of carbon dioxide. Moss discussed the possibility that it might be glycollate.

A discrepancy between Moss's results and those of earlier workers is of interest. Orchard, who also used *Pelargonium* [247, 151], and Gabrielsen [111] found almost the same rate of output of carbon dioxide into nearly carbon dioxide-free air in light as in darkness (Fig. 5.16). Moss also obtained this result in one of twelve experiments with *Pelargonium* and three out of five with sugarbeet, showing that the output in light was dependent on some uncontrolled factor. Such differences might be associated with water stress, for Moss used air dried by silica gel and the rate of flow relative to the leaf area must have been high to maintain not more than 15 p.p.m. of carbon dioxide in the emerging air. Zelitch [336] has proposed that the substrate for photorespiration is glycollate and Goldsworthy [123] has obtained evidence to support this.

iii) *Oxygen effects on* Γ

Orchard [247, 151] found that Γ was linearly related to oxygen concentration in *Pelargonium* and *Hydrangea*; light intensity between 4,500 and 22,000 lux made little difference to the relation, suggesting that photoxidation [96] was not mainly responsible. Tregunna *et al.* [299] and Forrester *et al.* [92] also found a linear relation, for tobacco and soybean respectively; the estimated value of Γ at zero oxygen concentration was not significantly different from zero (Fig. 5.26) and this was taken as evidence that dark respiration, which was but little affected by oxygen concentration, was completely inhibited in the light. Evidence to suggest that in maize some at least of the dark respiratory output of carbon dioxide continues in light is discussed in the next chapter (page 179).

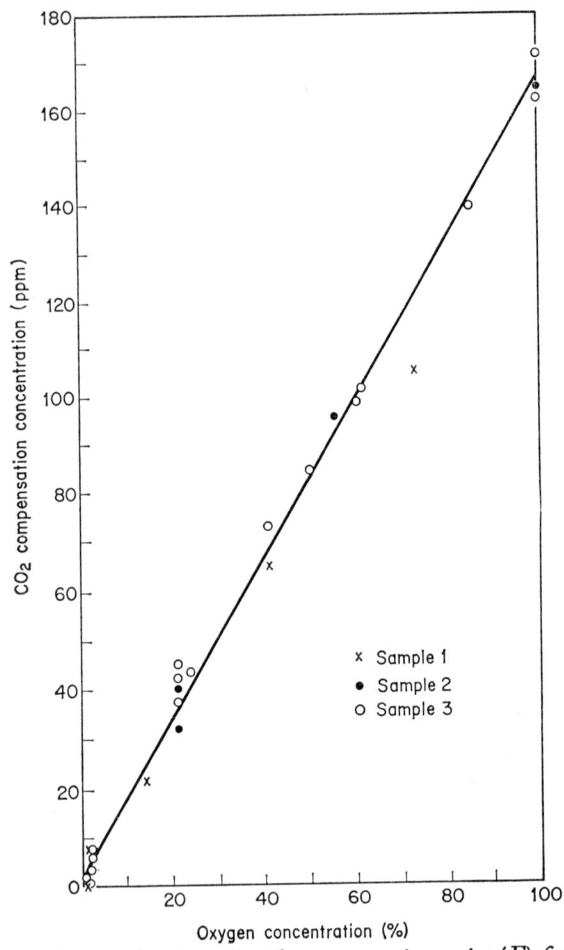

FIG. 5.26 Relation of carbon dioxide compensation point (Γ) for soybean leaves to oxygen concentration. After Forrester *et al.* [92].

iv) *Oxygen effects on apparent assimilation*

Tregunna *et al.* [299] investigated the effects of different oxygen concentrations up to 41 per cent on apparent assimilation by tobacco leaves in a closed circuit, in which the continuously changing carbon dioxide concentrations (page 76) prevented direct comparison of the rates in the different oxygen treatments. This was repeated, with soy-

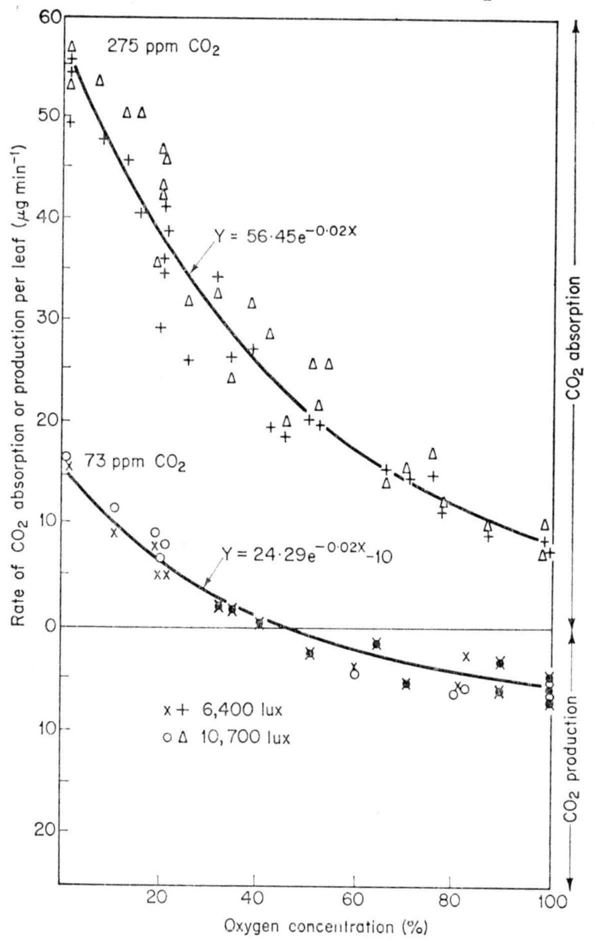

FIG. 5.27 Relations of apparent assimilation of carbon dioxide by soybean leaves to oxygen concentration, at two light intensities and with two concentrations of carbon dioxide in the gas stream entering the plant chamber. After Forrester *et al.* [92].

bean leaves and a wider range of oxygen concentrations, by Forrester *et al.* [92], who used an open circuit apparatus and could therefore measure steady-state assimilation. However, the carbon dioxide content of the air entering the system was maintained constant at 275 or 73 p.p.m. and thus the mean concentration over the leaf must have changed with the rate of photosynthesis—no correction seems to have been made for this and the data and fitted curves shown in Fig. 5.27 are therefore subject to increasing systematic errors the further they are above or below Γ. The stomata must also have opened wider and had a much lower diffusive resistance with 73 p.p.m. carbon dioxide than

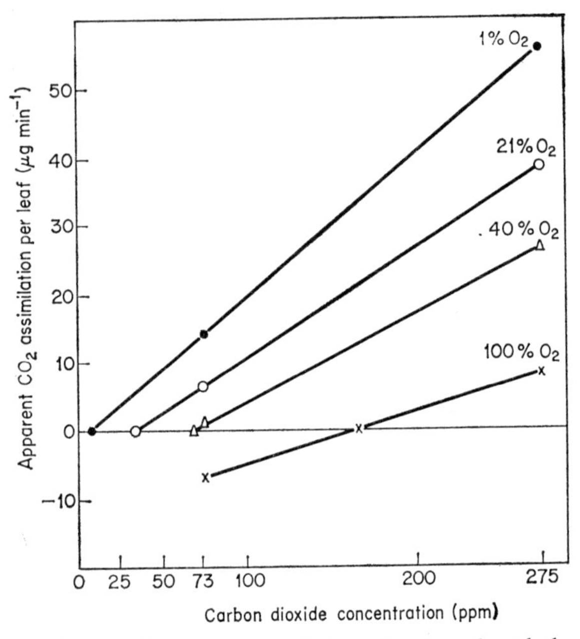

FIG. 5.28 Relation of apparent assimilation of carbon dioxide by soybean leaves to carbon dioxide concentration, at various oxygen concentrations; derived from the curves in Fig. 5.27; values of Γ obtained by linear extrapolation or interpolation. After Forrester *et al.* [92].

275 p.p.m., and these resistances would change with changing concentration [153]. Nevertheless the results clearly demonstrate the increasing inhibition of apparent assimilation with increased oxygen concentration. Fig. 5.28 was derived from the curves in Fig. 5.27 on the assumption that, for each of the four oxygen concentrations shown, apparent assimilation was linearly related to carbon dioxide concentration. The values of Γ plotted were obtained by extrapolation or interpolation where the straight lines cut the x-axis; the values of Γ

read off from the regression line in Fig. 5.26 appear to agree well with these, suggesting that the relations were in fact practically linear within the range shown.

The rate of photorespiration at various oxygen concentrations was next calculated by a method which amounted to that mentioned on page 157; it is there pointed out that in this method it is assumed that there is no effect of carbon dioxide concentration on light respiration

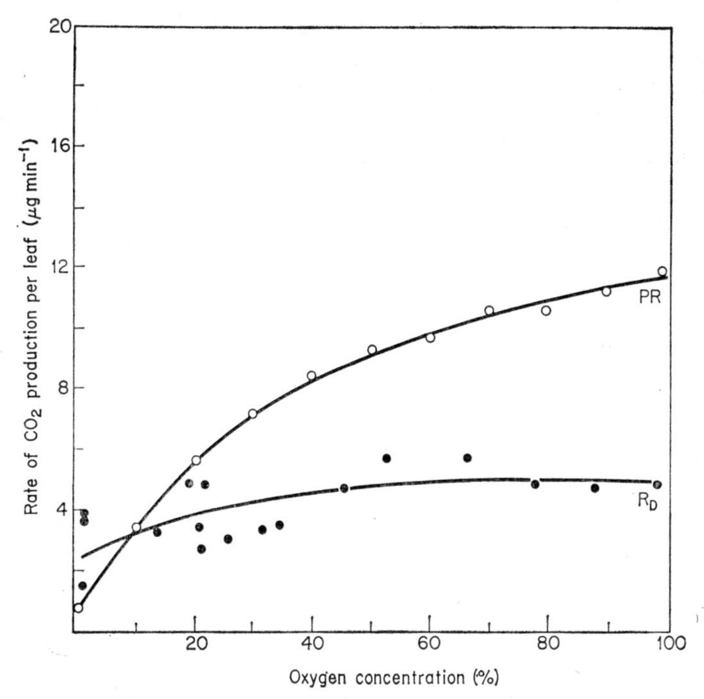

FIG. 5.29 Relations to oxygen concentration of dark respiration (R_d) of soybean leaves and also of photorespiration (PR) calculated on the assumption that it was unaffected by carbon dioxide concentration. After Forrester *et al.* [92].

and it may again be noted that Goldsworthy's [123] data suggest that it had in fact an appreciable effect, though he also mentions other possible interpretations. The effects of oxygen concentration on photorespiration so calculated, as well as on dark respiration, are shown in Fig. 5.29.

If, as assumed, photorespiration at any given oxygen concentration was constant throughout the carbon dioxide concentration range explored, and if the inhibition of apparent assimilation by oxygen was solely due to such photorespiration, then the lines in Fig. 5.28 should be parallel; the inhibition would arise from the increase in Γ making less carbon dioxide available for net uptake. However, the slope of the lines, which represents the net uptake per unit of external carbon dioxide above Γ or 'carboxylation efficiency' [299], also falls with increasing oxygen concentration. This implies either that photosynthesis itself is inhibited by oxygen or that photorespiration changes with carbon dioxide concentration at a rate depending on the oxygen concentration; Forrester *et al.* discussed only the first alternative and calculated inhibition of true photosynthesis as about 20 per cent in normal air of 21 per cent oxygen concentration and 65 per cent in pure oxygen.

Support for the view that one effect of high oxygen concentration was to inhibit true photosynthesis came from the experiments of Forrester *et al.* [93] on maize. Even in 100 per cent oxygen, maize gave a Γ value of zero and as already mentioned there was no carbon dioxide burst in any oxygen concentration. It was therefore concluded that maize entirely lacked photorespiration and hence any effects of oxygen concentration on carbon dioxide uptake in light could be assumed to act upon photosynthesis itself. It was found that this was progressively inhibited by increasing concentrations of oxygen and further that such inhibition was only partly reversible—the first determination of assimilation in 21 per cent oxygen always gave a higher value than a final one after the leaf had been exposed to high oxygen concentrations (Fig. 5.30). It was suggested that such damage might be due to photoxidation, but in the figure the effect is larger at the lower light intensity.

If the values for photorespiration obtained by the rather indirect method described above are accepted, they show that below 10 per cent oxygen concentration respiration in terms of carbon dioxide production was less in light than in darkness (Fig. 5.29). This is the result found by Brown and Weis (page 149) for a green alga and by Ozbun *et al.* for bean leaves at low oxygen concentrations, though at the same time they found an increased oxygen uptake in bright light. Such an increase was also found, except at very low light intensities, by Hoch *et al.* [168], for *Anacystis* and *Scenadesmus*, apparently with higher oxygen tensions.

The results described in Section D of this chapter leave little doubt that carbon dioxide output from leaves in light into nearly carbon dioxide-free air, with 21 per cent oxygen, can be as high as in darkness [111, 151] or higher [236] and this implies greater actual production; whether production is also higher in light at an external carbon dioxide concentration of 300 p.p.m. is more problematical. The increases in

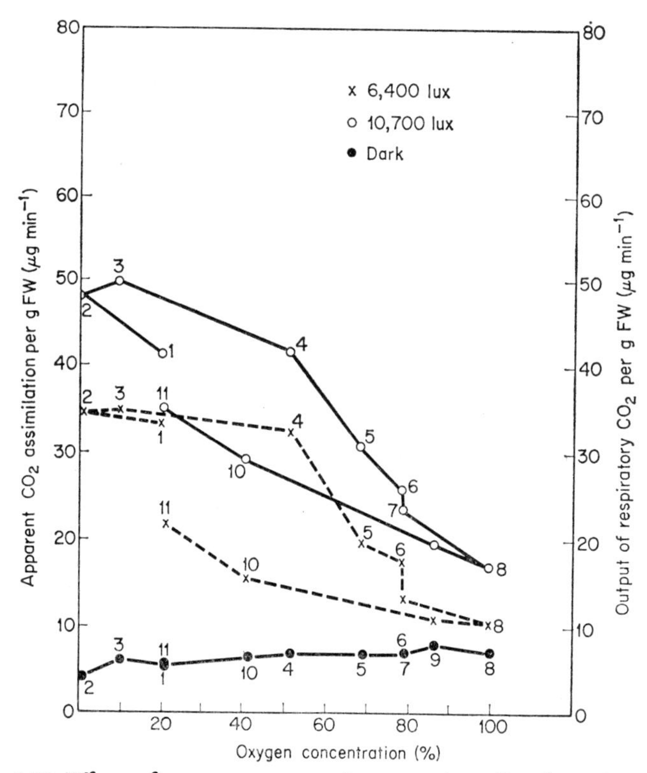

FIG. 5.30 Effects of oxygen concentration on carbon dioxide exchange by maize leaves at two light intensities and in darkness. The points are numbered in chronological order. After Forrester *et al.* [93].

oxygen uptake just referred to were, however, found in the presence of carbon dioxide: Brown and Weis [44] started their experiments with about 0·5 per cent in the gas phase and Ozbun *et al.* [249] about 2 per cent. It may be taken as well established therefore that respiration in terms of oxygen uptake is increased by light.

6: Water Supply and Leaf Water Content

IT is self-evident that leaf water content must depend upon the external factor of water supply but it is too often forgotten that it equally depends upon water loss and therefore upon the humidity of the air, the external resistance to diffusion and that due to the stomata. Except in a saturated atmosphere the deficit of leaf water content below full turgidity must therefore change with every stomatal movement. Further, this deficit itself affects the movements of the stomata. Severe wilting in the long run causes stomatal closure, but very rapid lowering of leaf water content, resulting in local differences of turgor within the leaf, can bring about a temporary opening; conversely, a rapid increase of water content can cause temporary closure [143].

Many workers, since the end of the last century, have recorded reduction in apparent assimilation rate with wilting but in most cases the picture has been confused by the likelihood (or certainty) of concurrent stomatal movements. Scarth and Shaw [273] avoided such confusion by taking advantage of the temporary opening on wilting. They subjected detached *Pelargonium* leaves without water supply to severe drying conditions and followed apparent assimilation of carbon dioxide in open circuit with an infra-red gas analyser. The stomata showed the usual temporary opening followed by closure and when the porometer reading had returned to the initial value (usually after about 20 minutes) assimilation was in each case considerably lower than at the start. This fall could not, therefore, be attributed to stomatal closure.

Heath and Meidner [148] evaded the stomatal factor by making measurements in closed circuit of Γ, at which concentration there should be no net diffusion through the stomata (see below). They subjected detached wheat leaves to much milder and more controlled water deficits than those in the experiment of Scarth and Shaw. Results for one of the eight leaves used are shown in Fig. 6.1. The solid line shows the carbon dioxide concentration and this was taken to be Γ when a steady state had been maintained for 20 minutes. The experiment started with the leaf-sheath in water. Mannitol solutions of 0·2 M

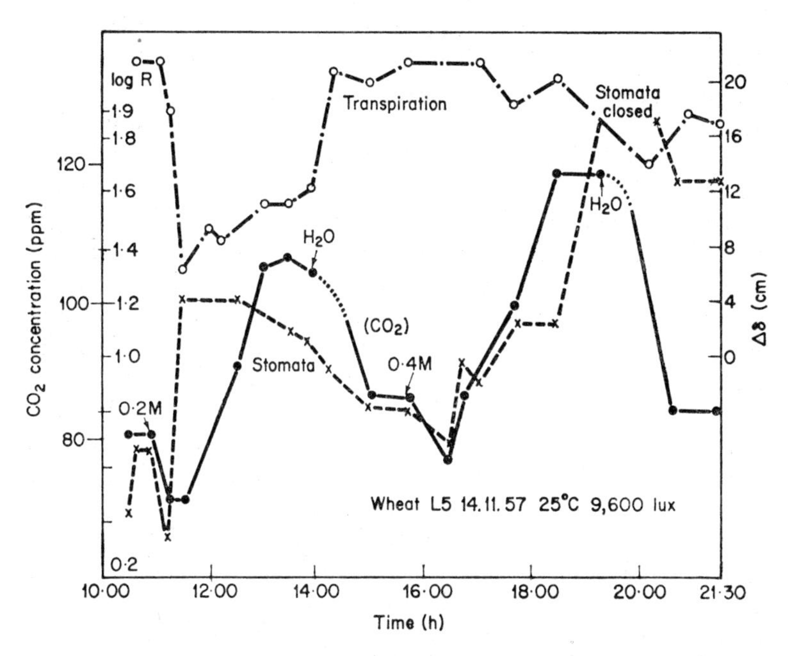

FIG. 6.1 Time course of transpiration ($\Delta\delta$ cm), stomatal resistance (log R) and carbon dioxide concentration in a closed circuit with a detached wheat leaf illuminated with 9,600 lux at 25°C. Water deficits induced by applying mannitol solution to the leaf-sheath where shown (0·2 M and 0·4 M) and relieved by replacing this with water (H_2O). Horizontal portions of the carbon dioxide curve taken as Γ. After Heath and Meidner [148].

and 0·4 M concentrations (about 5 and 10 atmospheres osmotic pressure) replaced the water, and water the mannitol, at the times marked by arrows. In every case an initial *fall* of about 10 per cent of the carbon dioxide concentration was observed when the leaf was treated with mannitol, followed by a rise to a new and higher Γ value which was usually reached after 1 to 2 hours and which was significantly greater with 0·4 M than with the 0·2 M solution. On treatment of the leaf with water the carbon dioxide fell again to a Γ similar to the earlier value. The record for stomatal resistance, plotted as log R, shows changes very similar to those in carbon dioxide and at the Γ for the 0·4 M mannitol treatment (about 120 p.p.m.) the stomata closed completely. The transpiration showed the converse changes and was presumably to a large extent controlled by the stomata.

The cause of the transient initial increase of apparent assimilation, indicated by the *fall* in carbon dioxide concentration with mannitol treatment, is quite uncertain. The stomatal opening could not have been responsible unless in fact a considerable quantity of respiratory carbon dioxide was escaping from the leaf directly through the cuticle (instead of into the intercellular spaces)—this would mean that at Γ there was an appreciable gradient inwards through the stomata. This is thought improbable; the authors suggest the possibility of increased light absorption (perhaps due to wrinkling of the epidermis) for it was noted that the leaves appeared darker at these times, but light must have been nearly saturating at so low a level of carbon dioxide. The result, whatever its cause, somewhat resembles a finding by Brilliant in 1924 [38] that water deficits increased the apparent assimilation rates of leaves of *Hedera helix* and *Impatiens parviflora*. For the former species a water deficit of 5–15 per cent resulted in some increase in the rate, which reached a maximum with a deficit of 15–25 per cent; even more severe deficits caused a marked reduction in apparent assimilation and, above a certain value, an abrupt fall to zero. Her statement that removal of pieces of epidermis 'n'a rien changé aux resultats' suggests that stomatal movements were not responsible for these effects.

Although it is thought that the initial stomatal opening with mannitol treatment was unlikely to have caused the increased assimilation observed, the converse could be true and the reduction in carbon dioxide concentration may have been a contributory cause of the stomatal opening. This also seems rather improbable at carbon dioxide levels below 100 p.p.m. (page 135) and most if not all of the initial opening in this experiment may be attributed to local differences of turgor. However, with the leaf in ordinary air such an increase in apparent assimilation would be more effective in controlling the stomata and the usual preliminary opening on wilting may therefore be supposed in part due to carbon dioxide changes, at least in wheat leaves.

Heath and Meidner argue that the very marked subsequent closing effects of mannitol treatment, as shown in Fig. 6.1, were unlikely to have been due to the increases in intercellular space carbon dioxide—though again, in ordinary air, such increases would be from a level considerably above Γ and therefore have a much more important effect. They point out that at 8,500 lux, Heath and Russell [153] found maximal opening with air of 170 p.p.m. carbon dioxide content forced through the leaf and that even with 290 p.p.m. the closure was not great. In the experiment now under discussion complete closure occurred with a Γ value of about 120 p.p.m. However, more recent work [146] with another species (*Xanthium pennsylvanicum*) has shown that loss of turgor very greatly increases the sensitivity of the stomata to carbon dioxide. It is probable, therefore, that stomatal closure with water deficit is to a very large extent carbon dioxide-operated.

The mean value for Γ with the wheat leaf-sheath in water and the highly significant increases due to mannitol treatment are shown in Table 6.1*a*. Results of similar experiments with a palm, *Phoenix reclinata*, and maize are presented in Table 6.1*b* and *c*. The wheat data showed a slight increase in Γ, as measured with the leaf-sheath in water, for all determinations after the initial reading in the morning (Fig. 6.1); this could be attributed to an after-effect of the mannitol treatment or perhaps to a diurnal trend, though the latter has not been found in other experiments with wheat. Such an after-effect was absent with the palm leaves but was extremely marked with maize (Table 6.1*c*). In a few check experiments, maize leaves from the 0·25 M mannitol treatment were kept with their sheaths in water for 48 hours and again tested; although the after-effect was slightly reduced, Γ never again approached zero. This persistent after-effect of water deficit was also shown in a comparison of leaves from plants which previously had experienced slight wilting out of doors with those from greenhouse-grown plants which had always been well watered. At a leaf temperature of 30°C the latter gave a Γ of 0·0 p.p.m. as usual and the former gave 3·9 p.p.m. (see Table 5.1). This effect may well explain the relatively high value of 9 p.p.m. found for maize by Moss (Table 5.1). Glover [119] in E. Africa found that although a short period of drought caused failure of maize stomata to open and a reduction of apparent assimilation, these effects did not persist on removal of the drought conditions; more severe drought, however, caused permanent damage to the stomata and restriction of photosynthesis even though the leaves had regained their normal appearance. Kaffir corn (*Sorghum*) leaves did not show these after-effects. Such differences between species in recovery from the effects of severe water deficit appear to be common but unpredictable. Thus the stomata of wheat leaves open fully on the day after being wilted and rewatered but those of dandelion do not recover completely for several days [146].

In none of the experiments on water deficit and Γ just discussed

TABLE 6.1

Effects of water deficit on Γ

a. Mean Γ values (maintained for at least 20 minutes) in p.p.m. of CO_2 from eight leaves of *Triticum sativum*. Leaf temperature 25°C; light intensity 9,700 lux; V.P.D. in air entering chamber 12 mm Hg.
From Heath and Meidner [148].

Mean Γ with leaf sheath in water	80·6
Mean change in Γ due to 0·2 M mannitol	+13·5
,, ,, ,, ,, ,, ,, 0·4 M ,,	+35·7

b. Mean Γ values (maintained for at least 30 minutes) in p.p.m. of CO_2 from four leaves (each on three occasions) of *Phoenix reclinata*. Leaf temperature 30°C; light intensity 15,000 lux; V.P.D. in air entering chamber 12 mm Hg for water treatment and during 2 hours in mannitol, zero during actual reading of Γ for mannitol treatment.
From Meidner [216].

Mean Γ with leaf sheath in water	78·9
Mean change in Γ due to 0·7 M mannitol	+43·2

c. Mean Γ values (maintained for at least 30 minutes) in p.p.m. of CO_2 from four leaves of *Zea mays*. Leaf temperature 30°C; light intensity 6,450 lux; V.P.D. in air entering chamber 12 mm Hg for water treatment and during 2 hours in mannitol, zero during actual reading of Γ for mannitol treatment.
From Meidner [217].

Successive treatments	Γ
Water	0·8
0·25 M mannitol	16·0
Water	5·3

were measurements of dark respiration rates made. More recent work on rates of photosynthesis and respiration of maize leaves (Meidner [218] has provided direct evidence of a decrease in dark respiration rate with increasing water deficit. When treatment with 0·6 M mannitol had caused a leaf water deficit of about 3 per cent (as measured on parallel samples), carbon dioxide uptake in light of 9,000 lux was much reduced but the respiration rate as measured in darkness had also fallen somewhat. On the assumption that respiration continued in light at the same rate as in darkness, this would imply that the increase in Γ with water deficit (Table 6.1*c*) must be attributed entirely to a reduced rate of photosynthesis and occurred in spite of the slight reduction in respiration. However, it has been postulated (page 161) that dark respiration is completely inhibited by light and that maize is without photorespiration. Since Γ rose above zero there must have been some respiratory production of carbon dioxide, and Meidner obtained the following evidence that this was not due to photorespiration: first, that a Γ value above zero for maize leaves under water stress was still found in the virtual absence of oxygen (commercial nitrogen) and Γ did not increase between 21 per cent and 100 per cent oxygen (Table 6.2);

TABLE 6.2

Effects of oxygen concentration and water deficit on Γ of maize leaves

Mean Γ values (maintained for at least 30 minutes) in p.p.m. of CO_2 from four leaves of *Zea mays*. Leaf temperature 20°C; light intensity 9,000 lux; V.P.D. in air zero during reading of Γ.
From Meidner [218].

	0·6 M Mannitol	*Water*
Commercial N_2	7·5	0·6
21% O_2	13·7	0·4
100% O_2	14·0	0·4

secondly, no carbon dioxide burst occurred when the light was extinguished. It seems, therefore, that in maize some at least of the dark respiration continues in light, and this is also suggested by the Γ values above zero produced by the combination of low light intensity and high temperature (see Table 7.3). If so, the photosynthetic mechanism must be extraordinarily efficient to prevent any escape of carbon dioxide under normal conditions in light, especially (as mentioned earlier) that produced by the epidermal cells.

Of other species some have shown an increase in dark respiration rate with water deficit and others a decrease. Thus Heath [141] found that if he passed a stream of dry air through the intercellular space system of a *Pelargonium* leaf there was a net *output* of carbon dioxide even under an illumination of 26,460 lux and the dark respiration rate was about four times the normal; with humidified air all the carbon dioxide except a nearly constant 100 p.p.m. (Γ) was assimilated, as already noted.

Iljin [177] found a progressive rise in the dark respiration rate of *Ranunculus repens* with fall of water content: a 7 per cent water deficit increased respiration by 7 per cent and a 39 per cent deficit increased it by 74 per cent. At this stage he stated that the leaves 'looked like rags'! Of the other species he investigated, six, including wheat and *Rumex confertus*, showed a rise and seven a fall. An effect of dehydration in reducing photosynthesis but not affecting respiration was found for *Chlorella* by Greenfield [129]. He observed that immersion in sucrose-buffer solution had no effect on respiration up to a concentration of 1·0 M, which would subject the cells to a water stress equivalent to about 23 atmospheres, but above 0·4 M the photosynthetic rate was progressively reduced. More recently, Nir and Poljakoff-Mayber [243] found that chloroplasts isolated from flaccid leaves of Swiss chard (*Beta vulgaris*) had much lower photochemical activity, as shown by cyclic photophosphorylation and the Hill reaction, than those from turgid leaves. Chloroplasts from turgid leaves which had been subjected to six cycles of wilting and recovery were only slightly less active in the Hill reaction than those from leaves kept turgid throughout; photophosphorylation was less effectively restored on recovery from wilting.

While the mechanism by which water deficit reduces apparent assimilation rate and increases Γ is a subject for further research, the practical importance of these effects is already manifest. The two mannitol treatments in the experiment of Heath and Meidner (Fig. 6.1 and Table 6.1*a*) would correspond to about one-third and two-thirds of the usually accepted permanent wilting point for soil (a total soil moisture stress of about 15 bar). Owing to the absence of roots, which normally have a high resistance to water movement [196, 197], the water deficit in the leaves might well be less than this comparison suggests, though this would also depend on the atmospheric component (page 174). By agricultural standards therefore, the degrees of water deficit were probably not very severe, but the higher concentration of mannitol produced complete stomatal closure. Such a drastic effect upon stomatal resistance would be even more important in reducing dry matter production than the fall in photosynthetic efficiency indicated by the rise of Γ. These results are in line with recent findings in agricultural research that *any* appreciable reduction in water supply below free availability (field capacity) has some effect in reducing growth and yield, unless there is a body of wet soil lower down into which the roots can grow [174].

7: Light, Temperature and Compensation Points

THE chemical resistance has been described as a function of light and temperature (page 124) but in fact these control very different parts of the photosynthetic chemical system. It is a characteristic of photochemical reactions in general that they proceed at rates almost independent of temperature and that the amount of reaction is directly proportional to the total amount of light received, that is, to time × intensity. On the other hand ordinary chemical reactions, sometimes called dark reactions because they are unaffected by light, proceed at rates which change markedly with temperature, usually increasing in a ratio of about 2 to 3 for every 10°C rise (the same is true of reactions catalysed by enzymes, though over a limited range which depends upon the characteristics of the enzyme concerned). This ratio, (Rate at $\theta° + 10 \div$ Rate at $\theta°$) is called the temperature coefficient or Q_{10}. It may be noted in passing that physical processes, such as diffusion, usually have a Q_{10} of about 1·1 photochemical reactions, being unaffected by temperature, have a Q_{10} of 1·0. The magnitude of the Q_{10} thus gives a clue to the type of process concerned—*if* one can be sure of not measuring an average Q_{10} for a number of processes of different types. Such confusion, however, commonly occurs with photosynthesis, unless one stage in the process has such a high resistance relative to the others as effectively to limit the rate of the whole.

The theoretical relation of the rate of a dark chemical reaction to temperature is exponential (Arrhenius equation) and can be written in the form

$$\frac{k_2}{k_1} = e^{-\frac{E}{R}\left(\frac{1}{T_2} - \frac{1}{T_1}\right)}$$

whence
$$\ln \frac{k_2}{k_1} = \frac{E}{R}\left(\frac{T_2 - T_1}{T_1 T_2}\right)$$

where k_2 and k_1 are the specific reaction rates at temperatures T_2 and T_1°K and E/R is a constant for the reaction concerned [118, 165].

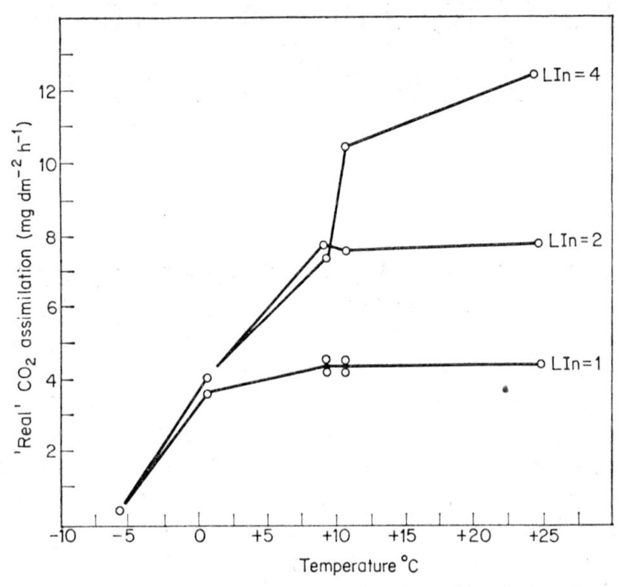

FIG. 7.1 Relation of 'real' assimilation of carbon dioxide by cherry laurel leaves to temperature, at three different light intensities (L.In = 1, 2 and 4). After Matthaei [215].

Over the limited temperature range concerned in biological processes, $T_1 T_2$ is not far from constancy and therefore

$$\ln \frac{k_2}{k_1} \simeq c(T_2 - T_1)$$

If T_2 and T_1 are 10°C apart

$$\ln Q_{10} \simeq c \times 10°$$

and
$$Q_{10} \simeq \text{constant}$$

The interaction of light and temperature in controlling rate of assimilation was the first to be studied experimentally, by Matthaei [215] in F. F. Blackman's laboratory, and it seems to have been on the basis of some of her results for cherry laurel leaves (Fig. 7.1) that he first enunciated his 'principle of limiting factors' [19]. These results he interpreted as showing an exponential (Arrhenius) relation between rate of assimilation and temperature, with the rate becoming suddenly 'light limited' when the light energy supplied (L.In. = 1, 2 or 4) was insufficient to fix any more carbon dioxide. The supposed relation was therefore as in Fig. 7.2, but this only held for temperatures up to about 25°C. At higher temperatures Matthaei found a continuous falling off in rate of photosynthesis with time and the higher the temperature the more rapid was the fall. As she used a gas flow method, with the

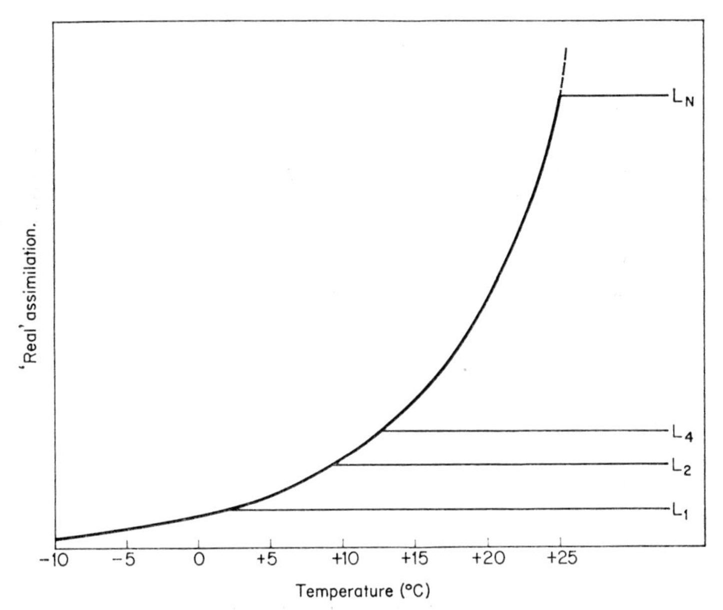

FIG. 7.2 Hypothetical limiting factor relation between assimilation and temperature up to 25°C, at different light intensities (L_1, L_2, L_4 . . . L_n) based on Matthaei's data (Fig. 7.1).

carbon dioxide uptake from carbon dioxide-enriched air estimated by absorption and titration (page 78), she could only determine assimilation over 1-hour, or more often 2-hour periods. In order to estimate the assimilation rates at the times of first exposure of the leaves to the various high temperatures Blackman carried out a remarkable feat of extrapolation (Fig. 7.3): first, the curve was extrapolated above 25°C by assuming the same Q_{10} (namely 2·1) as was obtained experimentally between 9° and 19°C; in this way the part of the curve above B in

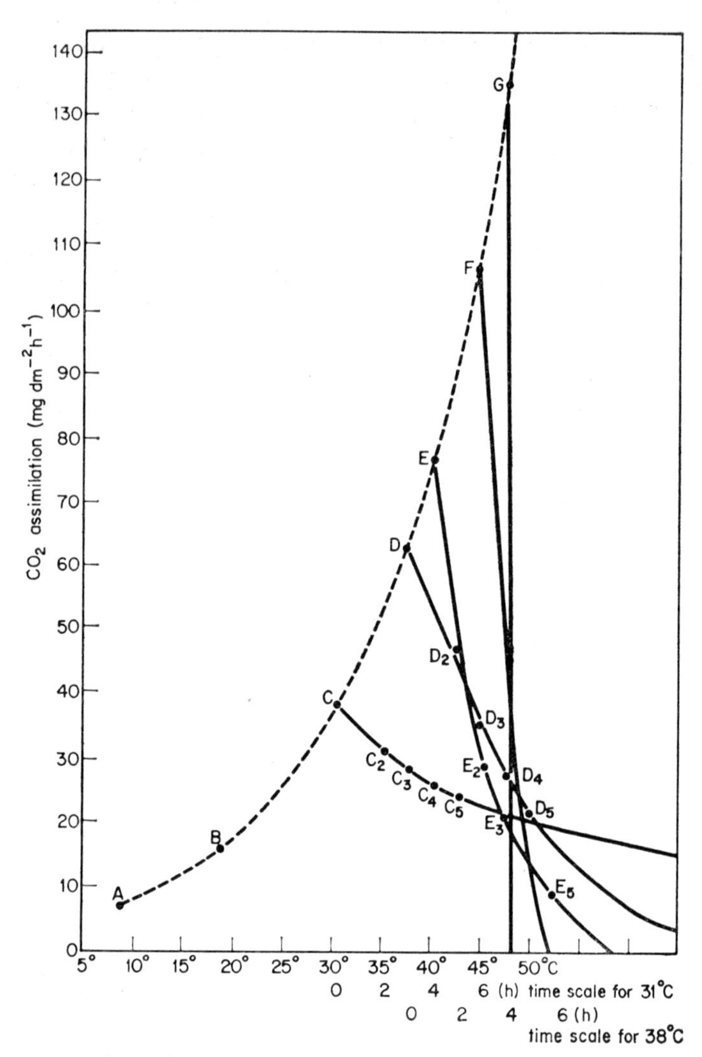

FIG. 7.3 Relation between 'real' assimilation of carbon dioxide by cherry laurel leaves and temperature: A and B, observed values at 9° and 19°C, giving a Q_{10} of 2·1; C_2 to C_5, successive observed values at 31°C plotted on a time scale with zero time coinciding with 31°C; D_2 to D_5, successive observed values at 38°C, similarly plotted; E_2 to E_5, successive observed values at 41°C, similarly plotted; C, D and E, hypothetical values at zero time, on an extrapolated curve (dotted line) with Q_{10} constant at 2·1. Curves F and G are entirely hypothetical. After Blackman [19].

Fig. 7.3 was derived. Secondly, the observed decline of assimilation at each high temperature used was plotted against a time scale on the temperature axis, with zero time coinciding with the temperature concerned in each case. These 'decline curves' were extrapolated upwards to meet, at the zero times and the appropriate temperatures, the extrapolated assimilation *v.* temperature curve. Since 'the calculated initial value and the observed subsequent values fall into one fairly harmonious curve for each temperature', Blackman considered this 'satisfactory evidence for a preliminary acceptance of the theory that such [initial] values actually occur' [19]. Curves F and G are not based on data but are purely hypothetical, the latter being for an 'extinction temperature' (here placed at 48°C) at which the rate would become *nil* almost at once.

Such extremes of extrapolation can be valuable in suggesting new hypotheses to be tested experimentally, though they should only be accepted as evidence with the utmost caution. On the additional basis of similar (unpublished) findings for respiration of leaves, Blackman put forward the suggestion that 'making suitable changes in the coefficients of temperature and time, this schema may possibly exhibit the hypothetical primary relation of all metabolic processes whatever to temperature'. He postulated that for photosynthesis the relation was really exponential with a Q_{10} of about 2, but that above 25°C a 'time factor' came into force with greater intensity the higher the temperature. This he supposed to be an injury effect which was at first reversible so that the system could recover if the exposure was not too long.

The existence of the time factor means that, in the special case of temperature, the concept of an optimum level (page 114) of a factor controlling the rate of a physiological process is particularly unsatisfactory. Not only does the optimum temperature depend upon the levels of the other factors operating but, owing to the time factor, the rates measured at various high temperatures will depend on the durations of the exposure to those temperatures preceding and during the measurements. In a temperature experiment with three different times of exposure (1, 2 and 3) a set of curves as in Fig. 7.4 might be expected, all coincident up to about 25°C but then separating to give lower optima, and these at lower temperatures, the longer the exposure. Jost [184] derived a set of curves something like this from Blackman and Matthaei's data (Fig. 7.5).

As noted, Matthaei found for the photosynthesis of cherry laurel

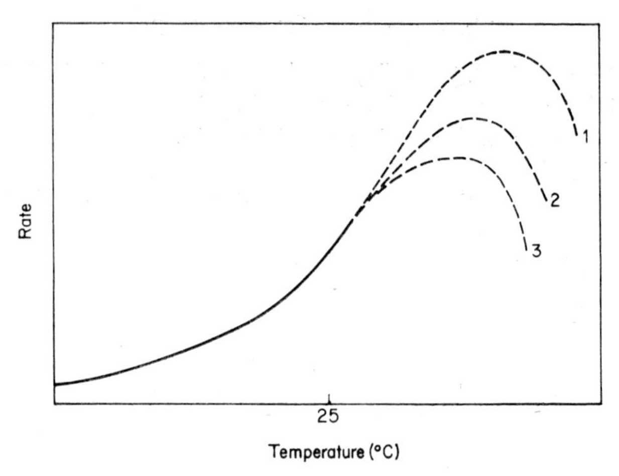

FIG. 7.4 Hypothetical relation to temperature of the rate of a biological process, such as photosynthesis, with three durations of exposure (1, 2 and 3) to the various temperatures.

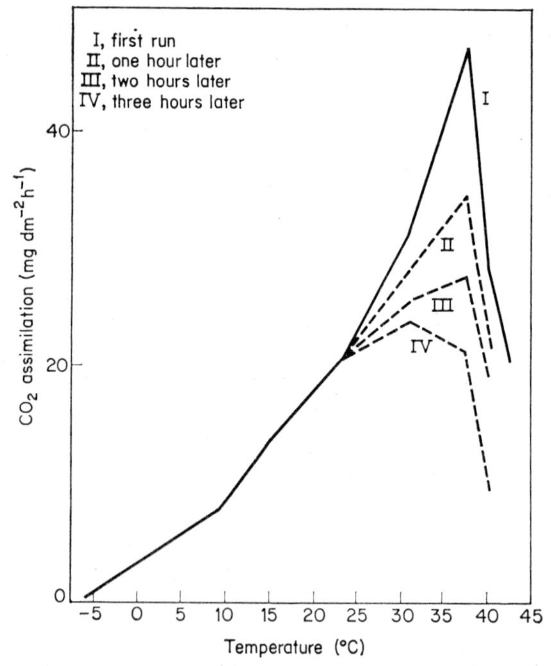

FIG. 7.5 Relation to temperature of 'real' assimilation of carbon dioxide by cherry laurel leaves, with four durations of exposure to the various temperatures. After Jost [184], derived from Blackman and Matthaei's data.

(*Prunus laurocerasus* var. *rotundifolia*) a Q_{10} of 2·1 between 9° and 19°C; Blackman and Matthaei [20] found one of 2·5 for *Helianthus tuberosus* and Blackman and Smith [21] one of 2·05 for *Elodea*. These high values of Q_{10}, with high carbon dioxide concentration and high light intensities, indicated that in addition to a photochemical stage in photosynthesis (which should have a Q_{10} of 1·0) there must be at least one ordinary chemical stage or dark reaction; later Warburg, who inferred the existence of such a dark reaction in the photosynthesis of *Chlorella*, called this the Blackman reaction. With high light Warburg found Q_{10} values as high as 4·7 (Table 7.1) at a low temperature range; here

TABLE 7.1

Temperature coefficients (Q_{10}), *assuming a* linear *relation, for* Chlorella, *with dense suspensions (complete light absorption) and high carbon dioxide concentration* (4 *per cent in gas phase*). *From Warburg* [306]

Relative light intensity	*Temperature range* °C	Q_{10}
1	25–32	1
1·8	15–25	1·06
16	5–10	4·7
16	16–25	2·0
45	5·4–10	4·3
45	10–20	2·1
45	20–30	1·6

the temperature-controlled dark reaction (or reactions) would be the main rate-limiting component of the whole process. At higher temperatures the dark reactions could proceed more rapidly and the Q_{10} fell as light became more limiting even at the highest intensity used; at low light intensity control of the rate was entirely photochemical with a Q_{10} of unity.

For an analysis of the interaction of light and temperature in controlling rate of photosynthesis we need estimates of the chemical resistances both in the dark reactions (with a Q_{10} of 2–4) and in the light reactions (with a Q_{10} of unity). With a small unicellular alga such as *Chlorella*, if the medium is strongly stirred, the diffusive resistances are probably negligible. The carbon dioxide concentration at the chloroplast surface (P_s) will thus be little lower than that in the external medium (P) and attention may be confined to the chemical resistances. With a higher plant and especially with a land leaf the picture is more complex. Maskell, who was concerned with interactions of light and carbon dioxide concentration, did not include temperature in his analysis. Monteith [231] adopted 'a conventional two-stage model in which carbon dioxide unites with an acceptor molecule A, forming a compound that is reduced to carbohydrate at a rate proportional to light intensity' (see p. 927 of ref. [265]). Combining the relation at the chloroplast derived from this model with a diffusion equation of the sort already discussed, he obtained for his field crop of beans (by making certain assumptions) an equation resembling (4.5). He used resistances $r = L/K$ s cm^{-1} per unit area instead of effective lengths and the final chemical phase was split into $r_t + r_l$. Here $r_t = 1/k$ was the resistance and k the velocity constant for the carboxylation of A, a non-reversible dark reaction and hence temperature-dependent; $r_l = P_s/k^\star$ was a light-dependent excitation resistance, $k^\star$ being the velocity constant of the reduction process; $P_s/k^\star$ could be written $P_s/\varepsilon I$ where I was the light intensity and ε a photosynthetic efficiency factor for the particular leaf type. ε was estimated assuming that the variation through the day both of r_{aq} (the aqueous diffusive resistance in the mesophyll cells) and of r_t with temperature were much smaller than the variation of r_l with light intensity, which appeared to hold at least for one of the two days of the experiment and would be expected in the bright summer weather; moreover, $(r_{aq} + r_t)$ was found by difference so that changes in r_t with temperature would be masked by the low Q_{10} for diffusion in r_{aq} (about 1·1).

In discussing the type of relation he had developed (page 125), Maskell wrote:—'. . . the analysis of the interaction of external factors must take a form determined by the special conditions of the process concerned. The particular equation developed in this paper is only a first approximation to such a form and its limitations are fully realised.' Our knowledge of the 'special conditions of the process concerned' (photosynthesis) has, in terms of its chemistry, been completely revolutionized since that time (1928). As noted previously (page 125), far from the process being a single enzyme system, many enzymes are now known to be involved; nor is it simply a two-stage carboxylation–reduction process in linear series, with a single dark reaction followed by a single light reaction. We have in the chemical stages of photosynthesis something reminiscent of a phrase quoted scornfully in a letter to *The Times*: 'a vicious circle of inter-dependent bottlenecks'. The compounds are kept circulating through the various reactions by energy derived originally from light but stored (independently of carbon dioxide) as reducing power and in high-energy complex organic phosphates which may also take part in other (non-photosynthetic) reactions; at one point in the cycle carbon dioxide enters; at another point a corresponding amount of the product is flung out but the rest of the reacting materials go round again to regenerate the compound which accepts the carbon dioxide. The rate of circulation, and hence the rate at which carbon dioxide can enter and the product be ejected, is controlled by:

(1) the carbon dioxide concentration at the chloroplast: this is dependent on the external concentration and the diffusive resistances already discussed, but also on the rate at which it is removed (that is, on (2) and (3) below);

(2) the availability of stored energy, which enters the cycle at two points—this depends on the external light intensity and the efficiency with which it is used;

(3) the sizes of all the bottlenecks (dark reaction resistances): these become less restrictive at higher temperatures so that their conductances are approximately trebled for every 10°C rise.

It will be seen that the light reaction itself is not really in series with the dark reactions [231] but constitutes part of a completely separate system; the two systems have in fact been separated *in vitro* experimentally [296].

Maskell's equation and others of similar type have been found to fit a great variety of experimental results reasonably well; an invaluable synthesis was made in showing that the many different shapes of curve, obtained for the relation of assimilation rate to carbon dioxide concentration and to light intensity, could be reconciled in terms of resistances. However, if an equation is to be used in testing a scientific hypothesis, rather than merely to fit a smooth curve to data, it must be of a form consistent with current knowledge and with the hypothesis to be tested; if it then fails to fit the data the hypothesis may have to be rejected, but if it fits well this is to be regarded as failure to disprove rather than as proof. It is possible that a single dark resistance and a single light resistance in series [231] constitute a reasonable first approximation to the complex system of reactions which is now known at least in outline. If so, the first would represent an average value for the many dark reactions in the cycle and the second an average of the resistances for the production of reducing power and high-energy phosphates, the supply of these being proportional to light intensity and to the efficiency of light utilization and it being assumed that they were all used in the photosynthetic cycle. It seems likely, however, that a new and more elaborate model is needed; it would appear only possible to test experimentally an hypothesis appropriate to such a model using a small unicellular organism and even then the difficulties of evaluating the constants would be formidable.

An important example of the interaction of light and temperature is provided by the temperature effects upon the light compensation point. This is the name given to that light intensity which gives a rate of assimilation just high enough to balance respiration. Its value will tend to fall with increasing carbon dioxide concentration, which is thought to have little or no effect on respiration rate but which increases assimilation; we will return to this aspect later. The carbon dioxide concentration should therefore always be specified; in the absence of such information it is usual to assume that ordinary air (300 p.p.m.) was used. It is even more important to specify the temperature. We have seen that with severe light limitation the Q_{10} for photosynthesis is near to unity; the Q_{10} for dark respiration, on the other hand, is usually between 2 and 3: that for respiration in light is not known. If a leaf illuminated at the compensation point has its temperature raised, the respiration rate may be expected to increase while the rate of photosynthesis will increase much less or not at all. In order to balance the new respiration rate the photosynthetic rate must be raised by increasing the light intensity. Thus the compensation point rises with temperature (Table 7.2). This has important bearings on glasshouse management. In seasons of very low light intensity, as in the English winter, it is essential to keep temperatures at the safe minimum for the crop concerned—too much heat is likely to raise the compensation point above the prevailing light intensity so that the plants lose dry weight and in fact literally respire themselves to death. During sunny intervals it is safe to raise the temperature and thus reduce temperature-limitation of photosynthesis—fortunately this occurs automatically if the ventilators are not opened.

Shade leaves have lower compensation points than sun leaves of the same species, for example 150–200 lux for the former and 500–750 lux for the latter in *Fagus sylvatica*, both at 20°C and 300 p.p.m. carbon dioxide [34]; this may be attributed partly to the shade leaves having a lower (dark) respiration rate (0·2 as compared with 1·0 mg CO_2 dm^{-2} h^{-1}) and partly to the fall in intensity of the light as it penetrates to the lower parts of the thicker sun leaves.

Owing to shading within crops or close natural communities, or in the crowns of large trees, it is quite possible for the light falling on inner leaves to be at or below the compensation point when those on the outside of the canopy are in bright light. Within the crowns of trees small branches die off as the crown grows, probably because even the shade leaves are not self-supporting for a large enough proportion

TABLE 7.2

Temperature effect on light compensation points

Species	*Temperature* °*C*	*Compensation point, lux (m/candles)*	*Carbon dioxide concentration*	*Author*
Entheromorpha compressa	10°	299	Uncontrolled	Ehrke [69]
	16°	457	,,	,,
Fucus serratus	10°	270	Uncontrolled	Ehrke [69]
	16°	408	,,	,,
Lactuca sativa (var. 'Grand Rapids')	15°	150	300 p.p.m.	Heath & Miedner [149]
	25°	400	,, ,,	Heath & Meidner [149]
Pelargonium zonale	15°	195	300 p.p.m.	Meidner (unpublished data)
	25°	490	,, ,,	

of the time to survive. Wiesner (326) measured the light intensity in such self-pruning regions and called the ratio of this to full daylight outside the crown the 'minimum light requirement'; this gave some measure of the compensation point in terms of average daylight during the growing season but included the effects of respiration at night. Comparisons of these ratios for different species at the same latitude were of interest, for example in Vienna: 1/108 for *Buxus sempervirens*, 1/55 for *Acer platanoides*, 1/9 for *Betula alba* and 1/5 for *Fraxinus excelsior*. The ratio rose in higher latitudes, as the total light intensity fell, in spite of the generally lower temperatures and longer days. Thus for *Betula nana* the values were 1/3·4 at Christiania in Norway (lat. 59° 55′ N) with a midsummer midday intensity of 1·15 units, 1/2·2 at Tromsö (lat. 69° 38′ N) with an intensity of 0·85, and 1/1 at Advent Bay in Spitzbergen (lat. 78° 12′) with an intensity of 0·75. These values also give a numerical measure of the change in habit of the tree.

Values of net assimilation rate (NAR, page 93 and Table 5.2) obtained in the field or glasshouse by the methods of growth analysis are also average values, covering day and night during the periods between successive samples. Such values are more difficult to interpret in terms of variation in individual external factors than measures of photosynthesis obtained (for example by gas analysis) under controlled laboratory conditions; on the other hand they are directly relevant to growth under the conditions actually experienced by the crop. G. E. Blackman [22] and co-workers made an extensive study of the effects of applying artificial shade by means of perforated screens, over small plants widely spaced in order to avoid mutual and self-shading, using twenty-two herbaceous species in southern England. Even in summer with mean light intensities of about 20,000 lux, in eighteen of the species including some normally found in shady habitats the NAR of the unshaded plants was light-limited. With the lower light intensities there was always some compensation for the reduced NAR by an increase in the leaf area per unit total dry weight (leaf area ratio or LAR). In spite of this, shading to 50 per cent of full daylight reduced the relative growth rates of twenty-one of the twenty-two species, leaving only *Geum urbanum* as deserving to be called a shade plant.

Where plants are growing close together, shading of the lower leaves greatly reduces the mean NAR values (compare values in Table 5.2 A and B, especially for *Lolium multiflorum*). Monteith [232] gives light transmission coefficients for leaves (kale and barley) varying between 1 per cent and 12 per cent with wave length in the range of the visible spectrum (400–700 mμ). He concludes, therefore, that light that has passed through more than one leaf is negligible in amount for photosynthesis by the leaves below. It has been recorded [171] that in a dense crop of *Vicia faba* at the flowering stage the light intensity at ground level was only 3 per cent of full daylight; at this stage it was estimated that 38 per cent of each plant was at or below the compensation point. A similar figure, of 2–3 per cent of daylight at ground level, has been found in a number of Japanese herbaceous natural communities [230], and in a sward of *Trifolium subterraneum* it is said that light absorption can be 'complete' [16]. Blackman and Black suggested

[22] that for maximal dry matter production the total leaf area per acre (leaf area index or LAI [311]) must be such that the most shaded leaves receive light at just above the compensation point; also that 'save for extremes of environment the vegetation in any region reaches a dynamic equilibrium when there is the maximum exploitation of in-coming radiation to produce the greatest production of dry matter'. If these suggestions are well founded it follows that if water supply, nutrients (especially nitrogen) and temperature are at adequate levels, light intensity, however high, must always limit the average NAR of natural vegetation.

It has been suggested (page 139) that the high NAR found for maize

TABLE 7.3

Minimum intercellular space carbon dioxide concentration (Γ) in p.p.m. (maintained for at least 30 *minutes), for isolated leaves of* Zea mays. *Interaction of light and temperature. Means for four leaves. From Meidner* [217]

Light intensity, lux	*Temperature °C*			
	25	30	35	40
1,600	0·5	2·8	12·7	38·0
3,200	0·0	2·5	4·0	7·0 .
9,700	0·0	1·5	2·2	5·5
13,000	0·0	0·2	1·5	3·8

Significant difference between means for 5% *probability* = 2·4
" " " " " 0·1% " = 4·4

(Table 5.2) may at least in part be attributed to the very low values for Γ that this plant can achieve and consequently to the steep maxi-mum gradient for carbon dioxide diffusion into the leaf. It may be assumed that both light and temperature were generally at high levels at Ibadan and the interaction of these two factors in controlling Γ is of some interest. This interaction for maize is shown in Table 7.3; although Γ tends to increase at high temperature and low light inten-sity, even at 40°C and 1,600 lux (perhaps 1 or 2 per cent of the intensity of direct tropical sunlight) the value is still lower than any shown in Table 5.1 for other species except sugar cane and *Chlorella*.

The same type of interaction but for a species having more usual values of Γ is illustrated in Fig. 7.6, in which values of Γ for *Pelargonium zonale* at two temperatures are plotted against light intensity on a log scale. There was very little effect of light on Γ between 15,000 and 3,800 lux at either temperature. Obviously photosynthesis was very severely limited by carbon dioxide concentration and above

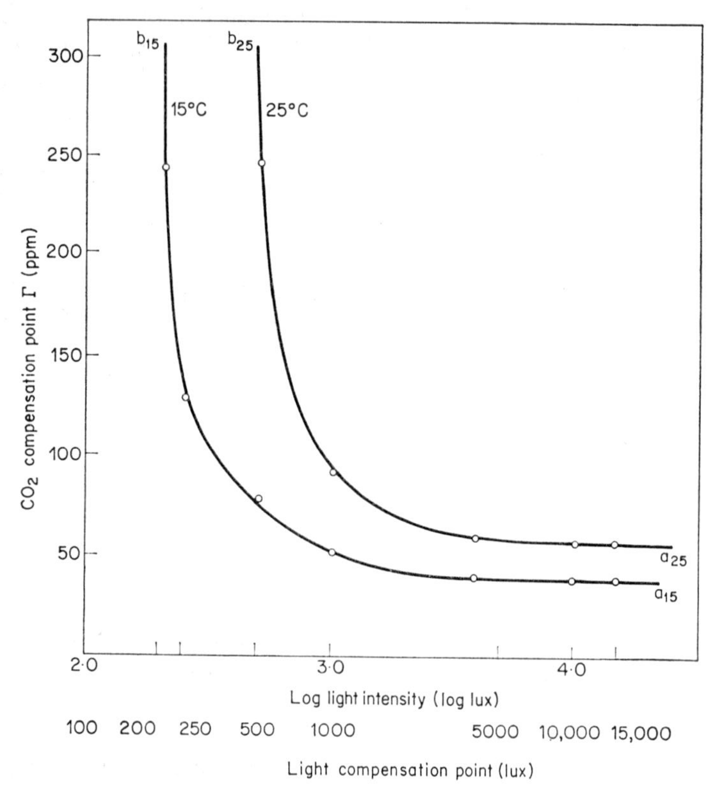

FIG. 7.6 Relation of the carbon dioxide compensation point (Γ) to log light intensity, at two leaf temperatures. Each point is a mean for four leaves of *Pelargonium zonale*; freehand curves. a_{25} and a_{15} indicate approximate positions for the carbon dioxide asymptotes, at the two temperatures, below which a steady state cannot be maintained at any light intensity; b_{25} and b_{15} similarly indicate the approximate asymptotes in terms of log light intensity. Data from Meidner and Glinka (unpublished).

10,000 lux little effect of light intensity would be expected, unless indeed photoxidation (page 144) were important, in which case a marked *increase* of Γ would be expected at high light intensity.

The curve for each temperature shows, at any light intensity, the carbon dioxide concentration (Γ) at which production (whether light stimulated or not) is just balanced by uptake and which we may here discuss as the 'carbon dioxide compensation point'. Alternatively we may say that at any carbon dioxide concentration the curve shows the light intensity at which production is just balanced by uptake, that is the light compensation point. Each curve in fact shows the relation, for that temperature, between the carbon dioxide compensation points and the light compensation points. At concentrations below the curve and light intensities to the left of it there will be a net output of carbon dioxide; at concentrations above and intensities to the right of the curve there will be a net uptake. The curve thus divides an area of loss in terms of plant dry weight from an area of profit. With data for more temperatures we could bring in the temperature compensation point as the third variable in a three-dimensional diagram.

The curves illustrate the fall in light compensation point with increasing carbon dioxide concentration mentioned earlier (page 190) and also the fall in the carbon dioxide compensation point with increasing light intensity. A comparison of the 25°C curve with the 15°C curve to the left of it at the same Γ value indicates the fall in the light compensation point with temperature at that concentration; the comparison with the 15°C curve below at the same light intensity shows the fall in carbon dioxide compensation point with temperature.

The fact that Γ increases with temperature even at high light intensity shows that the Q_{10} for photosynthesis is still lower than that for respiration. This might be expected with carbon dioxide so severely limiting, for although we should expect no net diffusion through the stomata at Γ, diffusion ($Q_{10} \simeq 1{\cdot}1$) from the mitochondria to the chloroplasts, both direct and via the intercellular spaces, would lower the average Q_{10} for photosynthesis below that of the dark reactions involved in respiration.

As the light intensity is reduced Γ rises, sooner at the higher temperature, and ultimately each curve approaches a vertical asymptote. Below this light intensity a steady state cannot be maintained, at the given temperature, as respiratory output will always exceed photosynthetic uptake whatever the carbon dioxide concentration. Similarly, as the carbon dioxide concentration is reduced the curve approaches a horizontal asymptote below which a steady state cannot be maintained whatever the light intensity. The two asymptotes (*a* and *b* in Fig. 7.6), in terms of carbon dioxide and log light intensity respectively, may be calculated from fitted curves [149, 280] and should then provide useful numerical (inverse) measures of net photosynthetic efficiency for different species and at different temperatures; carbon dioxide is completely limiting and light saturating at *a*, while the converse applies at *b*. For maize, *a* approximates to zero up to 35°C (Table 7.3) if the leaves have not experienced wilting. The value of *b* includes the proportion of the light not absorbed by the leaf; thus as an inverse measure of photosynthetic efficiency *b* includes the efficiency of the leaf as an absorber of light.

Rather than make the considerable extrapolations to infinite light intensity for *a* and to 'infinite carbon dioxide concentration' for *b*, it is better and easier to choose arbitrary levels of carbon dioxide and light at which the light and carbon dioxide compensation points may be specified as measures of plant performance. As already noted, it is conventional to specify the light compensation point at 300 p.p.m. of carbon dioxide, and 10,000 lux may be suggested as a convenient light intensity at which to specify Γ.

Horticulturists have found beneficial effects upon the growth of glasshouse crops, especially lettuce, of supplying extra carbon dioxide in winter and at rather higher temperatures than are normally used without such carbon dioxide enrichment. These effects seem at first sight paradoxical, for with light intensity severely limiting, as it usually is under glass in winter, little response to carbon dioxide might be expected; further, it might be thought that the higher temperatures would raise the light compensation point above the prevailing light intensities as suggested earlier (page 190). Fig. 7.6 may provide a clue to this paradox [149]. The enrichment of the air with carbon dioxide lowers the light compensation point and thus makes available a greater range of light intensities for net gain in dry weight. This might appear to be a rather marginal advantage but it seems likely that leaves which remain below the light compensation point for periods of days on end become senescent; even a rather small lowering of the compensation point may therefore prevent the yellowing and death of partly shaded leaves. This would be best achieved by keeping the temperature rather low, as in the 15°C curve of Fig. 7.6, but growth rate depends not only on photosynthesis per unit leaf area but also on other processes such as cell enlargement and leaf expansion which proceed more rapidly at higher temperature and in turn lead to more photosynthesis. The lowered light compensation point makes it safer to take advantage of the accelerating effect of higher temperature on such processes.

For *Pelargonium* Γ has been found to rise smoothly with temperature, at 9,670 lux, with a relative rate of increase of about 1·8 per 10°C rise between 10°C and 35°C [150]. In some species, however, the relative rate of increase suddenly rises when the temperature exceeds about 30°C (Fig. 7.7). This occurs in leaves of onion (*Allium cepa*) [150], coffee (*Coffea arabica*) [150] and a palm (*Phoenix reclinata*) [216], all of which exhibit midday closure of their stomata, but not in maize, wheat or *P. zonale*, which apparently do not. It has therefore been suggested

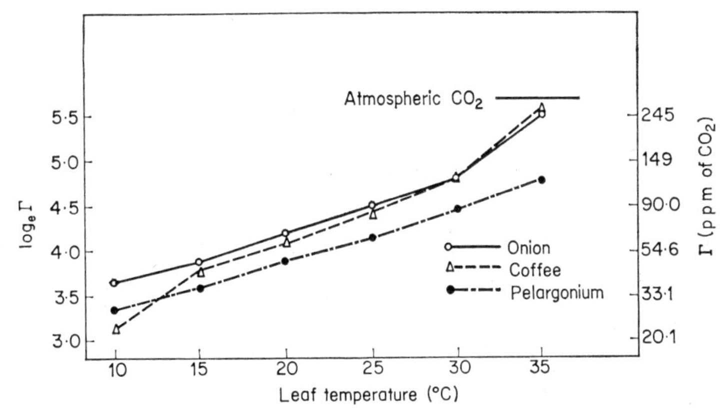

FIG. 7.7 Log Γ as a function of leaf temperature for three species, at a light intensity of 9,670 lux. The slope of the line (relative rate of increase of Γ) steepens above 30°C for onion and *Coffea* but not for *Pelargonium*. After Heath and Orchard [150].

[143] that midday closure is a high-temperature effect operating via the intercellular space carbon dioxide concentration; such an effect could be reinforced by one due to water deficit (page 174), especially as the latter increases the sensitivity of stomata to carbon dioxide [146]. It may be noted that in these experiments on the effects of high temperatures upon Γ the operation of the time factor (page 185) was not observed. Even at 35°C, or 40°C for *P. reclinata* and 45°C for maize, the value of Γ did not rise during the 30-minute period of the readings.

8: Chlorophyll Content and Light (Light Absorption)

THE history of the investigation of the relation between chlorophyll content and rate of photosynthesis provides an excellent example of the difficulties of interpretation encountered when attention is confined to a single factor. The first observations suggested a deceptively simple state of affairs. Weber [315] in 1879 had found differences between species, grown under the same conditions, in the dry matter production per unit leaf area; several years later Haberlandt [133] made counts of the numbers of chloroplasts, also per unit leaf area, in the same species as used by Weber, and found a truly remarkable correlation (Table 8.1).

Haberlandt attributed the slight divergencies between the figures in the last two columns to differences of chloroplast size and leaf structure but in fact, in view of the way the data had been assembled, the surprising thing was the good agreement obtained. He concluded that the rate of photosynthesis in leaves was proportional to their chlorophyll content. Later work did not appear to substantiate this conclusion.

Willstätter and Stoll, who had studied the chemistry of the chlorophylls in a long series of researches [328], used their methods of analysis in comparing chlorophyll content with rate of photosynthesis for a great variety of leaves [329]. In order to ensure that photosynthesis was controlled by internal factors only (including chlorophyll) they worked with a light intensity of about 48,000 lux, a gas stream containing 5 per cent carbon dioxide by volume and, in most experiments, a temperature of 25°C. They calculated their results as 'true' assimilation in g CO_2 per hour per g of chlorophylls ($a + b$) and called this the assimilation number (A_c). Constancy of A_c would indicate direct proportionality of assimilation to chlorophyll, while variation in A_c would indicate that other internal factors besides chlorophyll were having an

TABLE 8.1

Relation between dry matter formation (Weber, 1879) and number of chloroplasts (Haberlandt, 1882). After Haberlandt, 1914 [134]

Species	*Dry matter formation g per m² of leaf area in 10 hours*	*Number of chloroplasts per mm²*	*Relative photosynthetic activity per unit leaf area*	*Relative number of chloroplasts per unit leaf area*
Tropaeolum majus	4·47	383,000	100	100
Phaseolus multiflorus	3·22	283,000	72	74*
Ricinus communis	5·29	495,000	118·5	129
Helianthus annuus	5·57	465,000	124·5	122

* Haberlandt [133] gives 64, but also gives the figures shown in the third column.

important effect. They compared A_c values for leaves of different sorts, which may be grouped under five heads:

1. mature, but not old, leaves of different species having fully green leaves;
2. leaves of green and yellow (*aurea*) varieties of the same species;
3. leaves of the same species of different ages;
4. etiolated leaves greening under the influence of light;
5. chlorotic leaves suffering from nutrient deficiency.

Some of the results of Willstätter and Stoll are presented in Table 8.2. For fifteen species in class 1, the values of A_c at 25°C ranged from 5·2 in *Acer pseudoplatanus* to 16·7 in *Helianthus annuus*, while the chlorophyll content varied only from 3·1 to 6·6 mg dm^{-2}. Nine of these fifteen species gave A_c values between 5·2 and 7·7, and these they called 'normal'; here the range of chlorophyll content was rather small, from 3·1 to 5·2 mg dm^{-2}, though this appeared larger (1·3–4·0 mg) when expressed per g fresh weight of leaf as in their book. Unless these 'normal' A_c values could legitimately be considerd as a class apart, and there appears to be no justification for this, the data indicated that chlorophyll content was not an important factor controlling assimilation under high light and with high carbon dioxide concentration.

TABLE 8.2

Chlorophyll content and assimilation (under saturating light and carbon dioxide concentration). Light intensity 48,000 *lux; carbon dioxide in air stream* 5 *per cent; temperature* 25°*C. From Willstätter and Stoll,* 1918 [329]

Species	*mg Chl* g^{-1} *F.W.*	*mg Chl* dm^{-2}	*Assimilation rate mg* CO_2 dm^{-2} h^{-1}	*Assimilation number* (A_c) *g* CO_2 h^{-1} g^{-1} *Chl*	*Assimilation time* (T_A) *s* mol^{-1} *Chl*
(1) *Fully green leaves*					
Acer pseudo-platanus	4·00	5·22	27	5·2	33·8
Acer negundo	2·48	5·17	40	7·7	22·9
Helianthus annuus	1·50	4·50	75	16·7	10·5
Tilia cordata	2·81	4·19	28	6·7	26·3
Laurus nobilis	1·27	3·22	19	5·9	29·8
(2) *Green and yellow vars.*					
Ulmus sp.	1·62	3·06	21	6·9	25·5
Ulmus sp. *aurea*	0·12	0·29	24	82	2·1
Sambucus nigra	2·35	5·19	32	6·2	28·4
S. nigra aurea	0·081	0·175	21	120	1·5
(3) *Age of leaf*					
Sambucus nigra (1st May)	1·17	3·76	46	12·2	14·4
Sambucus nigra (8th May)	2·31	5·80	57	9·8	18·0
Sambucus nigra (14th July)	2·35	5·19	32	6·2	28·4

This conclusion was greatly reinforced by the results for class 2 (Table 8.2). Yellow leaves with one-tenth of the normal chlorophyll content or less could assimilate carbon dioxide at about the normal rate and thus gave very high A_c values.

Willstätter and Stoll concluded that some other internal factor, controlling the rate of dark reactions in photosynthesis, was concealing the effect of the chlorophyll factor. This is exactly what might be expected at high (saturating) light intensity if the effect of the chlorophyll content is to control the amount of light absorption; by confining their attention almost entirely to high light they ensured that even the yellow leaves absorbed enough light for some other factor to be severely limiting. They did, however, carry out two experiments on the assimilation rates of green and yellow leaves at a series of different light

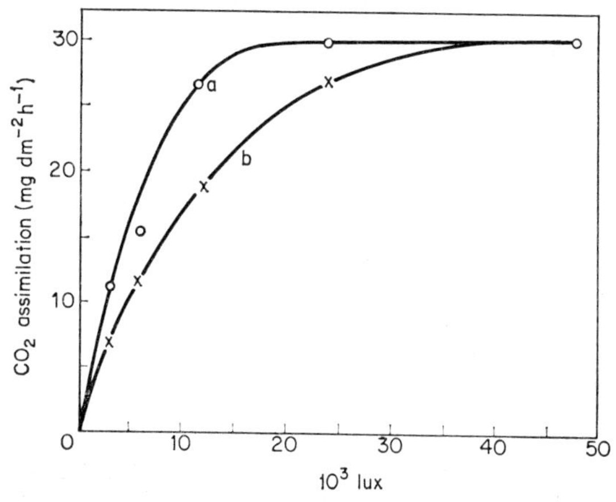

FIG. 8.1 Relation of assimilation of carbon dioxide by leaves of *Ulmus* sp; to light intensity. *a*, green leaves with 3·1 mg dm^{-2} of chlorophyll ($a + b$). *b*, yellow leaves with 0·40 mg dm^{-2} of chlorophyll ($a + b$). After Gabrielsen [112], who recalculated the data from Willstäter and Stoll [329].

intensities and for these Gabrielsen [112] recalculated the data on a basis of leaf area instead of leaf fresh weight, since light absorption depends on chlorophyll content per unit area. The results are shown in Figs. 8.1 and 8.2. In the former it is seen that to attain any given assimilation rate about twice as much light was needed for the yellow *Ulmus* leaves as for the green ones; this implies that a reduction in chlorophyll content from 3·1 to 0·4 mg chlorophyll ($a + b$) dm^{-2} halved the amount of light absorbed; at 48,000 lux the rates are indistinguishable and the curves are almost horizontal. The results in Fig. 8.1, if the drawing of the curves is accepted, imply that all the non-photochemical factors were the same for both green and yellow leaves, for the curves tend towards the same maximum rate (asymptote). The results for

Sambucus in Fig. 8.2 were interpreted by Gabrielsen as showing that the non-photochemical processes went on more slowly in the yellow than in the green leaf, since the curves tended towards different asymptotes. (Rabinowitch [266] concluded for these same data: 'The extrapolated saturation value of the yellow variety appears, however, to be equal to, or only slightly below, that of the green one.') By comparing the initial slopes, in fact by taking the ratio between the rates for the two leaves at a low light intensity, Gabrielsen concluded that light absorption was in the ratio 0·36 : 1, although the chlorophyll contents were

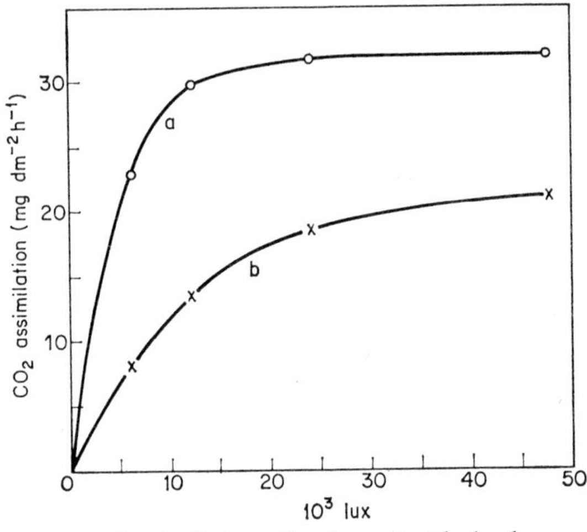

FIG. 8.2 Relation of assimilation of carbon dioxide by leaves of *Sambucus nigra* to light intensity. *a*, green leaves with 5·2 mg dm^{-2} of chlorophyll ($a+b$). *b*, yellow leaves with 0·18 mg dm^{-2} of chlorophyll ($a+b$). After Gabrielsen [112], who recalculated the data from Willstäter and Stoll [329].

as 0·18 : 5·2. Such a less than proportional increase in absorption with increasing chlorophyll content would be expected from the law governing light absorption in coloured solutions (Beer's Law) according to which the logarithm of the relative amount of light transmitted decreases linearly with increasing solute concentration. This means that at low concentrations a small difference in chlorophyll content will have a large effect but at high concentrations it will make very little difference (compare Fig. 8.3). It should be remembered, however, that the distribution of chlorophyll in leaves is by no means as homogeneous as in a solution and Beer's Law, which in any case only applies to dilute solutions, would not be expected to apply exactly.

To check his interpretation of Willstätter and Stoll's results, Gabrielsen measured photosynthetic rates, in normal air and at light intensities from 0 to 9,000 lux (0 to 29·0 J m^{-2} s^{-1}) for seventeen leaf types whose chlorophyll contents ($a + b$) were found to range from 0·21 to 8·7 mg dm^{-2}. The leaf types are listed in Table 8.3. Since the chlorophyll factor was expected to be most important at low light intensity, when

TABLE 8.3

Leaf material and maximum energy yield. From Gabrielsen 1948 [112]

No. of category	*Species*	*Colour*	*Chlorophyll* $(a + b)$ *(mg dm^{-2})*	E_m
1	*Sambucus canadensis aurea*	pale yellow	0·21 ± 0·04	1·7
2	,, ,, ,,	yellow	0·47 ± 0·05	3·7
3	*Ulmus glabra lutescens*	,,	0·76 ± 0·15	6·0
4	,, ,, ,,	yellowish green	1·1 ± 0·1	7·7
5	*Ulmus glabra lutescens*	pale green	1·8 ± 0·2	10·0
6	*Atriplex hortensis chlorina*	,, ,,	2·5 ± 0·3	10·4
7	*Sambucus nigra aurea*	,, ,,	4·2 ± 0·3	11·7
8	*Fraxinus excelsior* (shade leaves)	,, ,,	4·8 ± 0·2	9·4
9	*Populus canadensis aurea*	,, ,,	5·1 ± 0·7	13·1
10	*Corylus maxima*	green	6·0 ± 0·4	14·1
11	*Atriplex hortensis*	,,	6·4 ± 0·6	15·4
12	*Frazinus excelsior* (sun leaves)	,,	6·6 ± 0·4	12·1
13	*Triticum sativum*	,,	6·7 ± 0·8	13·4
14	*Sinapis alba*	,,	7·0 ± 0·4	13·4
15	*Ulmus carpinifolia*	dark green	8·4 ± 0·8	14·4
16	*Sambucus nigra*	,, ,,	8·6 ± 0·5	13·7
17	*Populus canadensis*	,, ,,	8·7 ± 0·7	14·7

light was most severely limiting, he expressed its effect as maximum energy yield (E_m) which was calculated as the percentage of an incident radiation of 4·7 J m^{-2} s^{-1} (about 1,500 lux) which was bound as chemical energy. E_m for each leaf type was thus proportional to the slope of the initial part of the curve relating its photosynthetic rate to light intensity. The values for all his seventeen leaf types are given in Table 8.3 and Fig. 8.3; they suggest that even at such low light intensity increasing the chlorophyll concentration above about 5 mg dm^{-2} has little further effect.

Gabrielsen's results suggest that, to a first approximation, light intensity and chlorophyll content make up a single factor, namely light absorption. At high light intensity even a small chlorophyll content can entrap enough of the incident quanta for light to remain saturating; it is only when light intensity is very low that the extra light absorbed by a high concentration of chlorophyll becomes important. This might seem to make the high chlorophyll contents of many shade leaves biologically advantageous, but in Table 8.3 the leaves corresponding to numbers 1 to 9 are characterized as 'pale yellow' to 'pale green', while 10 to 17 were 'green' to 'dark green'. It appears therefore that any fully

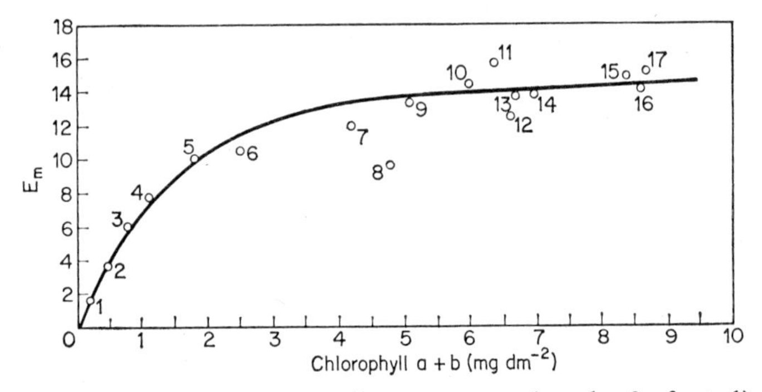

FIG. 8.3 The relation to chlorophyll concentration (mg dm^{-2} of $a+b$) of 'maximum energy yield' (E_m) for leaves with 17 different chlorophyll contents as listed in Table 8.3. E_m is the percentage of incident radiation at 4·7 J m^{-2} s^{-1} fixed as chemical energy. After Gabrielsen [112].

green leaf has an excess of chlorophyll even for light intensities approaching the compensation point, but this may be important for absorbing green and far-red light remaining after passage through other leaves.

Rabinowitch [266] seems to accept that Willstätter and Stoll's 'normal' leaves constituted a separate class in which rate of photosynthesis was more or less directly proportional to chlorophyll content in spite of both light and carbon dioxide being presumably almost saturating. He concludes that in these leaves the rate of a dark reaction was limited by a catalyst present in amount proportional to that of chlorophyll. If this is well founded it may be that in such 'normal' leaves the variation in chlorophyll is mainly due to variation in the *number* of chloroplasts of nearly constant chlorophyll content. If it is accepted that the whole enzyme system for photosynthesis resides in the chloroplasts [6], the enzymes, including whichever one was most rate-limiting, might then be expected to increase more or less proportionately with the chlorophyll. On the other hand, in those leaves where lower chlorophyll content is accompanied by a great increase in A_c it may be suggested that there is a decrease in the chlorophyll content of the individual chloroplasts without a corresponding decrease in the rate-limiting enzymes. These speculations require testing by means of data for chloroplast numbers—an aspect that seems to have been neglected since Haberlandt's work (page 198). It is possible that differences in the *size* of the chloroplasts, with constant chlorophyll and enzyme content per unit volume of plastid, might act in a similar way, though in general chloroplasts do not vary much in size—from about 3 to 10 μ across in different species [264].

Although the distribution of chloroplasts in the leaf is markedly non-homogeneous, this should not have important effects under light-saturated conditions, and variation in their number or size should then somewhat resemble in its effects variation of the number or size of cells in a suspension of unicellular algae, while variation in the chlorophyll content of individual plastids should simulate variation of that in the algal cells. An experiment carried out by Emerson [72] with *Chlorella*, in high carbon dioxide concentration and strong light for 16 hours, showed effects of the kind postulated. At the end of the experiment the total volume of cells had increased by a factor of 2·8, which could have been due to increase in cell number or cell size or a combination of the two, but the total chlorophyll content was virtually unchanged so that the concentration in each of the cells was reduced almost to one-third. Nevertheless, the measured rate of photosynthesis had more than doubled giving a similar increase in A_c. Thus the increase in cell number or cell size gave an almost proportional increase in total photosynthesis, presumably because the enzymes responsible for the dark reactions were multiplied by a similar factor.

The comparison as between different species of the relation of assimilation rate to chlorophyll content, or even as between different varieties of the same species, is obviously likely to be confused by the effects of differences in many other internal factors. An alternative and better approach is to attempt to vary chlorophyll content experimentally in a single species, though here again (as in all experiments) it is impossible to confine the effects of treatments to the desired internal factor and many other changes are inevitably produced—it thus becomes a matter for judgement or further experimentation to decide whether such unwanted changes may safely be neglected for the purposes of the experiment.

Perhaps the least drastic treatment for modifying chlorophyll content is simply to allow leaves to age, though even here many other things such as water content, nitrogen content and doubtless enzyme complement change also. These simultaneous changes make the choice of a basis for expressing chlorophyll content particularly critical. Both fresh and dry weight change continuously as the leaf ages and once it has expanded fully it is probable that leaf area is the most constant basis to use (Table 8.2(3)). It will be noted that as the *Sambucus* leaves aged the chlorophyll content increased continuously on a fresh weight basis, but on an area basis both chlorophyll and assimilation first rose and then fell and A_c declined throughout.

The development of the capacity to photosynthesize in etiolated (dark-grown) leaves becoming green under the influence of light was also studied by Willstätter and Stoll. They found very high A_c values for such leaves, for example for *Phaseolus vulgaris* after 6 hours pre-illumination A_c was 133, with a chlorophyll content of only 0·07 mg per g fresh weight; even with no illumination until the measurements began, when the leaves were 'pure yellow' and had less than 0·02 mg of chlorophyll per g fresh weight, the A_c value was greater than 70. This contrasts with the findings of Irving [178] in Blackman's laboratory that leaves of 5–6-day-old etiolated barley seedlings continued to give out carbon dioxide at the dark rate during 14 hours of continuous illumination. Both in barley and *Vicia faba*, shoots which had developed a considerable green colour could still fail to show any photosynthetic activity, which in barley seedlings at the coleoptile stage only became detectable after two day periods and two night periods. Briggs [36], using the palladium black method for measuring photosynthesis (page 90), found that no further chlorophyll appeared to develop in leaves while they were exposed to light under the practically anaerobic conditions it necessitated. He could therefore produce any desired degree of green coloration in light and ordinary air before an experiment and maintain this for days by only illuminating the leaves when they were in hydrogen. By bringing young plants of *Phaseolus vulgaris* into the light at various ages he found that the initial rate of assimilation was determined by the number of days from sowing rather than by the apparent chlorophyll content. Seven days from sowing, leaves of 'grass green' colour showed no oxygen output, but 'green-yellow' leaves nine days old had an appreciable photosynthetic rate. Whatever the rate initially, it increased on subsequent days, almost reaching the value for normal green leaves in spite of a persistent low chlorophyll content. Subsequently Briggs [37] found that this delay only occurred with seedlings in which the first specialized photosynthetic organs were separate from the storage organs (*Phaseolus*, *Ricinus*, *Zea*). In *Helianthus*, *Acer* and *Cucurbita*, where the storage organs (cotyledons) were themselves the first assimilating organs, photosynthetic activity appeared directly after germination. This suggested that in the former class the delay was due to the need for some factor which had to be translocated from the storage organs.

In both Irving's and Briggs' work the rate of photosynthesis was found to increase with successive periods of light and dark. Smith [277] investigated this with the phosphorescence-quenching method for measuring oxygen evolution (page 90), using leaves from 9–11-day-old etiolated barley seedlings. Ten minutes' illumination, at about 2,500 lux with the leaves in carbon dioxide-free hydrogen, converted 85 per cent of the dark-formed protochlorophyll initially present into chlorophyll *a*, with virtually no oxygen production. Illumination in air for 10 minutes also failed to induce appreciable oxygen production in an immediately subsequent 10-minute period in hydrogen, although nearly all the initial protochlorophyll content was converted to chlorophyll *a*; if two 5-minute periods of illumination in air were separated by dark periods of various durations, photosynthesis was increased far more than in proportion to the new chlorophyll formed but almost in direct proportion to the length of the dark period (up to 110 minutes).

The effect of alternating light and dark was demonstrated in a much more striking manner by Gabrielsen *et al.* [113]. They used intact etiolated wheat seedlings 6–8 days old, the diaferometer method (page 80) with 3 per cent carbon dioxide in the air stream and a higher light intensity (10,000 lux) than any of the other workers mentioned except Willstätter and Stoll. Under continuous illumination for 20–30 minutes, the longest period found possible owing to zero drift of the diaferometer at the high sensitivity needed, the seedlings apparently gave out carbon dioxide continuously at a rate in excess of the dark respiration rate. This was evidence that photoxidation (page 144) was occurring, especially as the output increased with light intensity when this was varied. In Fig. 8.4 are shown the effects of alternating light and darkness. In (a) and (b) the photoxidation effect is seen to have been less in each successive light period, with definite evidence of photosynthesis (carbon dioxide uptake) appearing in the fifth one, that is, after 48 minutes from first illumination, or four of the 12-minute cycles used. It will be noted that the total *amount* of light (8 minutes in (a) and 2 minutes in (b)) had no effect on the time taken to reach this stage; in a further experiment, with cycles of 3-second light periods followed by 10-minute dark periods, photosynthesis was again detected (in a single 2-minute light period) 50 minutes after first illumination, that is, after a total of about 15 seconds' light in 5 cycles. In the

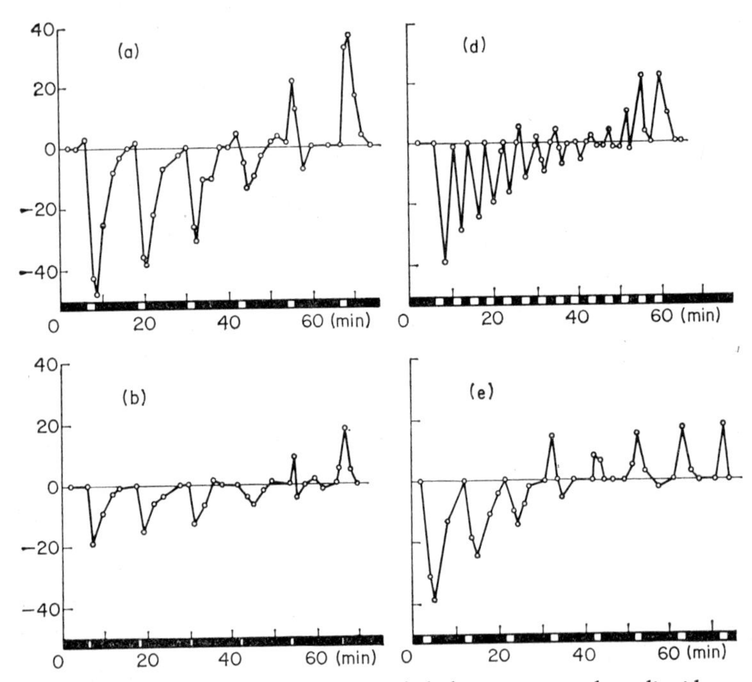

FIG. 8.4 Effects of alternating light and darkness upon carbon dioxide exchange by intact etiolated wheat seedlings. Ordinates, carbon dioxide output (negative) and uptake (positive) in mm galvanometer scale deflection. Abscissae, periods of dark and light (10,000 lux): (a), 12-min cycles of 2 min light and 10 min dark; (b) 12-min cycles of 30 s light and 11 min 30 s dark; (d), 4-min cycles of 2 min light and 2 min dark; (e), preliminary long cycle of 2 min light and 40 min dark (not shown) followed by 10-min cycles of 2 min light and 8 min dark. After Gabrielsen *et al.* [113].

experiment shown in Fig. 8.4 (d) the cycles were shortened to 4 minutes (2 minutes' light and 2 minutes' dark) and the authors state that the period before carbon dioxide uptake could be reliably demonstrated was reduced to 36 minutes or perhaps less; in (e) a preliminary long cycle of 2 minutes' light and 40 minutes' dark was followed by cycles of 2 minutes' light and 8 minutes' dark and photosynthesis was first detected 72 minutes after first illumination. The authors attribute these effects to the length of the dark period but it is rather suggestive that in (a), (b), (d) and (e), as well as in a fifth of their experiments not reproduced here (f), photosynthesis first appeared after four or five light–dark cycles whatever the lengths of either light or dark periods.

It appears that for the development of the ability to photosynthesize a light reaction must be followed by a dark reaction, that each successive light–dark cycle carries the process a step further and that the minimum periods for either light or dark were not reached in the experiments just described. The work of Smith [277] suggests that this process is not related at all closely with the production of chlorophyll *a*, and although he found a considerable quantity of the latter formed before photosynthesis (oxygen evolution) began at all the increase in the one appeared almost independent of increase of the other; chlorophyll *b* was not necessary for oxygen evolution. Continuous light would appear to tend to inhibit the dark reaction, especially perhaps in very young seedlings of species in which the storage organ is separate from the first assimilating organ. Briggs [36] attributes the discrepancy between the results of Irving on the one hand and Willstätter and Stoll on the other to the former having used seedlings 5–6 days from sowing and the latter, 14 days. Although Gabrielsen *et al.* found less photoxidation at lower light intensity this last does not appear to be the most critical factor for inhibition of the dark reaction, since Irving, Briggs and Smith all used very low light intensities.

Another method of varying chlorophyll content experimentally is to subject plants to deficiency of nutrients, especially of iron, magnesium or nitrogen, though this may again be expected drastically to alter other internal factors also. Willstätter and Stoll found that severe chlorosis, induced in plants of *Zea mays* and *Helianthus annuus* by growing them in water culture without iron, resulted in very low assimilation rates and in A_c values similar to those of 'normal' leaves. The small quantity of chlorophyll formed was thus used much less efficiently than in leaves of *aurea* varieties, probably indicating a failure in some other part of the system. Emerson [70, 71] obtained $2\frac{1}{2}$- to 5-fold ranges of chlorophyll content in *Chlorella vulgaris*, in different experiments, by varying the iron supply in the nutrient solutions. With carbon dioxide supplied from a carbonate–bicarbonate buffer (page 87) and even at high light intensity (10^5 lux) photosynthesis increased almost in proportion to chlorophyll concentration. For a normal and a chlorotic culture the relations to light intensity (Fig. 8.5) somewhat resembled Fig. 8.2 but with maximal photosynthetic rates in the ratio 3·4 : 1, that is, not much less than the ratio of 4 : 1 for their chlorophyll contents. Fleischer [87] obtained similar results for the effects on *Chlorella* of iron deficiency and also of nitrogen deficiency—in each case an almost linear relation of maximum photosynthesis to chlorophyll content. These results would not support the hypothesis of a direct control of photosynthesis by chlorophyll content acting simply through light

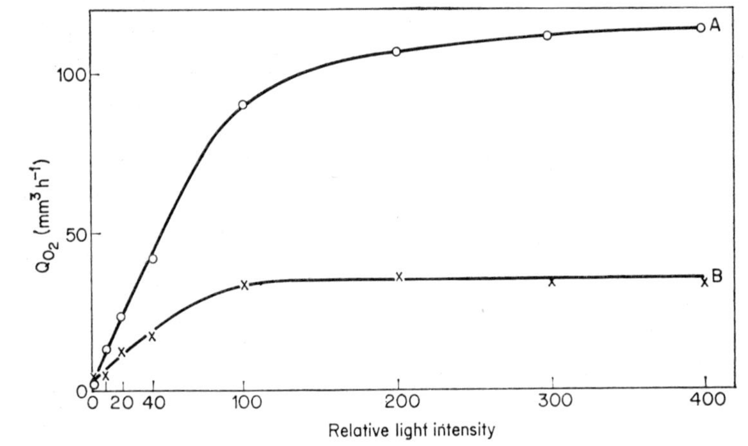

FIG. 8.5 'True' photosynthesis (Q_{O_2}, oxygen output) by *Chlorella vulgaris* as a function of light intensity. A, normal culture, extinction coefficient $\varepsilon = 0{\cdot}099$ for pigment solution determined at 670 nm as a measure of chlorophyll content; B, chlorotic culture, $\varepsilon = 0{\cdot}026$. Temperature 20°C. Data from Emerson [71].

absorption, for then the light curves should resemble Fig. 8.1, but taken by themselves they would be consistent with the chlorophyll itself being the rate-limiting catalyst, so that photosynthesis was at all light intensities limited by chlorophyll content. This possibility is rejected by Rabinowitch [266] on the basis of the evidence from leaves (with the tacit assumption that the mechanism is there the same as in algae) and this rejection is confirmed by more recent biochemical studies. It seems that iron or nitrogen deficiency must inhibit the formation of an enzyme or enzymes in the same proportion as chlorophyll.

In connection with nitrogen deficiency it is interesting to note that van Hille [163] found a steady decline in photosynthesis with *age* of culture in *Chlorella pyrenoidosa*, even though chlorophyll content increased; this was attributed to increasing nitrogen deficiency, for it could be cured by periodically adding nitrate to the nutrient solution. These changes are rather reminiscent of those in leaves (Table 8.2(3)) and it is suggestive that the nitrogen content of leaves often falls as they age [270].

Unlike iron and nitrogen deficiencies, magnesium appears to have a direct effect on photosynthesis quite independent of chlorophyll [87, 188]. Magnesium deficiency affects the chlorophyll content only when it is very severe and in a range where it has little effect on rate of assimilation; it affects the latter most at a higher range, where chlorophyll content is almost constant.

In addition to calculating assimilation numbers (A_c), Willstätter and Stoll also estimated 'assimilation times' (T_A) as the average time that each chlorophyll molecule took to reduce one molecule of carbon dioxide, with saturating light and carbon dioxide concentration. Since the average molecular weight of chlorophyll is about 900 and that of carbon dioxide 44

$$T_A = (44 \times 3{,}600)/900A_c = 176/A_c \text{ s}$$

Values of T_A for 'normal' leaves were 25–30 seconds (Table 8.2), but clearly this cannot be a fundamental attribute of the chlorophyll molecule, for in *aurea* leaves the value was found to be as low as 1·5 seconds.

The reciprocal of T_A gives the mols of carbon dioxide fixed per mol of chlorophyll per second and this varies (for the values in Table 8.2) from about 1 to 0·03. As we shall see in the next chapter, it was found by Emerson and Arnold [76] that if very short flashes of light (about 10 μs) were separated by much longer dark periods of the order of 100 ms, the amount of photosynthesis *per unit of light* was very greatly increased, although of course the total photosynthesis was much reduced. Using this technique and *Chlorella* cultures with various chlorophyll contents produced by growth under different light conditions, they found [77] that the maximum yield of oxygen per flash of light was linearly related to chlorophyll content (Fig. 8.6). The number of mols of oxygen liberated per mol of chlorophyll by each flash was of the order of 4×10^{-4}, equivalent to about 40 mols per second of illumination (compare 1 to 0·03 for continuous light). This corresponded to about 2,500 mols of chlorophyll for every mol of oxygen released and of carbon dioxide fixed. Similar values, from 2,000 to 4,000 were obtained for a number of different species in subsequent work [77, 3]. As light intensity was saturating it was assumed that all these chlorophyll molecules were activated. In view of the linear relation with chlorophyll content the suggestion was made that the chlorophyll was associated in 'photosynthetic units' of about 2,500 molecules. This topic and the interpretation of other flashing light experiments will be discussed in the next chapter. As far as I am aware flashing light experiments have not been described for *aurea* varieties, or

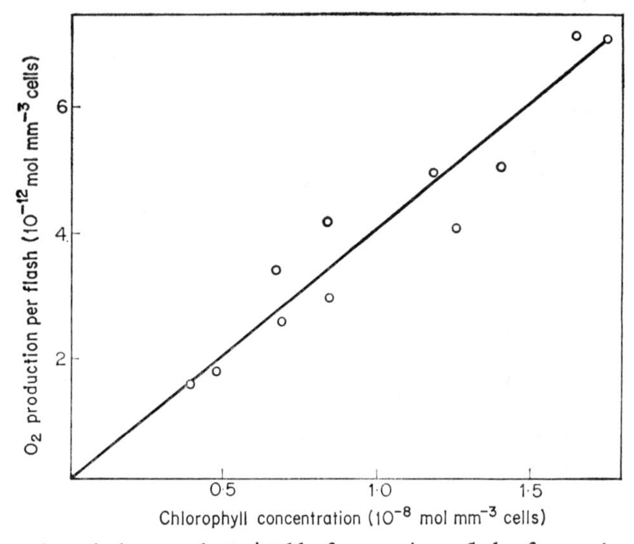

FIG. 8.6 'True' photosynthesis (yield of oxygen) per flash of saturating light as a function of chlorophyll content of *Chlorella pyrenoidosa* cultures. Temperature 25°C. After Emerson and Arnold [77].

very young leaves, which give very high A_c values in continuous light; nor do the questions of plastid number and size, or cell number and size (as distinct from total cell volume in a suspension) appear to have been considered.

9: Light Quality and Duration; Stages in Photosynthesis Postulated from Physiological Investigations

A. *FLASHING LIGHT EXPERIMENTS*

In their assimilation experiments described in 1905, Brown and Escombe [47] used a rotating disc with adjustable sectors to reduce the amount of light falling on the leaf. F. F. Blackman [20] had criticized such a method on the good theoretical grounds that it reduced the time of illumination and not the intensity, but in an *addendum* to the same paper he wrote: 'Consideration of a number of experiments, in which this method was employed by Brown and Escombe, allows one to see that the leaf behaves in assimilation as the eye does in vision, and that halving the light *in time* has, with all strengths of illumination, the same effect as halving it in intensity. The interest of the matter therefore shifts to the determination of the induction period and to finding at what slowness of rotation of the sectors the leaf begins to distinguish between the two methods of reduction of illumination.'

Some fifteen years later, Warburg [306, 307], using the same method for illuminating suspensions of *Chlorella* in a Warburg apparatus, showed that the interest in fact shifted in the opposite direction. He used sectors giving equal periods of light and darkness, thus reducing the total light to half, and found that as the rate of rotation was increased the amount of assimilation (estimated from oxygen output) approached that in continuous light of the same intensity. In one experiment with periods of 7·5 s the reduction in assimilation amounted to 30 per cent instead of 50 per cent, while periods of 3·8 ms gave a reduction of only 4 per cent so that the efficiency of the light was almost doubled (1·92×). Warburg suggested that *either* reduction of carbon dioxide continued in the dark periods *or* that it went on twice as fast in the brief light periods as in continuous light. He preferred the latter alternative, now generally believed to be the wrong one, and supposed that certain steps in the photosynthetic process continued in the dark so that a higher concentration of reactive substance would be available to take part in the decomposition of carbon dioxide in the subsequent light flash.

In 1932 Emerson and Arnold [76] showed that it was possible to improve the efficiency of light utilization in *Chlorella* by a factor of four, by making the dark periods five times as long as the light periods (Fig. 9.1). This result was obtained with a duration of light of 3·4 ms, almost the same as in Warburg's most rapid alternation, but with only 50 flashes per second as compared with his 133, giving dark periods of 16·6 ms. The 'continuous' light was supplied from a neon tube running on alternating (50 cycle) current so that the light was actually on for about 95 per cent of the time. For the flashing light, alternate half-cycles of current were shunted through a rectifier so that the tube flashed 50 times a second; the duration of the flash was controlled by a variable resistance in series with the rectifier. The improvement in efficiency was found with high light intensity and high carbon dioxide concentration, at 25°C; at low light intensity, with a filter passing only 5 per cent of the light, flashing light was no more efficient than continuous light and the rate of photosynthesis was proportional to the product of light intensity and duration, that is, to the total light given.

Emerson and Arnold [76] next investigated the effect of varying the dark periods, with a constant very short (10 μs) intense light flash obtained by discharging a condenser through a neon tube. The frequency of the flashes was controlled by a commutator driven by a synchronous motor. They found (Fig. 9.2) that at 25°C the shortest dark period used, 40 ms, was long enough to give maximum efficiency. At 1·1°C, however, the yield was much lower with such short dark periods but rose as they became longer and equalled the yield at 25°C with a dark period of 400 ms. This result showed that while the dark process was sensitive to temperature the light process was not, otherwise the curve for 1·1°C would not have reached the same level as that for 25°C. Emerson and Arnold suggested that the light process or photochemical reaction produced an intermediate product and this was converted to other intermediates or final products of photosynthesis through the operation of the 'Blackman reaction' (page 187), which could proceed in darkness and was temperature-sensitive.

Some of their other conclusions, though entirely reasonable in the

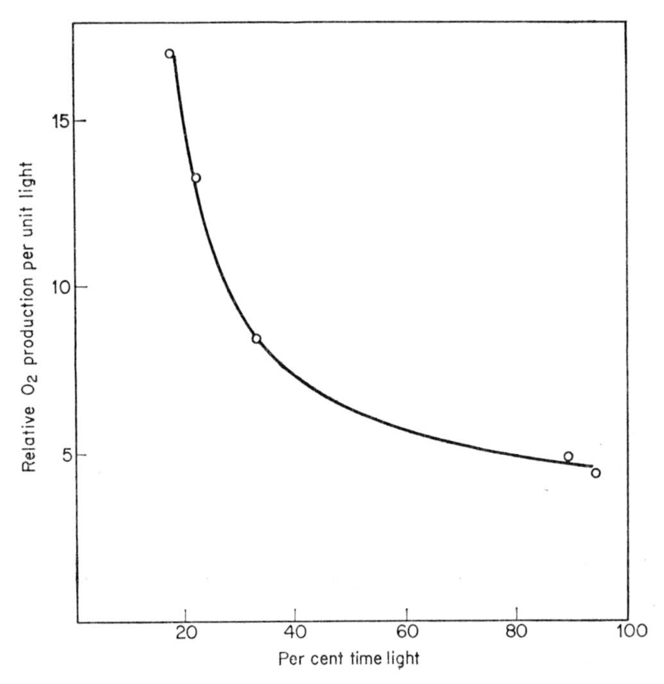

FIG. 9.1 'True' photosynthesis by *Chlorella pyrenoidosa* per unit of light received, in flashing light experiments, as a function of the percentage time in light. High carbon dioxide concentration; temperature 25°C. After Emerson and Arnold [76].

light of existing knowledge, have seemed less enduring. Thus they argued that if chlorophyll alone was involved in the light reaction and the absorbed energy was transferred later to a carbon dioxide compound (as is now generally thought to be the case) reducing the carbon dioxide concentration should not decrease the maximum yield per flash but only require a longer dark period. An experiment to test this showed (Fig. 9.3), however, that at a lower carbon dioxide concentration increasing the dark period would not increase the yield to that reached with a higher concentration. They therefore concluded that 'the carbon dioxide enters the process of photosynthesis either before or coincident with the photochemical reaction'. (Compare Warburg's recent conclusions discussed on page 254 below.) Because 10 μs seemed too short a time for the carbon dioxide molecules to move into position on and react with the chlorophyll, they supposed

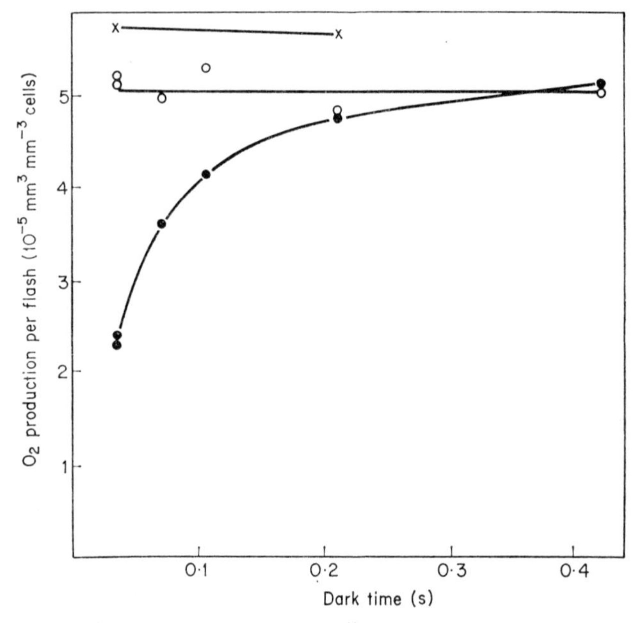

FIG. 9.2 'True' photosynthesis by *Chlorella pyrenoidosa* per intense short flash of light (10 μs) as a function of the time in darkness. Open circles, temperature 25°C; solid circles, 1·1°C; crosses, subsequent check determinations at 25°C. After Emerson and Arnold [76].

that carbon dioxide first combined with chlorophyll in darkness and that the light flash then activated the 'chlorophyll-CO_2 molecules' and left them ready to undergo the Blackman reaction. This, they supposed, released the chlorophyll which could then react with more carbon dioxide, so that a cyclic process was concerned.

Emerson and Arnold's interpretation seemed to be further supported by another experiment. If a chlorophyll-carbon dioxide complex was formed, low chlorophyll contents and low concentrations of carbon dioxide should have similar effects, that is, it should not be possible to compensate for a lower chlorophyll content by a longer dark period. Cells of *Chlorella pyrenoidosa* grown with neon lamps were found to have about one-quarter the chlorophyll content of those grown with mercury lamps. When these were compared in a flashing light experiment it was found, as expected, that the curves much resembled those at two carbon dioxide levels shown in Fig. 9.3; if the values at the low chlorophyll concentration were multiplied by a constant they agreed

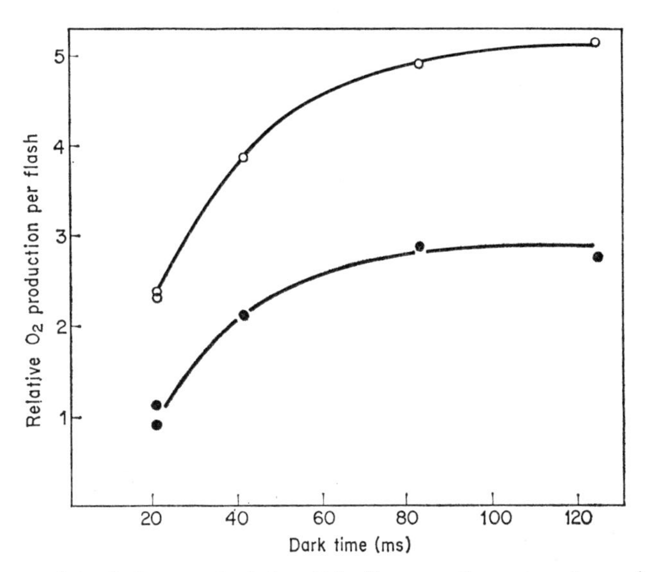

FIG. 9.3 'True' photosynthesis by *Chlorella pyrenoidosa* per unit number of intense short flashes of light (10 μs) as a function of the time in darkness, at two carbon dioxide concentrations. Open circles, carbon dioxide 71 μmol l^{-1}; solid circles, carbon dioxide 4·1 μmol l^{-1}; temperature 5·9°C. After Emerson and Arnold [76].

well with those at the high concentration, implying that only the light reaction and not the dark reaction was affected by the chlorophyll content of the cells. This was also of course consistent with the first and rejected hypothesis that chlorophyll alone was involved in the light reaction, but in any case in a later paper, Emerson and Arnold [77] attributed the result to chance; they then had evidence that cultures grown with neon and mercury light might have different capacities for the Blackman reaction, as well as different chlorophyll contents, owing perhaps to their very different growth rates. The difficulties of experimentation on the effects of different chlorophyll contents, arising from associated differences in other characteristics of the plant material, have been stressed in the previous chapter.

B. *QUANTUM EFFICIENCY AND WAVE LENGTH OF LIGHT*

Warburg and Negelein [309] stated in 1923 that photosynthesis (oxygen production) by *Chlorella* could attain a value indicative of an efficiency of 0·25 molecule of carbon dioxide reduced per absorbed quantum of visible light, this being the (then) theoretical maximum. Emerson and Lewis [81] in 1939, with similar use of Warburg's manometric method (page 85), were able to obtain even higher apparent

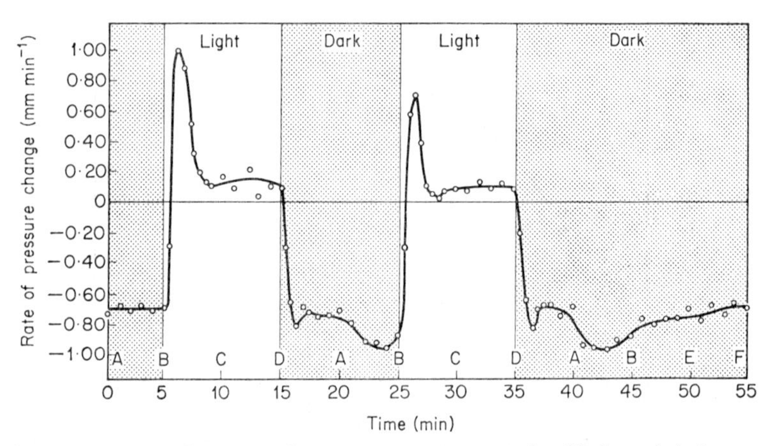

FIG. 9.4 Rates of pressure change in successive periods of light and darkness. *Chlorella pyrenoidosa* in Warburg apparatus; light intensity 1·4 J m^{-2} s^{-1}. For further explanation see text. After Emerson and Lewis [81].

efficiencies by modifying various conditions. They attributed these high values to artifacts due to transient changes in the ratio of exchange of oxygen and carbon dioxide during the short periods of light and darkness used. By taking readings at less than one-minute intervals they were able to detect a sudden increase in rate of pressure change at the beginning of each period of illumination as well as a low rate at the beginning of each dark period (Fig. 9.4). Warburg and Negelein had only taken readings of pressure at the times corresponding to A and B in the figure. Their method of estimating oxygen evolution in 10 minutes of photosynthesis was to subtract three times the mean pressure change measured from time A to time B (representing 15 minutes of oxygen uptake in respiration) from the pressure change found from B to A (representing 10 minutes of photosynthesis plus 15 of respiration). Emerson and Lewis found that this method gave much higher apparent quantum efficiencies than if they estimated photosynthesis

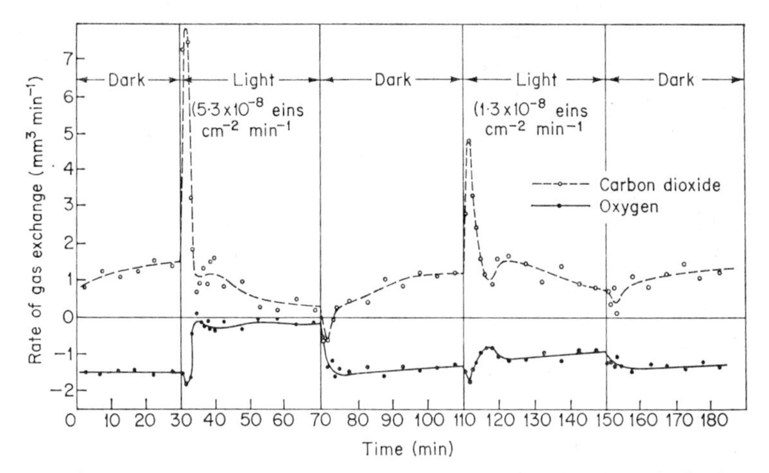

FIG. 9.5 Rates of exchange of oxygen (solid line) and carbon dioxide (broken line) in successive periods of light and darkness, calculated from rates of pressure change with *Chlorella pyrenoidosa* in modified Warburg two-vessel apparatus. Note carbon dioxide burst or Emerson effect on first illumination. After Emerson and Lewis [82].

from C to D and respiration from E to F when more or less steady states had been achieved.

Later, Emerson and Lewis [82] estimated both oxygen and carbon dioxide by the Warburg two-vessel technique (page 87); these results supported their suggestion that the fluctuations in rate of pressure change were due to differences in ratio of oxygen to carbon dioxide. In particular the rapid increase of pressure on first illumination was found to be due to a sudden output of carbon dioxide (Fig. 9.5). This 'CO_2 gush' came to be known as the Emerson effect; it was followed on darkening by a less sudden initial uptake of carbon dioxide or 'CO_2 gulp'.

Results obtained by manometric methods can never be unequivocal (page 88), but in 1955 Brown and Whittingham [45] published mass spectrometer data which confirmed beyond doubt that the strain of *Chlorella pyrenoidosa* used by Emerson gave both the CO_2 gush and CO_2 gulp, when grown and tested under his conditions; these included continuous bubbling of 5 per cent carbon dioxide in air through the cultures in order to ensure that carbon dioxide was not limiting. However, reducing the carbon dioxide content to 0·5 per cent during the experimental period eliminated the gush and in 1959 Hiller and Whittingham [164] found that only two out of five strains of *C. pyrenoidosa* tried gave off carbon dioxide on illumination even when supplied with high carbon dioxide concentrations. This illustrates the large part played by chance in scientific controversy (and indeed in scientific investigation generally) for the CO_2 gush loomed large in the controversy on quantum efficiency.

It is not intended to discuss this controversy, in which the Warburg technique was continually and ingeniously modified and attempts were made to use it far beyond its limits of accuracy. A maximum possible efficiency of 0·125 seems now to have been established on physicochemical grounds, and since 100 per cent efficiency cannot be obtained experimentally owing to inevitable losses of energy as heat, the values of 0·08 to 0·10 found by Emerson and many other workers appear to be well substantiated. The subject was reviewed by Emerson in 1958 [74]. More important, as seen in retrospect, his interest in quantum efficiency led him to investigate the effects of different wave lengths of light. During the period 1938–40 Emerson and Lewis [82] had measured the quantum efficiency for various algae, with sodium light of wave length 589 nm, and found similar values of about 0·09 for green algae, in which practically all the absorption was due to chlorophyll, and for the blue-green alga *Chroococcus*, in which a large part of the energy at this wave length must have been absorbed by phycocyanin (Fig. 1.13). This strongly suggested that phycocyanin was active in photosynthesis and led to attempts to determine the parts played by this and other accessory pigments. For this purpose quantum efficiencies were determined for different wave lengths and compared with estimates of the relative absorption at those wave lengths due to the various pigments [83, 84].

The distinction should be emphasized between graphs of quantum efficiency as a function of wave length (as in Fig. 9.8) and the more usual action spectra.

Action spectra. In these the rate of a light-dependent biological process, such as photosynthesis, for a constant incident light intensity, or alternatively the incident light intensity to produce a constant rate, is plotted against the wave length of the exciting light. For either type of action spectrum it is important that the rate should be nearly proportional to the light intensity at each wave length used (Fig. 9.6); the action spectrum obtained will not then vary appreciably with the value chosen for the constant intensity or constant rate. This implies

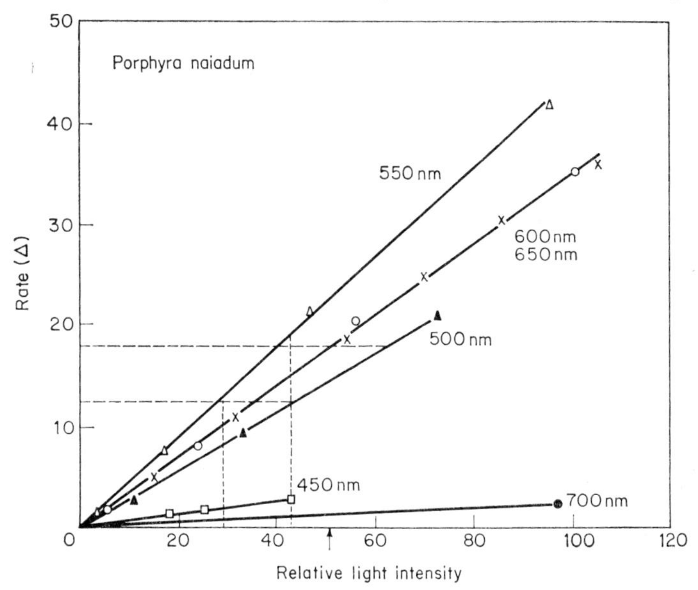

FIG. 9.6 Rate of photosynthesis (Δ) by *Porphyra naiadum* as a linear function of light intensity over the ranges shown for different wave lengths. Action spectra could appropriately be determined, for example, at the light intensities indicated by dotted lines or (except for long and short wave lengths) at the rates indicated by dashed lines. The arrow indicates a light intensity of approximately 0·8 J m^{-2} s^{-1}. After Haxo and Blinks [137].

that light intensity must be severely limiting the rate. Action spectra may be compared with absorption spectra measured for the same cells or plant organs and thus yield clues as to the pigments concerned in the process studied. If the two graphs run closely parallel in a range of wave lengths where a known pigment is mainly responsible for the absorption it is probable that light absorbed by that pigment is driving the process (Fig. 9.30).

Quantum efficiency graphs. As an alternative to comparing an action spectrum with the corresponding absorption spectrum, a graph of rate of the process studied per absorbed quantum may be plotted against wave length, i.e. a graph of quantum efficiency. Since the amount of a photochemical reaction depends upon the number of absorbed quanta and not on their energy content (as long as this is sufficient for the reaction to occur at all) one might expect the quantum efficiency to

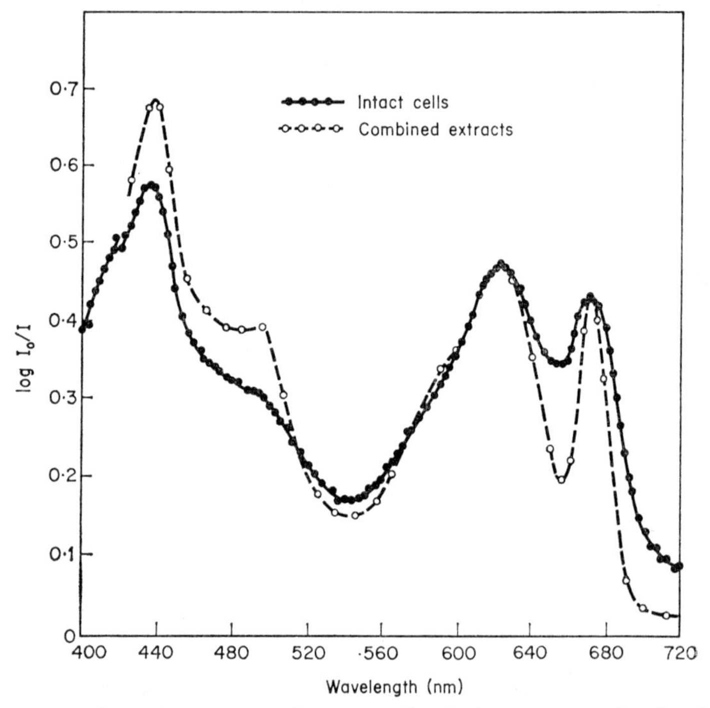

FIG. 9.7 Absorption spectrum for intact cells of *Chroococcus* sp.; also that for total extracted pigments, obtained by adding the curves for the individual components after shifting them towards longer wave lengths to give the best correspondence of individual maxima with those for intact cells. After Emerson and Lewis [83].

remain constant throughout the wave length range where a single active pigment was absorbing, even though the amount of light absorbed varied greatly (see, however, page 226 *et seq.*). The same would apply where two or more pigments of equal efficacy were absorbing. The quantum efficiency found at any wave length is an

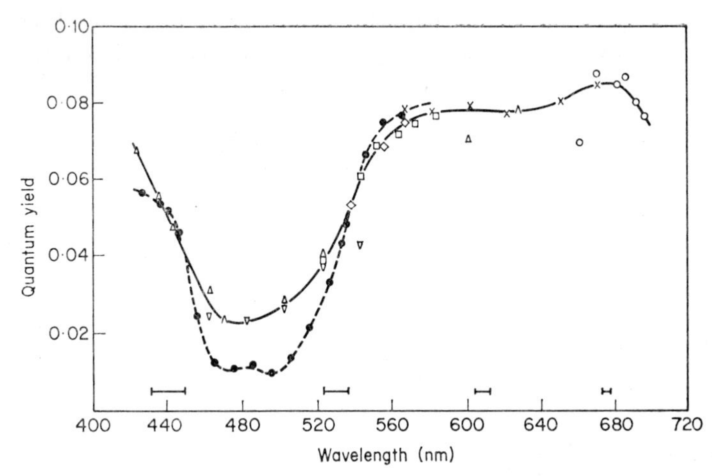

FIG. 9.8 Relation of quantum yield of photosynthesis by *Chroococcus* sp. to wave length of light. Solid line, freehand curve through experimental points (with different symbols for different runs); dotted curve, calculated on assumption that light absorbed by chlorophyll and phycocyanin gave a quantum yield of 0·08 at all wave lengths and that light absorbed by carotenoids was completely ineffective. After Emerson and Lewis [83].

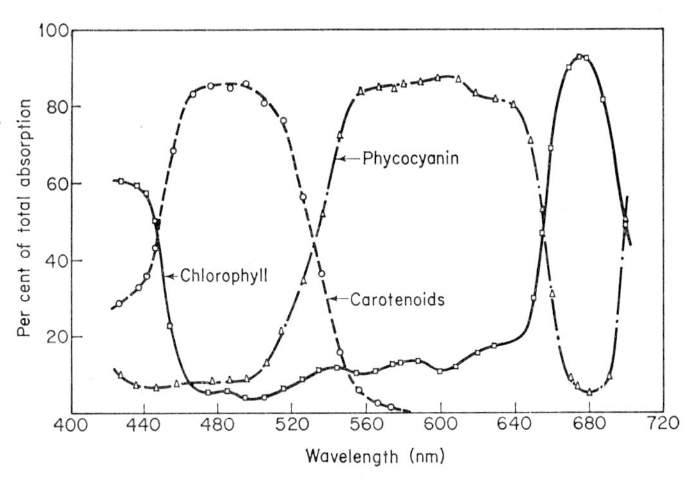

FIG. 9.9 Relation to wave length of the percentages of total light absorbed by *Chroococcus* sp. due to chlorophyll, carotenoids and phycocyanin respectively; estimates based on measurements with extracted pigments. After Emerson and Lewis [83].

average value for all the pigments absorbing at that wave length weighted by the proportions they absorb. A marked fall in quantum efficiency at wave lengths where an active pigment is still absorbing usually implies absorption of some of the light by an inactive screening pigment or by a less effective pigment. As for action spectra it is important that the light intensity should be so low as to be severely limiting.

In the experiments of Emerson and Lewis [83, 84] light of narrow

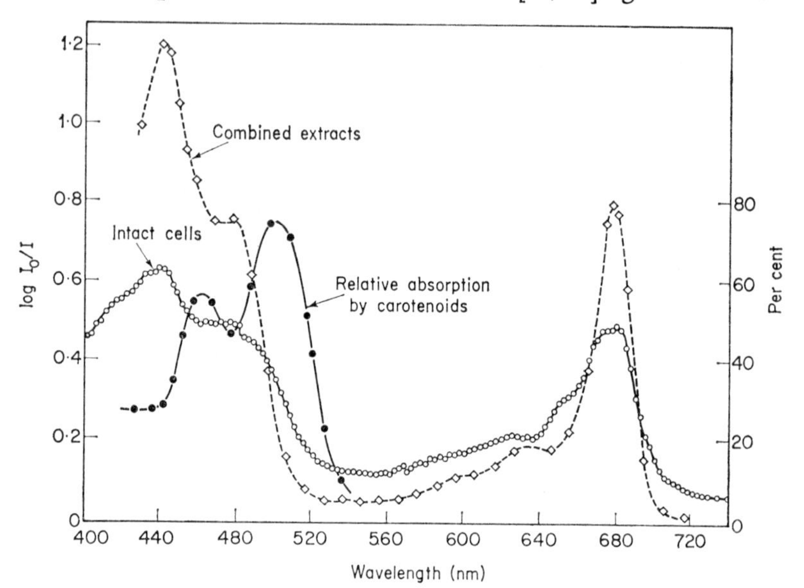

FIG. 9.10 Absorption spectrum for intact cells of *Chlorella pyrenoidosa*; also that for total extracted pigments, obtained by combining the curves for chlorophyll and carotenoids after shifting them to compensate for the wave length shift due to extraction; also relation to wave length of the percentage of total light absorption due to carotenoids, based on measurements with extracted pigments. After Emerson and Lewis [84].

wave bands was obtained from a large grating monochromator. Measurements of photosynthesis at a limiting light intensity in each wave band were made in a Warburg apparatus both with suspensions of the alga concerned thick enough to absorb all the incident light, and thin suspensions for which the absorption had to be estimated by also measuring the amount transmitted and scattered. The pigments were extracted and the absorption spectra of the extracts compared with those of the intact cells. For *Chroococcus* [83], in which the pigments are distributed throughout the cell, the agreement was remarkably good (Fig. 9.7), after certain reasonable adjustments had been made, but for *Chlorella* [84] there were large differences (Fig. 9.10) which were attributed to the pigments being concentrated in the chloroplasts.

In Fig. 9.8 the solid line shows the relation found between quantum yield and wave length for *Chroococcus*. If this is compared with the

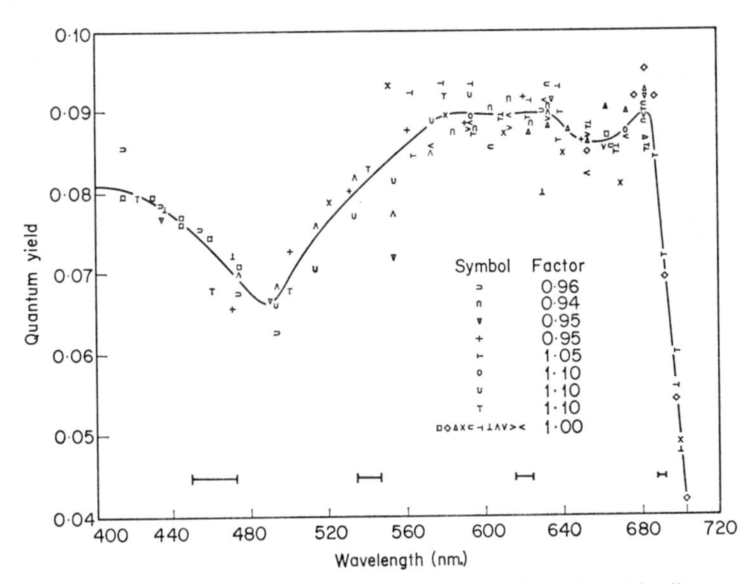

FIG. 9.11 Relation of quantum yield of photosynthesis by *Chlorella pyrenoidosa* to wave length of light. Different symbols for each of 19 separate runs; for 8 of these, arbitrary multiplying constants were used as shown. Horizontal lines indicate band half-widths used in various parts of the spectrum. After Emerson and Lewis [84].

estimated percentages of the total absorption due to the various pigments at the different wave lengths (Fig. 9.9) it is clear that absorption by phycocyanin was almost if not quite as effective as absorption by chlorophyll itself. On the other hand there was a marked fall in quantum efficiency in the region where most of the absorption was due to carotenoids, and obviously these were much less effective. The dotted line in Fig. 9.8 shows the quantum yields calculated on the assumption that at all wave lengths these were 0·08 for light absorbed by chlorophyll or phycobilin and zero for light absorbed by carotenoids. The last assumption has depressed the curve too far and Emerson and Lewis concluded that a small part of the light absorbed by carotenoids might be available for photosynthesis.

The quantum yield curve for *Chlorella* (Fig. 9.11) again shows a marked fall in the region of absorption by carotenoids (Fig. 9.10), and Emerson and Lewis again considered that some small photochemical activity of these pigments was indicated. At the red end of the spectrum the quantum yield curve fell abruptly above 685 nm in a region where chlorophyll *a* absorption was still appreciable and where no other pigment was known to absorb. A similar fall in efficiency above 685 nm was seen in *Chroococcus* (Fig. 9.8). This observation was later to prove of the utmost importance, as will be seen in the next section; at the time the authors suggested that it could be due to the quanta of far-red light having too low an energy content for the primary photochemical process. Later, Tanada [288] found a similar decline at about the same wave length for the diatom *Navicula minima*.

C. ENHANCEMENT EFFECTS

In 1956 Emerson [80, 79, 73, 75] returned to the problem of the decline in quantum efficiency in the far-red—the so-called 'red drop'. If it was connected with the small energy content of the quanta, the quantum yield might be expected to be further reduced at low temperature. He and his co-workers found however that at 5°C far-red light was *more* effective than at 20°C, both in *Chlorella* (Fig. 9.12) and in the red alga *Porphyridium cruentum* (Fig. 9.13). They at first [80] considered that this temperature dependence of the red drop was characteristic of low light intensities, for they found that in the presence of a supplementary beam of more intense light of shorter wave length the small increment in quanta given in the far-red beam was efficient further into the long wave lengths even at 26°C. However, this finding of the effect of supplementary shorter-wave light led to a new hypothesis so exciting that the temperature effect was not investigated further; this was that for maximum efficiency of photosynthesis light quanta must be absorbed by *two* pigments, namely chlorophyll *a* and one of the accessory pigments—chlorophyll *b* in green algae, a phycobilin in blue-green or red algae, and so on. As it was only in the far-red that chlorophyll *a* was the sole pigment absorbing, it was only here that monochromatic light failed to excite a second pigment also and hence only here that the quantum efficiency fell off. This two-pigment hypothesis was contrary to the current view, based on fluorescence studies [66], that photosynthesis was entirely due to the excitation by light energy of chlorophyll *a* and that the only function of the accessory pigments was to contribute the energy of quanta they had absorbed to the chlorophyll *a*, probably by inductive resonance (page 36).

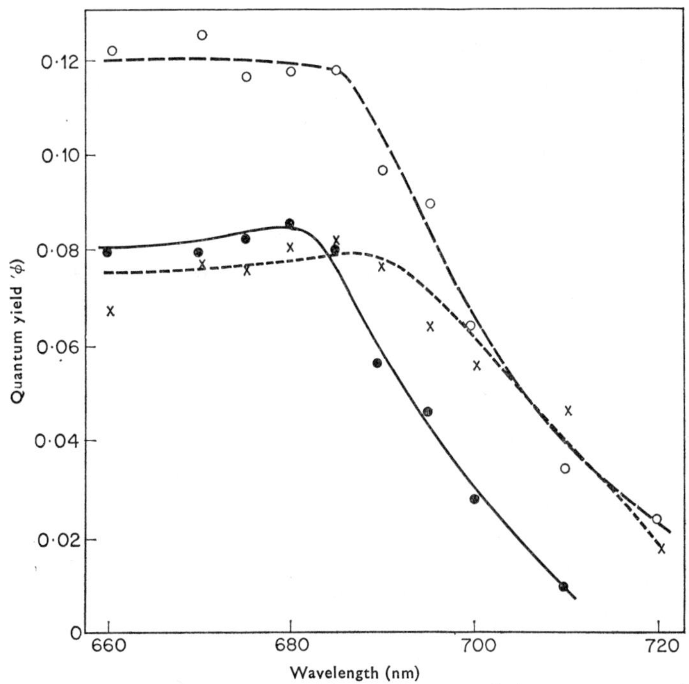

FIG. 9.12 Effects of temperature and supplementary light on quantum yield of photosynthesis by *Chlorella pyrenoidosa* at different wave lengths of red light. Solid line and solid circles, temperature 20°C and no supplementary light; dashed line and open circles, temperature 5°C and no supplementary light; short-dashed line and crosses, temperature 20°C and supplementary light from a mercury-cadmium lamp. After Emerson *et al.* [79].

Red algae were of particular interest in relation to this hypothesis, for Haxo and Blinks [137] had concluded from action spectra that in red algae light absorbed by phycoerythrin was used *more* efficiently than that absorbed by chlorophyll *a*; further Duysens [65] had found that fluorescence of chlorophyll *a* was more efficiently excited when light was absorbed by phycobilin than when directly absorbed by the chlorophyll (page 37). Emerson chose *Porphyridium cruentum* as a red alga that could conveniently be used in a Warburg apparatus. Brody and Emerson [40, 41] were able to vary the concentrations of chlorophyll *a* and phycoerythrin in *Porphyridium cruentum* and their relative proportions by growing the cells in light of different intensities and wave lengths. For cells grown in low-intensity 'white' light from fluorescent tubes (Fig. 9.13) or in blue light (Fig. 9.14), they obtained normal quantum efficiencies of about 0·09 with red light (644 nm) where chlorophyll *a* and phycocyanin absorbed about equally; with green light, however, where nearly all the absorption was due to phycoerythrin, the efficiency was much lower. Cells grown in green light, which contained the highest proportion of chlorophyll to phycobilin, gave quantum yields in green light almost as high as those in red (Fig. 9.14). However, later work by Brody and Brody [39] showed that any attempts to interpret these results in terms of pigment contents or ratios represented an over-simplification. With cells grown in blue light and then adapted by exposure to green light for 12 hours they obtained a quantum efficiency curve almost identical with the green light curve of Fig. 9.14, and after a further 12 hours in blue light (readaptation) these same cells gave a curve almost identical with the blue light curve of Fig. 9.14; no changes in absorption spectrum could be detected after adaptation suggesting that there had been no changes in pigment contents or ratios. Similar results were obtained with adaptation periods of only 2–3 hours.

In all these quantum efficiency curves for *Porphyridium* the red drop began at about 650 nm, in contrast to the value of about 685 nm found for *Chlorella*, perhaps because beyond this wave length chlorophyll *a* was the only pigment absorbing light, there being no chlorophyll *b* in red algae.

In the earlier experiments to compare quantum yields in red or far-red light with and without supplementary illumination, a beam of supplementary light of short wave lengths from a mercury–cadmium lamp was added to the red or far-red beam from the monochromator. This treatment gave greater rates of photosynthesis than the sums of those for the two light sources given separately; hence the quantum efficiency found for the two beams together was greater than the average of the two separate quantum yields and was maintained at the usual value of about 0·08–0·09 further into the far-red (Figs. 9.12 and

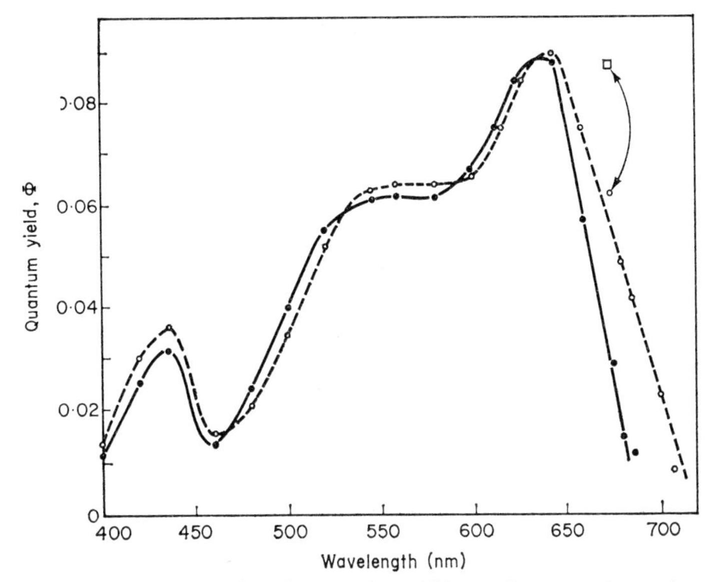

FIG. 9.13 Quantum yield of photosynthesis (Φ), as a function of wave length, for *Porphyridium cruentum* grown in low intensity 'white' light. Solid line and solid circles, measurements at 20°C; broken line and open circles, measurements at 5°C; square joined by arrow to corresponding open circle shows increase in quantum yield (enhancement) at 675 nm due to supplementary light of short wave length. After Brody and Emerson [40].

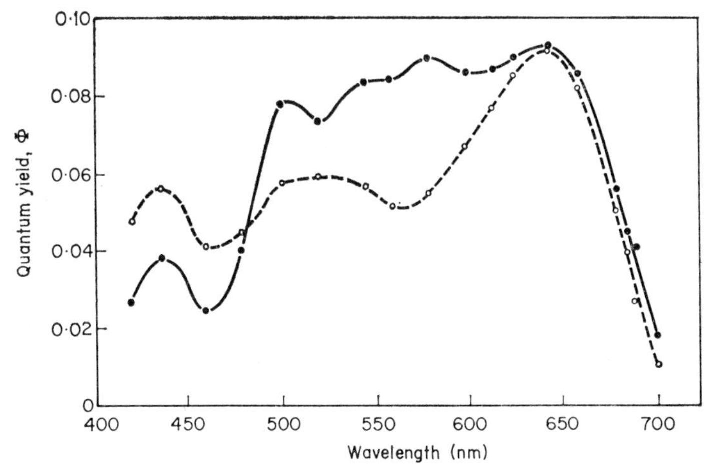

FIG. 9.14 Quantum yield of photosynthesis (Φ), as a function of wave length, for *Porphyridium cruentum* grown in green light (solid line and solid circles) or blue light (dotted line and open circles). After Brody and Emerson [40].

9.13). The next question was what wave lengths were most effective for the supplementary light and did they coincide with the absorption maxima of accessory pigments? Some preliminary results, shown in Table 9.1, indicated that they did. For *Porphyridium* the percentage increase in the yield attributed to the far-red, after subtracting the yield for the supplementary light when given alone, was greatest at 546 nm; this was approximately at the peak absorption of phycoerythrin, the principle phycobilin pigment of red algae. For *Chlorella* the pattern was more complex but the increases showed a rough correspondence with the absorption spectrum of chlorophyll *b*. Moreover,

TABLE 9.1

Increase in yield of photosynthesis from a band of long-wave red light (>690 nm) due to supplementary light of shorter wave length. From Emerson, R. [75]

Wave length of supplementary light	*Increase in yield (%)*	
	Chlorella	*Porphyridium*
644	28	0
578	19	15
546	12	100
508	5	85
480	60	0
468	40	
436	10	0
405	5	
365	7	0

as both red and blue light were effective (644 and 480 nm) it appeared that the important aspect was not the energy content of the absorbed quanta but their absorption by a second pigment (here chlorophyll *b*).

Emerson and co-workers also determined more complete action spectra for the supplementary light with a constant far-red beam and compared them either with the absolute absorption curves of the appropriate accessory pigments or with the curves for the estimated proportion of light absorbed by those pigments at the same wave lengths. Such data for algae from four main divisions were given in a paper by Emerson and Rabinowitch [85], prepared by the latter for a Congress in 1959, after Emerson's death. They are reproduced here in Figs. 9.15–9.18 and show the photosynthetic yield of oxygen for the combined beams, minus the yield for the supplementary light alone, as a percentage of the value for the far-red beam alone. (The intensity of the supplementary beam was adjusted to give, by itself, about the same yield as the far-red beam alone.) On the whole, these results strongly supported the two-pigment hypothesis. They suggested that for most efficient photosynthesis in the far-red, quanta absorbed there by chlorophyll *a* must be supplemented by quanta of shorter wave length absorbed by the principal accessory pigment—chlorophyll *b* in

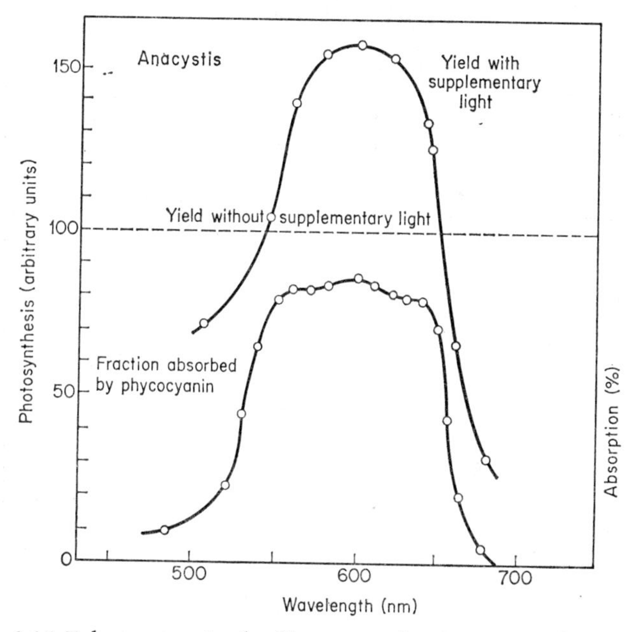

FIG. 9.15 Enhancement in the blue-green alga *Anacystis*. Abscissa: wave length of monochromatic supplementary light. Left-hand ordinate (upper curve): yield of photosynthesis attributed to far-red light in the presence of supplementary light and expressed as a percentage of the yield in its absence. Right-hand ordinate (lower curve): estimated percentage of total light absorption due to phycocyanin. After Emerson and Rabinowitch [85].

a green alga, phycocyanin in a blue-green, phycoerythrin in a red alga and the carotenoid fucoxanthol in a diatom. This has come to be known as the enhancement effect or Emerson effect; or sometimes as the second Emerson effect, the first being the gush of carbon dioxide output on first illumination of *Chlorella* (page 219).

In order to reconcile the enhancement effect with Duysens' observations of the high fluorescence yield of chlorophyll *a* from incident light absorbed by accessory pigments (page 36), Rabinowitch (*loc. cit.*) suggested that chlorophyll *a* must exist in at least two forms *in vivo*: one of these, a non-fluorescent form excited by far-red light, could not by itself carry out photosynthesis efficiently, thus accounting for the red drop; the other, a fluorescent form, could be excited either directly

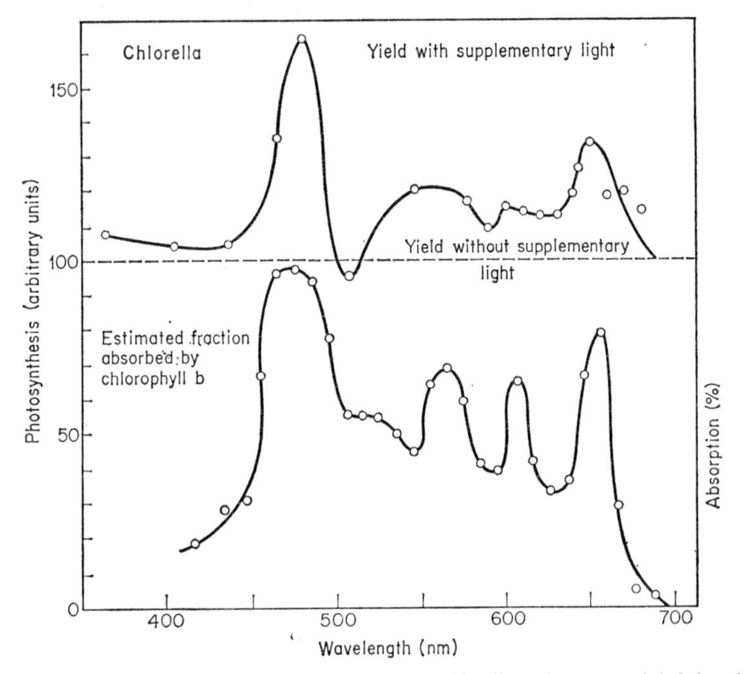

FIG. 9.16 Enhancement in the green alga *Chlorella*. Abscissa and left-hand ordinate (upper curve) as in Fig. 9.15. Right-hand ordinate (lower curve): estimated percentage of total light absorption due to chlorophyll *b*. After Emerson and Rabinowitch [85].

by near-red light or indirectly by resonance transfer of energy from an accessory pigment. It was suggested that there were two different primary photochemical processes required for photosynthesis (for example a photoreduction and a photoxidation) preferentially sensitized by the two different forms of chlorophyll *a*. Enhancement occurred when both forms were excited simultaneously. These suggestions were partly based on a theory of Franck's [95] (page 259) and supported by French's [99] observation that the red absorption band of chlorophyll *a* was complex.

A curious feature of Figs. 9.15–9.18 is that some regions of 'negative enhancement' are shown and obviously the interaction of different narrow wave bands is a complex matter which demands the use of orthogonal factorial experiments for its elucidation. Plants have in general become adapted to a wide continuous spectrum, though at

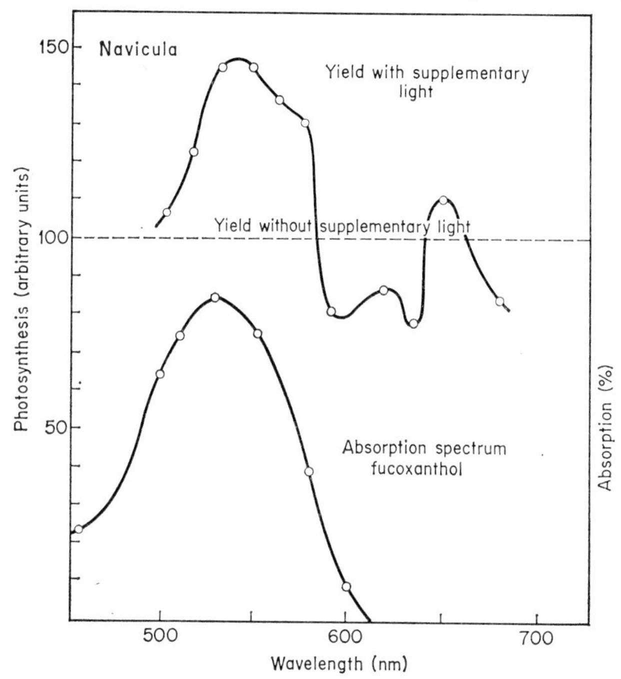

FIG. 9.17 Enhancement in the diatom *Navicula*. Abscissa and left-hand ordinate (upper curve) as in Fig. 9.15. Right-hand ordinate and lower curve: absorption spectrum of fucoxanthol. After Emerson and Rabinowitch [85].

depths inhabited by red algae the range of wave lengths is greatly reduced; interaction effects between different wave lengths might be expected therefore. Ordinary action spectra or quantum efficiency curves obtained with single narrow wave bands provide another example of the difficulties of interpretation posed by experiments in which only a single factor (here wave length) is varied, or to look at it another way, in which a number of factors (different wave lengths) are presented separately and at a single level (intensity).

Govindjee and Rabinowitch [127] later (1960) also found negative enhancement for *Chlorella* and *Anacystis*, with supplementary light of 680 or 690 nm added to far-red of 700 nm; for *Anacystis* this could be converted to positive enhancement by reducing the intensity of the far-red (700 nm) light. Their data suggested that light saturation occurred at lower intensities for longer wave lengths and even below

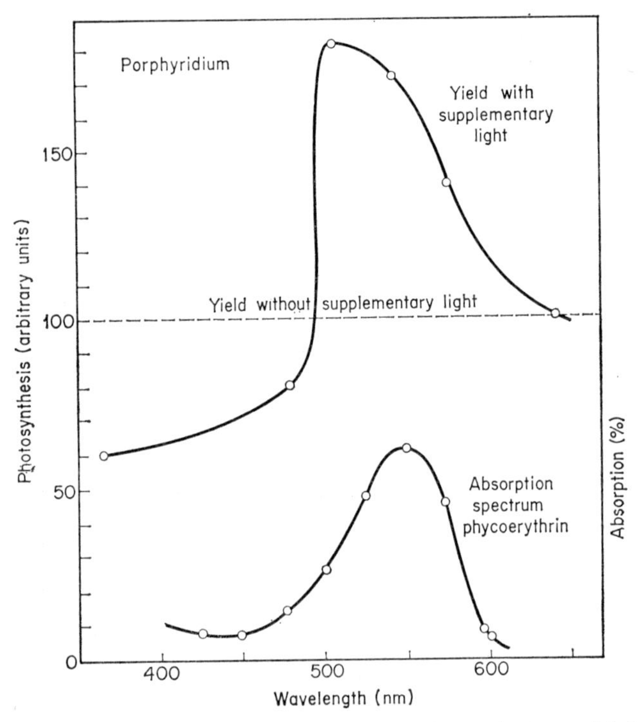

FIG. 9.18 Enhancement in the red alga *Porphyridium*. Abscissa and left-hand ordinate (upper curve) as in Fig. 9.15. Right-hand ordinate and lower curve: absorption spectrum of phycoerythrin. After Emerson and Rabinowitch [85].

the compensation point; thus the negative enhancements appeared to be associated with approaching saturation for the longer wave length (see page 261). This is the kind of hypothesis that would be most satisfactorily tested by orthogonal factorial experiments with different combinations of wave lengths and intensities. Some years later (1963) Jones and Myers [182] carried out a factorial experiment with *Anacystis*, with a series of intensities of 700 nm, either without background light

(A in Fig. 9.19) or with backgrounds at 700 nm (E) or 620 nm (G, F). This experiment showed that the curve for 700 nm alone (A) did not in fact become nearly horizontal but apparently parallel to that with a 700 nm background (E); they concluded that the Kok effect (page 145) was involved, and not light saturation. In E the Kok effect did not show because the total intensity, including the 700 nm background, was always too high. The initial slopes of A, F and G were similar and

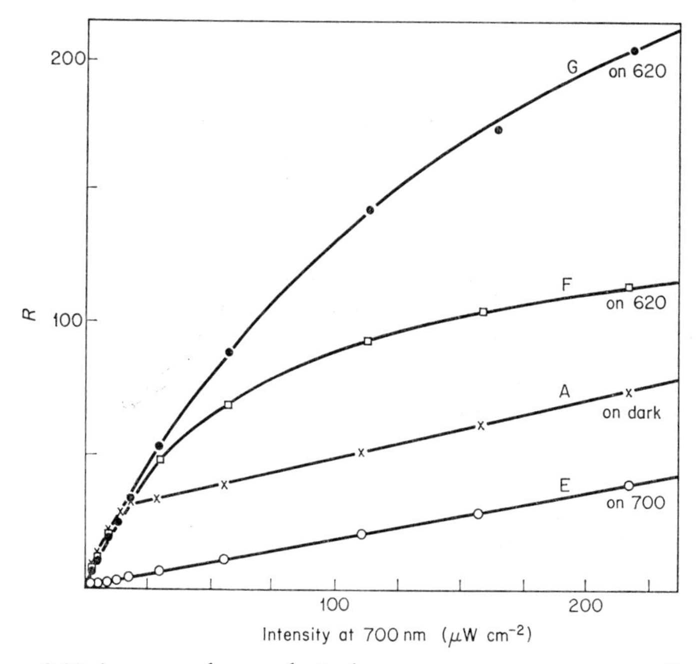

FIG. 9.19 Apparent photosynthesis (net apparent oxygen output, *R*, as measured with a platinum electrode polarograph) by *Anacystis nidulans* as a function of intensity of light of 700 nm wave length: A, without background light; E, with background of 180 μW cm^{-2} at 700 nm; F, with background of 35 μW cm^{-2} at 620 nm; G, with background of 145 μW cm^{-2} at 620 nm. Temperature 26°C. After Jones and Myers [182].

hence there was no enhancement at very low levels of 700 nm light; they suggested that this implied that at these levels light of 700 nm absorbed by chlorophyll *a* was efficient in suppressing oxygen uptake (Kok effect). At higher intensities of 700 nm light, curves F and G showed enhancement and F (at the lower intensity of 620 nm) became approximately parallel with A and E; this was interpreted as due to the supply of quanta at 620 nm becoming limiting so that further additions of quanta at 700 nm could not be enhanced. Jones and Myers [183] concluded that enhancement should always be measured by comparing the results of *increments* in light intensities over ranges high enough to avoid the Kok effect but low enough to avoid approaching light saturation. They also found the apparent Kok effect with another blue-green alga (*Anabaena variabilis*) but they noted that *Chlorella* showed only a minor non-linearity, suggesting that the effect was not general to all

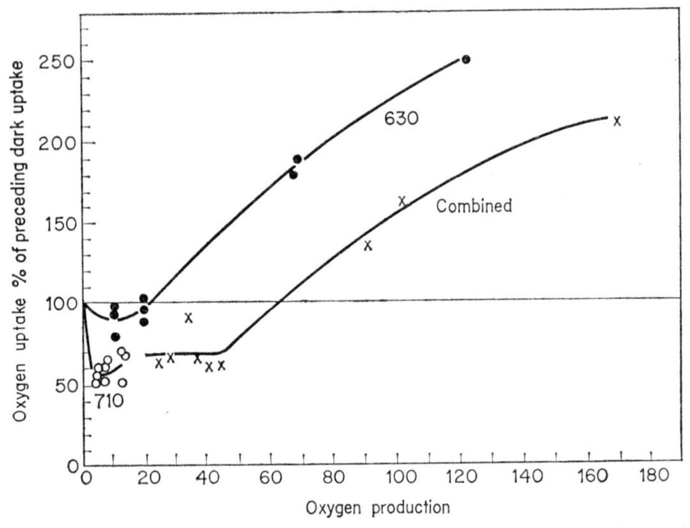

FIG. 9.20 Oxygen uptake by *Anacystis nidulans* in light, per cent of preceding dark uptake, as a function of oxygen production: solid circles, light of 630 nm alone; open circles, light of 710 nm alone; crosses, 630 nm and 710 nm combined. Mass spectrometer data. After Hoch and Owens [167].

photosynthetic systems (page 148). Govindjee [126] found a marked change of slope for the red alga *Porphyridium* with light of 700 nm.

The relevance of the Kok effect to enhancement studies was further indicated by the investigations of Hoch and co-workers [167, 168] who used a mass spectrometer as modified by Hoch and Kok [166] (page 91) and measured the uptake of marked oxygen by *Anacystis nidulans* concurrently with the photosynthetic output of ordinary oxygen. Some of their results, obtained with tungsten filament light, have already been discussed in Chapter 5 (see Figs. 5.14 and 5.15) but they also investigated effects of red and far-red light. Fig. 9.20 shows that with increasing oxygen production there was a great increase of uptake, above the dark rate, in light of 630 nm. On the other hand at 710 nm uptake was partially suppressed and when a high rate of photosynthesis was made possible by combining the two wave lengths such suppression apparently continued. Fig. 9.21 shows that the oxygen output in 710 nm light alone increased linearly with intensity while the uptake was greatly reduced; the relation to light intensity of net

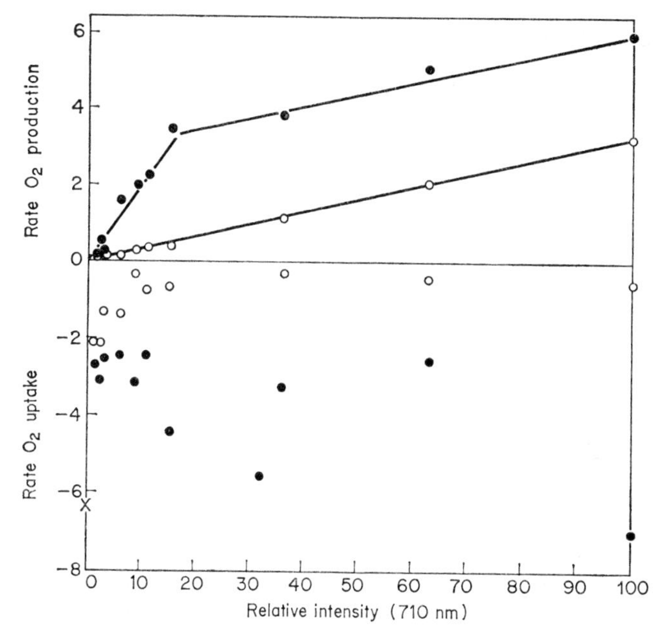

FIG. 9.21 Oxygen production and uptake by *Anacystis nidulans* as a function of intensity of light of 710 nm: open circles, without background light; solid circles, with background of 630 nm. Mass spectrometer data. After Hoch and Owens [167].

output would therefore be curved (Kok effect) as in Fig. 5.8 or curve A of Fig. 9.19. With the addition of background light of 630 nm wave length the oxygen production rose steeply due to enhancement and then became parallel with the 710 nm line as in Jones and Myers' experiment (Fig. 9.19); the oxygen uptake, which was greatly stimulated by the 630 nm light alone, was partly suppressed over the range of intensity of far-red light that gave the steep rise in output. Thus enhancement of oxygen production and suppression of uptake could occur together.

Net output of oxygen for three wavelengths of monochromatic light is shown in Fig. 9.22. In far-red light of 680 or 700 nm the relation shows the Kok effect, though the shape of the curve is in doubt owing to the wide scatter of the points; for 630 nm the relation is linear. Hoch *et al.* suggested that these results might explain the discrepancy between Emerson and Lewis's [83] finding of approximately equal efficiency for light absorbed by chlorophyll *a* or phycocyanin on the one hand and on the other that of Haxo and Blinks [137] and of

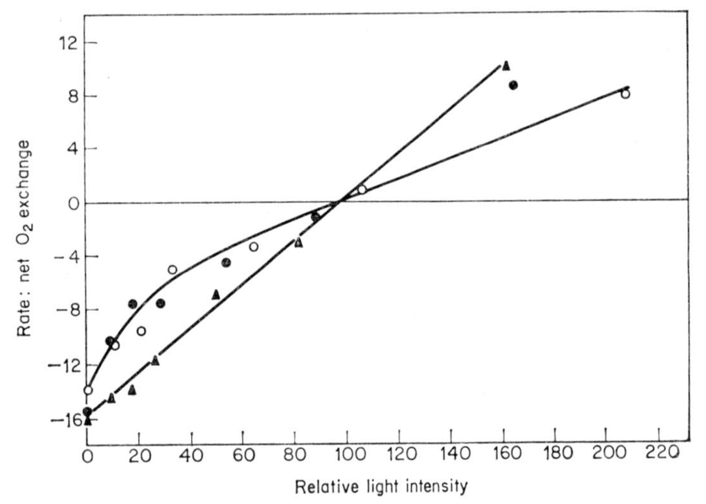

FIG. 9.22 Net oxygen exchange by *Anacystis nidulans* as a function of light intensity: triangles, 630 nm; solid circles, 680 nm; open circles, 700 nm. After Hoch *et al.* [168].

Duysens [65] that the chlorophyll was only half as efficient. Emerson and Lewis would have had low average intensities of monochromatic light in their Warburg vessels and might therefore have worked with the early part of the 680 nm curve where the yield per unit of light was highest; in the polarograph method of Haxo and Blinks the individual cells might have been exposed to higher light intensities where the 680 nm light was less efficiently used. Hoch and Owens [167] suggest that suppression of respiration also explains negative enhancement, such as that between 680 nm and 700 nm mentioned above (page 234). Either wave length alone results in both oxygen production and suppression of uptake, but the suppression is not increased when they are given together so that the second wave length is less efficient in combination than by itself. Hoch and Owens took the existence of negative enhancement in *Chorella* as evidence for suppression of uptake by light, which often could not be detected with this alga by their technique. Whether or not all their conclusions are ultimately substantiated, one must agree with them that 'respiratory changes can and do cause many anomalies when the net rate of gas exchange is taken as a measure of photosynthesis'.

There are several other suggestive findings, besides the negative enhancement, mentioned in the Emerson and Rabinowitch [85] paper. One is that the red drop can be shifted further towards the far-red by growing *Chlorella* with beef broth or earth extract, possibly because these may supply vitamin K, but not glucose or a wide range of different inorganic media. Another links the enhancement effect with the phenomenon of chromatic transients discussed below. In the enhancement experiments, both beams of light were completely absorbed by the thick cell suspensions used; the cells therefore moved between illumination and virtual darkness as the suspension was agitated by the movement of the Warburg vessel. The supplementary beam entered the vessel from below and enhancement was found to be considerably greater if the far-red beam entered from above than if it entered at the side of the vessel. This was attributed to the shorter average time needed for cells to travel between the bottom and the surface layer than between the bottom and side. This suggested that the cells were illuminated alternately by one beam and the other; also that the interval between the absorption of the two wave lengths might be important.

Emerson had adhered throughout his working life to a single research technique, that of the Warburg apparatus. Some of the uncertainties in the use of this method for photosynthesis have been mentioned in Chapter 3 (page 85) and we have just seen that in the enhancement studies, as in all other experiments with thick suspensions, questions of intermittency of illumination were involved. In another research group a quite different technique had been developed by Blinks and co-workers with different advantages and disadvantages; the use of the platinum electrode polarograph (page 89) with alternating illumination at two different wave lengths led to an independent confirmation of the enhancement effect.

With the platinum electrode method of Haxo and Blinks [137] there was little doubt about the times of illumination of the plant material, though there were unknown time lags in the readings due to the diffusion of oxygen, and, with some algae, some curious toxic effects (page 89). Blinks [26] found in 1957 that on first illumination of the red alga *Porphyra* with light at 560 nm, where phycoerythrin absorbed, the rate of oxygen output rose more rapidly than with 675 nm, absorbed by chlorophyll *a* only, and showed a characteristic cusp after the initial steep rise (Fig. 9.23). In 1960 [27] he interpreted somewhat similar results for the green alga *Enteromorpha* in terms of a transient increase in respiration occurring immediately on illumination with 700 nm and causing the slow initial rise, but being slightly delayed

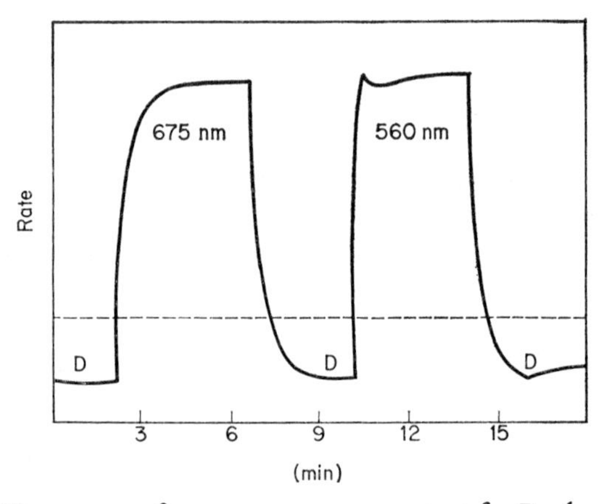

FIG. 9.23 Time course of net apparent oxygen output for *Porphyra perforata* in periods of red (675 nm) and green (560 nm) light with dark periods (D) between. Compensation point indicated by dotted line. Measurements with platinum electrode polarograph. After Blinks [26].

under 650 nm, absorbed by chlorophyll *b*, and so causing the depression after the cusp in the curve. Blinks [26] had also investigated the effects of alternating the longer and shorter wave lengths with no intervening periods of darkness. He found that if the steady-state rates of oxygen output were made equal for the two wave lengths by adjusting the light intensities, the change to the long wave length resulted in a sharp fall followed by a recovery to the steady-state level, while the change to the short wave length gave a sharp rise followed by a fall and recovery (Fig. 9.24). These effects have come to be known as chromatic transients.

Myers and French in 1960 [239] used the Haxo and Blinks platinum electrode as modified by Haxo [136] for unicellular algae and confirmed that the above results also applied with *Chlorella*, though in the chromatic transients the change in wave length 700→650 nm caused a rise, followed by a fall in oxygen output which scarcely passed below the steady-state value (Fig. 9.27). They noted that the upswing due to 700→650 nm was larger than the downswing due to 650→700 nm. They used the height of this upswing as a measure of the chromatic transients for various wave lengths following 700 nm and from three experiments obtained the action spectrum shown in Fig. 9.25(A); they also measured the steady rate of photosynthesis when the light of the various wave lengths was added to that at 700 nm and so obtained the action spectrum for enhancement in Fig. 9.25(B). These two almost identical action spectra strongly suggested that chromatic transients and

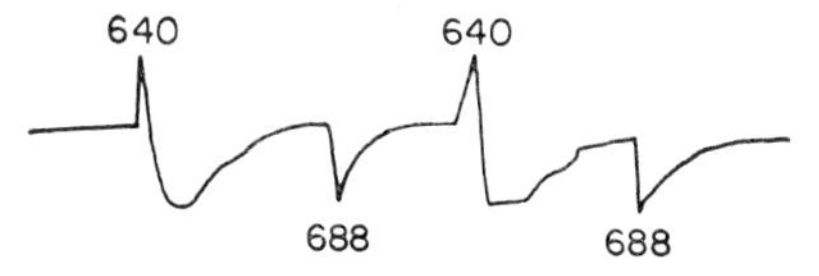

FIG. 9.24 Chromatic transients in the green alga *Ulva* illuminated alternately with 640 and 688 nm light. Measurements with a platinum electrode polarograph. After Blinks [28].

enhancement were basically the same phenomenon and related them both to absorption by chlorophyll *b*. In a second paper in 1960 the same authors [240] showed that the time courses of oxygen output for 650 and 700 nm differed greatly (Fig. 9.26) and they sought to explain the chromatic transients as the net effect of simultaneous rise and decay curves. Thus for 700→650 nm the rapid rise at 650 nm and slow decay on cessation of 700 nm would give a net rise; conversely for 650→700 nm. These very different time courses implied that the time constants for the reactions specifically associated with the two wave lengths differed in value by seconds. Hence it should be possible to obtain enhancement even when the 650 nm and 700 nm were separated in time. (We have seen that this probably occurred in Emerson's enhancement experiments—page 239.) Myers and French therefore tried a series of increasingly rapid alternations of the two beams (Fig. 9.27) and found that even with alternations as long as 15 seconds (not shown) there was some degree of enhancement, while with 6 seconds alternations the entire envelope of the curve was above the steady rate level for either beam alone. With the most rapid alternations (6×10^{-1} s ; period H) the amount of enhancement was exactly half that obtained when the two beams were superimposed (period L) giving double the total light; thus the relative enhancement was the same. When beams of 630 nm and 647 nm were alternated in the same way

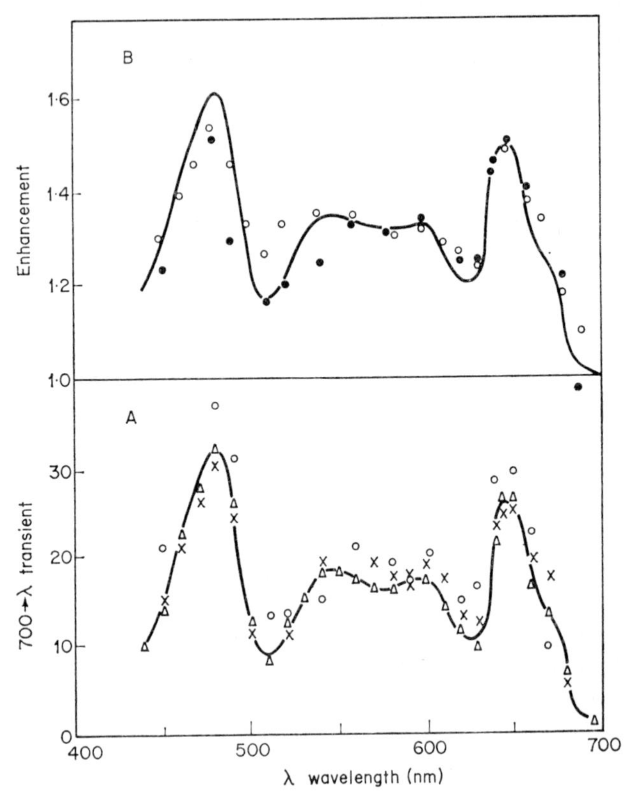

FIG. 9.25 Action spectra for *Chlorella pyrenoidosa*. A, chromatic transients: the transient rate given by the change from 700 nm to λ as a percentage of steady-state photosynthesis. B, enhancement: the increase in photosynthetic yield attributable to 700 nm due to the addition of λ calculated relative to the yield from 700 nm alone. The curve from A is repeated in B for comparison. Measurements with platinum electrode polarograph. After Myers and French [239].

as that shown in Fig. 9.27 they gave chromatic transients but no measurable enhancement. Experiments at 22°C and 2°C showed that the 700→650 nm and 650→700 nm transients were temperature-sensitive but with a rather low temperature coefficient.

Myers and French [240] were inclined to reject Blinks' hypothesis, that transients could be explained in terms of differences in respiration, for a number of reasons, one of which was the small temperature coefficient, but later work by French and Fork [103, 104] demonstrated differences in photo-stimulation of oxygen uptake measured in darkness immediately following illumination of *Porphyridium* with 570 nm or 695 nm. They found that the action spectrum for such respiratory

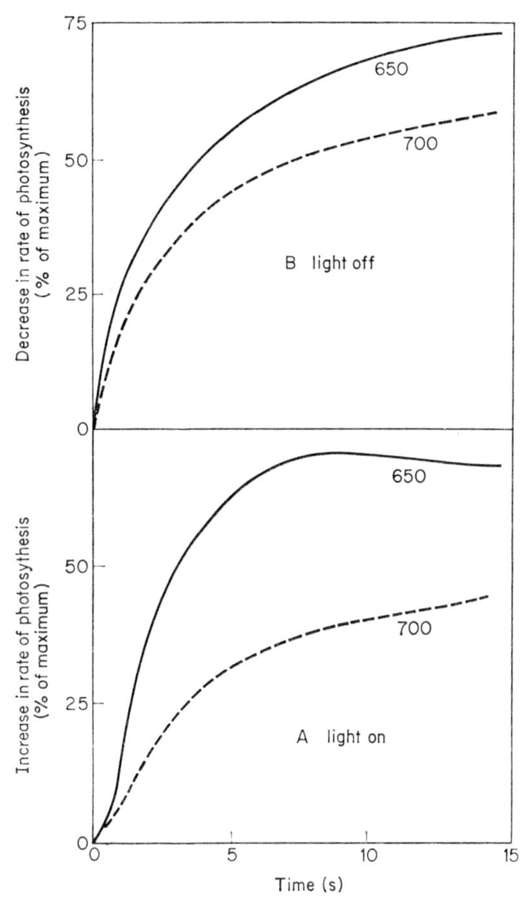

FIG. 9.26 Time course of net apparent oxygen output by *Chlorella pyrenoidosa*. A, rise in rate of photosynthesis, per cent of maximum rate, in 650 nm or 700 nm light following darkness; B, fall in rate of photosynthesis, per cent of maximum rate, in darkness following 650 nm or 700 nm light. Measurements with platinum electrode polarograph. After Myers and French [240].

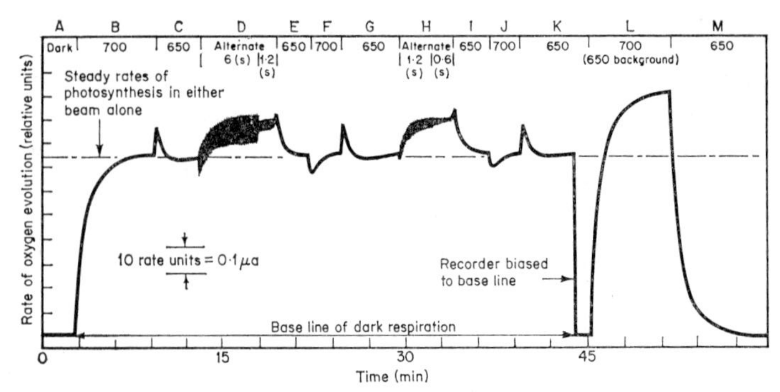

FIG. 9.27 Time course of net apparent oxygen output by *Chorella pyrenoidosa*, showing chromatic transients and enhancement of photosynthesis in alternating red (650 nm) and far-red (700 nm) light. In periods D and H and two wave lengths were alternated with the frequencies shown; in period L they were given together. Measurements with platinum electrode polarograph. After Myers and French [240].

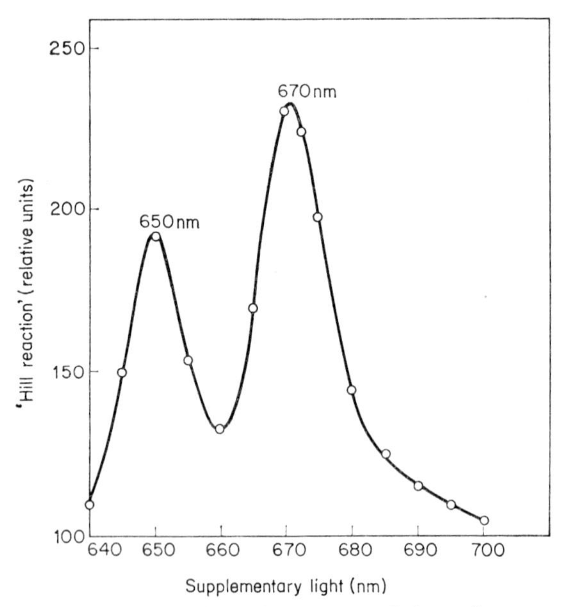

FIG. 9.28 Action spectrum for enhancement of the Hill reaction (photoreduction of quinone) with *Chlorella pyrenoidosa* cells. Rate of oxygen evolution attributable to far-red light (>680 nm) in presence of supplementary light expressed as a percentage of the rate in its absence. After Govindjee *et al.* [128].

stimulation resembled the absorption spectrum of chlorophyll *a*, and the action spectrum of oxygen evolution in light resembled the absorption spectrum of phycoerythrin. However, Govindjee, Thomas and Rabinowitch [128] found that they could obtain considerable enhancement in the Hill reaction with *Chlorella* cells poisoned with quinone. This treatment completely inhibited respiration as well as carbon dioxide fixation but allowed oxygen evolution to proceed with the photoreduction of the quinone. This showed that a large part if not all of the enhancement effect was independent of respiration and acted

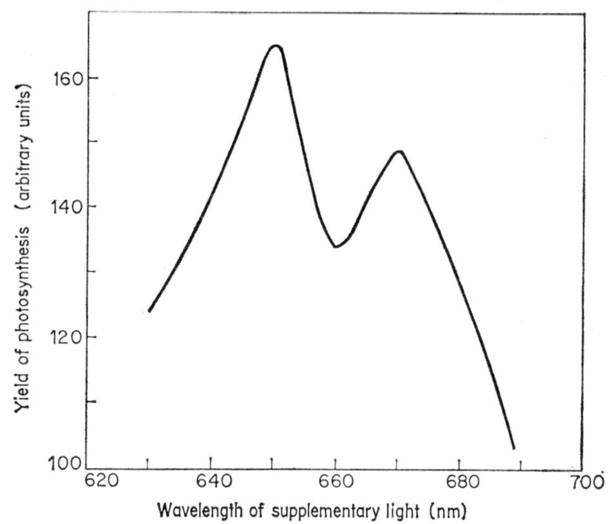

FIG. 9.29 Action spectrum for enhancement of photosynthesis in *Chlorella pyrenoidosa*. Yield of oxygen attributable to far-red light (>680 nm) in presence of supplementary red light expressed as a percentage of the yield in its absence. After Govindjee and Rabinowitch [127].

upon oxygen production rather than on carbon dioxide reduction. As discussed above (page 236) Hoch *et al.* later showed that enhancement in *Anacystis* acted upon oxygen production. In view of the common features of enhancement and chromatic transient phenomena it seems unlikely that the latter are mainly due to respiration changes.

In their experiments on the Hill reaction Govindjee *et al.* (*loc. cit*) found that the action spectrum for enhancement had two peaks in the red, one at 650 nm and the other at 670 nm (Fig. 9.28). A similar action spectrum for the enhancement of complete photosynthesis in *Chlorella* was found by Govindjee and Rabinowitch [127] (Fig. 9.29). They attributed the 650 nm peak to chlorophyll *b* but the 670 nm peak to absorption by an *in vivo* form of chlorophyll *a* which they called 'Chl *a* 670' and which showed as a shoulder on the absorption spectrum (page 32). This form of chlorophyll *a* could thus function as an accessory pigment in causing enhancement of the effectiveness of light absorbed by the long wave length absorbing form. Similar results were obtained with the diatom *Navicula minima.* These findings confirmed the speculations of Rabinowitch [85] and made possible a return to the view of Duysens [65] (page 38), based on evidence from fluorescence,

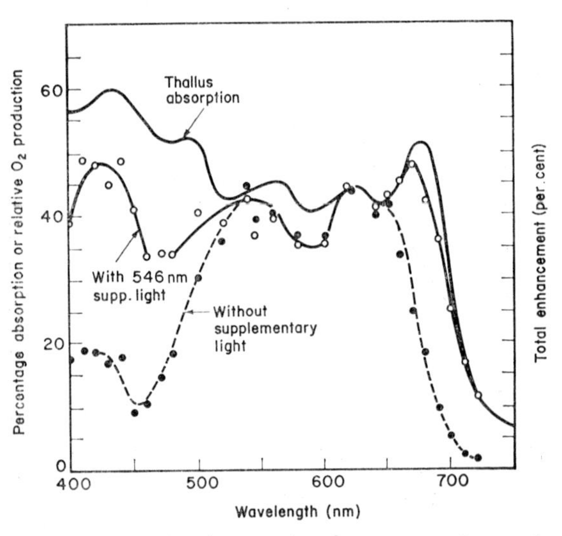

FIG. 9.30 Action spectra for photosynthetic oxygen production by *Porphyra perforata*, in the presence and absence of supplementary green light (546 nm), at a temperature of 20·5°C; also absorption spectrum. After Fork [91].

that all the photochemistry of photosynthesis was carried out by chlorophyll *a*, the other pigments merely collecting and transferring energy to it (compare the scheme proposed by Franck—page 259).

It may be noted in passing that if chlorophyll *a* 670 were to function as an accessory pigment in all plants, the red drop would not be expected to begin at about 650 nm in red algae, as found by Brody and Emerson (page 228) and shown in Fig. 9.30 from recent work by Fork [91]. In these algae it does appear as if excitation of chlorophyll *a* by either red or blue light requires that phycobilins should be excited also for maximum efficiency (Figs. 9.30 and 9.31). Conversely, the effect of green light, absorbed by phycoerythrin, can be enhanced by blue light absorbed by chlorophyll *a* (Fig. 9.31). As Fork remarks: 'The inactive chlorophyll of red algae could better be termed unenhanced chlorophyll.' The photosynthesis of red algae seems to show a number of discrepancies from that of green plants; careful comparative studies of these may be rewarding.

Myers and French [240] suggested that as transients were not confined to the use of a long wave length but could be found with changes such as 630→647→630 nm, 510→647→510 nm or 647→480→647 nm,

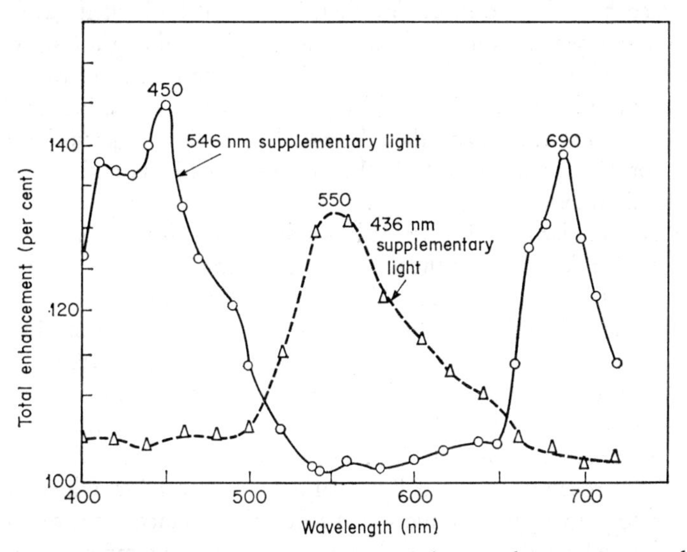

FIG. 9.31 Action spectra for enhancement of photosynthetic oxygen production by *Porphyra perforata*, calculated as the ratio of the yield for the combined light sources to the sum of the two separate yields. Circles and solid line, green supplementary light (546 nm); triangles and broken line, blue supplementary light (436 nm). After Fork [91].

they arose from a change in the ratio of supply of quanta to chlorophylls *a* and *b*; an upswing in rate of oxygen output resulted from an increase in *b/a* and a downswing from a decrease. *In vivo* the ratio of absorption *b/a* was always small in *Chlorella* and a large change in ratio could only be obtained when one of the two light beams had a wave length greater than 690 nm, so that *b/a* became zero. This might be the reason that enhancement could only be found under such conditions. They concluded that there were two kinds of events, one specifically associated with chlorophyll *b* or other accessary pigment. They also suggested that the intermediate(s) formed as a result of absorption of 650 nm light by chlorophyll *b* must reach rate-saturating concentrations in fractions of a second and have lifetimes measured in seconds. They called attention to the similar time constants found by Whittingham and Brown [324] for oxygen production by the green alga *Ankistrodesmus*, measured with the very sensitive Hersch oxygen cell (page 88), with a single flash or a single pair of flashes of 'white' light. The yield of oxygen from a single 35 ms flash was doubled if preceded by a flash so short (less than 5 ms) that by itself it gave no measurable yield. Some enhancement of yield, though less, was obtained when the two flashes were given in the reverse order. The degree of enhancement increased rapidly with the separation of the flashes up to about 1 s and then decreased more slowly, persisting up to 10 or 15 s (compare 6×10^{-1} s and 15 s in Myers and French's chromatic transient experiments).

In later similar experiments, Whittingham and Bishop [322] found that at 4°C the yield from the long flash alone was much decreased but if it was preceded by a short flash the maximum absolute yield was of the same order of magnitude as at 20°C; however, the optimal interval between the two flashes was now 16 s instead of about 1 s (Fig. 9.32). Oxygen production by isolated spinach chloroplasts with ferricyanide as hydrogen acceptor (Hill reaction) was also maximal with about 18 s between the flashes at 4°C. These results suggested two photochemical processes separated by a dark thermal reaction which was needed to convert the products of the first to reactants of the second. This thermal reaction proceeded more slowly at low temperature but its product had a greater half-life. Such longer persistence at low temperature of a product of a reaction following short wave length absorption might perhaps explain the finding of Emerson *et al.* [79] (page 226) that far-red light was more efficient at 5°C than at 20°C, for in their experiments they continually returned to a standard wave length of 660 nm, for which they knew the difference between rate of dark respiration and the net rate of photosynthesis and respiration. However, Govindjee [126] has more recently found that the red drop in *Chlorella*, which at 20°C begins at 680 nm, does not begin until 690 nm at 10°C.

Similar experiments but with monochromatic light of various wave lengths might be expected to yield further information as to the relations between the pigments concerned in the first and second photo reactions (see below). French [101] illuminated *Porphyridium* with pairs of flashes, one red (685 nm) and the other green (567 nm), mainly absorbed by chlorophyll *a* and phycoerythrin respectively and followed oxygen output or uptake with a teflon-covered gold electrode [102]. With a single pair of long flashes, each 3 s separated by 20 s dark, it was found that oxygen production from a green flash was enhanced by a preceding red flash, but the oxygen production from a red flash was somewhat reduced by a preceding green flash. When the separation of the flashes was varied (Fig. 9.33) the minimum yield of oxygen was obtained when a 2 s green flash and a 5 s red flash were started

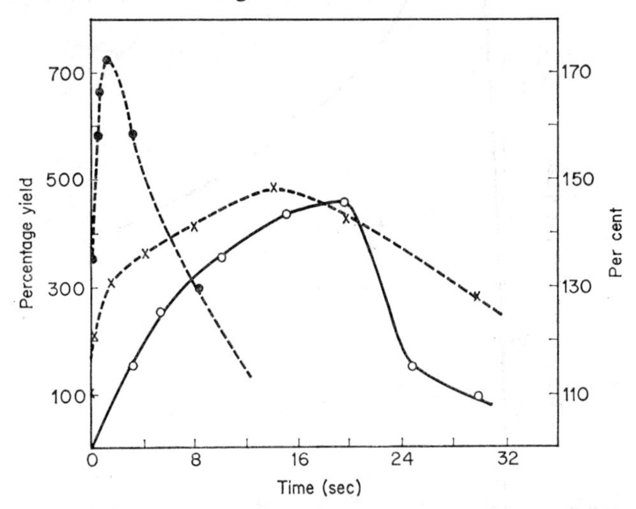

FIG. 9.32 Yield of oxygen from a short plus a long flash of 'white' light, as a function of the period of darkness between them, expressed per cent of the yield for the two flashes given together. Crosses and broken line, *Ankistrodesmus braunii* cells in bicarbonate buffer at 4°C (left-handed ordinate); solid circles and broken line, the same at 20°C (right-hand ordinate); open circles and solid line, isolated chloroplasts of spinach beet (*Beta vulgaris*) with ferricyanide at 4°C (right-hand ordinate). Measurements with a Hersch cell. After Whittingham and Bishop [322].

together and the maximum when the green flash followed immediately after the red flash. The yield then fell gradually as the separation was increased. The 'red product' which made the green more effective disappeared slowly with a half-life of about 18 s. It was thought that perhaps a 'green product' which might make red more effective might have such a short half-life as to be undetectable with the long time scales used. Further experiments [102] were carried out with 4 ms flashes in pairs with 10 ms between members of each pair and 430 ms between successive pairs. The steady-state oxygen evolution was measured when such pairs were given repeatedly. If a green product had a very short half-life it should disappear in 430 ms but not in 10 ms; in this case enhancement should be obtained when green

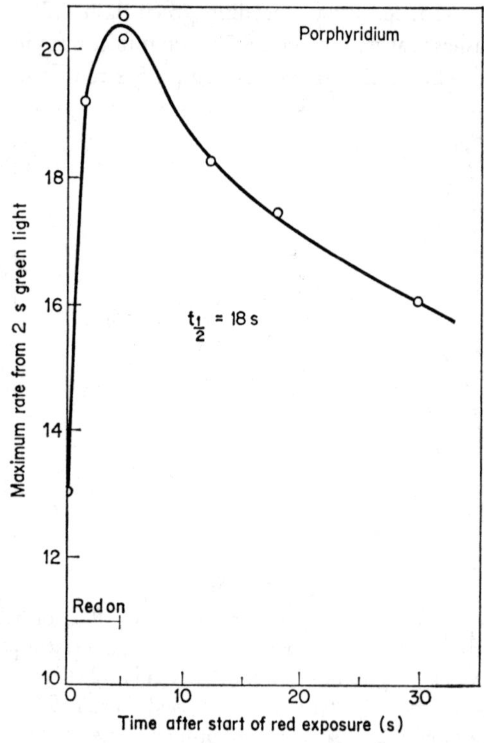

FIG. 9.33 Maximum rates of oxygen production by *Porphyridium cruentum* from a 2 s flash of green light (567 nm) given at various times after the start of a 5 s flash of weak red light (685 nm). Red alone gave a peak rate of 2·4 and green alone of 13·7. Measurements with a gold electrode polarograph. After French [101].

preceded red in each pair, as there would be plenty of red product surviving the 430 ms dark periods. No such effect was found. It was concluded that in *Porphyridium* material made by chlorophyll *a* enhanced the production of oxygen by green light; there was no evidence for production of a substance by green light which could subsequently enhance oxygen production in red light.

The results of further experiments by Whittingham and Bishop [323] on the Hill reaction of spinach chloroplasts at 4°C do not show this large effect of the order in which the flashes were given. With a single 35 ms monochromatic flash separated by various dark intervals from an earlier period of monochromatic light of another wave

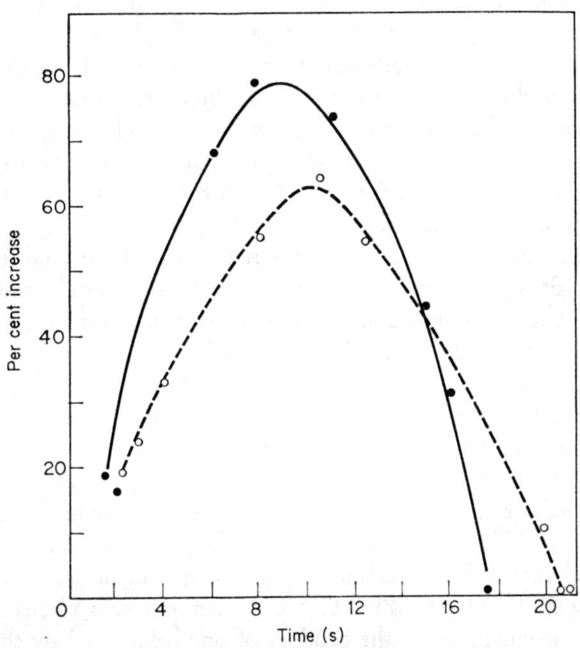

FIG. 9.34 Hill reaction of spinach chloroplasts reducing ferricyanide at 4°C, with phosphorylating cofactors present. Yield of oxygen from a short (35 ms) flash of monochromatic light, as a function of the dark interval after a preceding period of monochromatic illumination, expressed per cent of the yield from a similar flash given in prolonged darkness. Solid circles and solid line, yield from 644 nm flash after illumination at 700 nm; open circles and broken line from 697 nm flash after illumination at 653 nm. Measurements with a Hersch cell. After Whittingham and Bishop [323].

length, only slightly less enhancement was obtained for a 697 nm flash following illumination at 653 nm than for a 644 nm flash following 700 nm (Fig. 9.34). It should be remembered that respiration was involved in French's experiments and not in those of Whittingham and Bishop with isolated chloroplasts; also that the comparison is between a red alga with phycoerythrin and no chlorophyll *b* on the one hand, and higher plant chloroplasts with chlorophyll *b* but no phycobilin on the other. The former difference might well be crucial and the curious results obtained by French might be due to light-stimulated respiration at the shorter wave length oxidizing the green product.

It appears that there is opportunity for a great deal more physiological experimentation with short flashes of light of different wave lengths alone and in combination, in the further study of chromatic transients and enhancement, especially if photosynthesis and respiration can be measured concurrently. Biophysical and biochemical discoveries in this field have tended to follow physiological discoveries rather than *vice versa*. (This is in marked contrast to the later stages of the biochemistry of photosynthesis, involving the reduction of carbon dioxide, to which physiological experiments have made little contribution.) The idea of two separate photochemical reactions, driven by the far-red absorbing form of chlorophyll *a* and an accessory pigment, stemmed directly from Emerson's work in 1956–9; by 1960 this had resulted in a great number of papers confirming and extending his findings and some of these are mentioned above. These ideas quickly led biochemists to look for and find evidence at a chemical level.

The current views as to the photochemical stages of photosynthesis should perhaps be summarized, although really outside the scope of this book and likely to be modified with the passage of time. The most generally accepted view arises from a general scheme put forward in 1960 by Hill and Bendall [161]; it is that the two photochemical reactions occur in series, the product of one being used by the other (Fig. 9.35). System II is the accessory pigment activated by quanta of the shorter wave lengths of light. One of these quanta raises an electron in the pigment from +0·8 electron volts (the potential of the reaction $4OH^- = 2H_2O + O_2 + 4e^-$ at 25°C and pH 7) to zero potential. This enables the electron to be captured by an electron carrier (perhaps plastaquinone); it then falls back to + 0·4 eV in a dark reaction through a series of carriers (cytochromes and perhaps the modification of chlorophyll *a* termed P700 [195]) giving up energy with the formation of adenosine triphosphate (ATP) from adenosine diphosphate (ADP) and inorganic phosphate. It is then raised from +0·4 to −0·4 eV with the energy of a quantum absorbed by System I (the long wave length absorbing form of chlorophyll *a*, which can also make use of shorter wave lengths). Tagawa and Arnon [287] state that the electron from System I is captured by an iron-containing protein, which they call ferrodoxin, with a potential of —0·432 eV at pH 7·55; it is then passed to nicatinamide adenine dinucleotide phosphate, (NADP) to form NADPH. The 'hole' in the pigment of System II is supposed filled

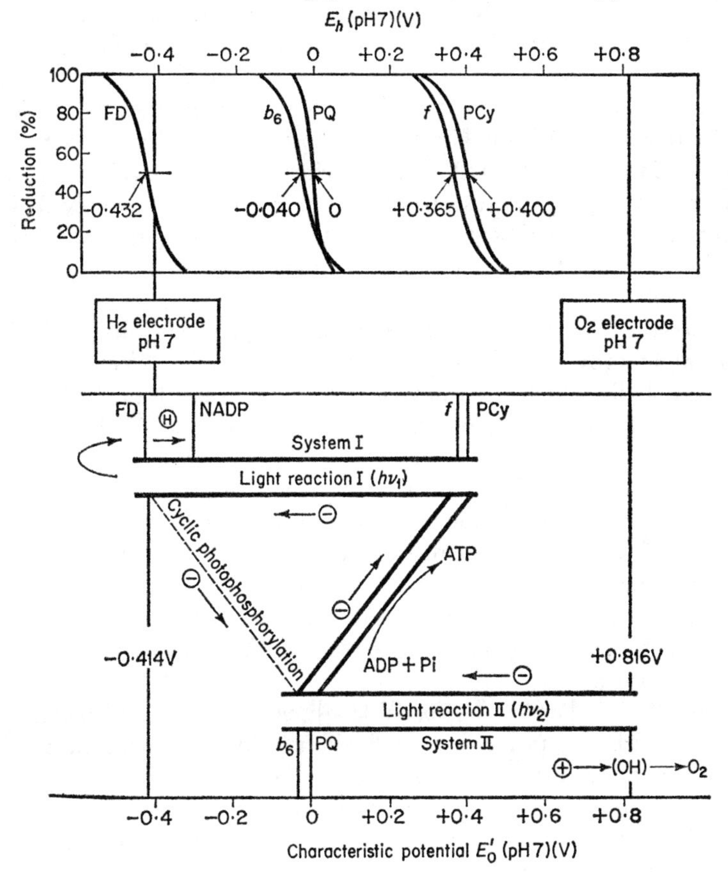

FIG. 9.35 Scheme for two photochemical reactions in photosynthesis related to the redox potentials of some components of chloroplasts. Circle with negative sign, electron; circle with positive sign, positive hole; PCy, plastocyanin; f, cytochrome *f*; PQ, plastoquinone; b_6, cytochrome b_6; FD, ferrodoxin. After Hill [160].

with an electron from a hydroxyl ion but the enzyme catalysing this is unknown; as the uncharged [OH] free radical which results cannot maintain an independent existence it at once unites with three others to form $2H_2O + O_2$ so that molecular oxygen is evolved; the corresponding hydrogen ion remains attached to the NADP where the electron from the chlorophyll has come to rest, forming NADPH.

In terms of this serial scheme, photosynthetic bacteria are considered to possess System I only. They have to be provided with electrons at about zero potential by a substrate. An historical account of the biochemical investigations leading to this scheme, many of them with isolated chloroplasts, has been given by R. Hill [160].

It will be seen that in this scheme System II operates to provide electrons at a sufficiently high potential in reduced carriers for System I to use in reducing NADP, thus oxidizing the carriers again. Further, unless System II absorbs quanta no oxygen should be evolved; in enhancement experiments far-red light alone always gives some yield of oxygen but this might be due to slight absorption by System II pigments.

Recently, Arnon *et al.* [7] have proposed a scheme in which a single quantum provides the energy for the transfer of an electron from chlorophyll at +0·8 electron volts to ferrodoxin at −0·43 eV. It may then return to the chlorophyll with the formation of ATP in cyclic phosphorylation; alternatively it may be transferred to NADP, in which case it is replaced in the chlorophyll from water (OH^-) and non-cyclic phosphorylation occurs between the OH^- and the chlorophyll. These are separate processes with the cyclic process driven by far-red quanta and the non-cyclic by shorter wave lengths. Far-red alone results in a shortage of $NADPH_2$, and shorter wave lengths cause enhancement by increasing the supply of reductant by non-cyclic phosphorylation.

In 1960 Warburg and Krippahl [308] put forward a fundamentally different scheme in which carbon dioxide is reduced in a photochemical reaction rather than as a dark process. They supported this by the observation that oxygen production in the Hill reaction is stimulated by the presence of carbon dioxide—a finding that has been confirmed in other laboratories. It is a remarkable and as yet unexplained fact that in the living plant it has not been found possible to separate the utilization of light from the assimilation of carbon dioxide, although isolated chloroplasts can first be illuminated and then fed with carbon dioxide in darkness to give some of the same photosynthetic products as when light and carbon dioxide are provided simultaneously [296]. Hill [160] points out that *in vitro* experiments with chloroplasts fall short in several ways of proof that carbon dioxide is fixed in a dark process; among others that much higher carbon dioxide pressures are needed than *in vivo*, that the activity of the reconstructed systems is lower and that the formation of glycollic acid by cells in low carbon dioxide concentrations and high light is not fully explained.

D. THE PHOTOSYNTHETIC UNIT

Just as the biochemistry and physical chemistry of modern studies of photosynthesis are dominated by ideas arising from Emerson's discovery of the enhancement effect, so the biophysics of photosynthesis is dominated by his discovery of the 'photosynthetic unit'.

Emerson and Arnold [76] in 1932 had concluded that photosynthesis (carbon dioxide reduction) consisted of a light reaction independent of temperature which could take place in one-hundred-thousandth of a second and a dark temperature-dependent reaction which at 25°C required about one-fiftieth of a second for completion. They next [77] considered the implication of light saturation under flashing light conditions, with dark periods long enough to complete the Blackman reaction between flashes. If the intensity of the flashes was increased until no further increase in photosynthesis was observed, this implied that each unit of chlorophyll capable of contributing to the photochemical reaction did so once in each flash—they discounted the possibility that it could undergo the light reaction more than once in a 10 μs flash because of the long time needed for the dark reaction. They therefore set out to determine experimentally the number of chlorophyll molecules in such a unit, defined as 'the mechanism which must undergo the photochemical reaction to reduce one molecule of carbon dioxide'. If, for instance, every chlorophyll molecule could absorb a quantum of light and was associated with the necessary enzymes to enable it to contribute to the reaction, the number of molecules in the unit would be equal to the quantum requirement for the reduction of one molecule of carbon dioxide; this they accepted as 4, following Warburg and Negelein [309], but later determined as 10 to 12—numbers which accord with the theoretical minimum of 8 now generally accepted, for owing to inevitable heat losses 100 per cent efficiency is not to be expected.

Light saturation was achieved (Fig. 9.36) under their flashing light conditions, by concentrating the light from the neon tube with mirrors. The maximum rate of oxygen production in moles per flash divided into the number of moles of chlorophyll found by analysis after the experiment gave the number of chlorophyll molecules in a unit which could fix one molecule of carbon dioxide, assuming a Q_p of unity. This determination was repeated with *Chlorella pyrenoidosa* cells having a wide range of chlorophyll contents, produced by growth with neon or mercury lamps, and light-saturated photosynthesis was approximately proportional to chlorophyll content (Fig. 8.6). The mean number of chlorophyll molecules in a unit, given by the slope of the straight line, was 2,480. Similar values, ranging from 2,500 to 4,200, were found by Arnold and Kohn [3] with six other species, namely

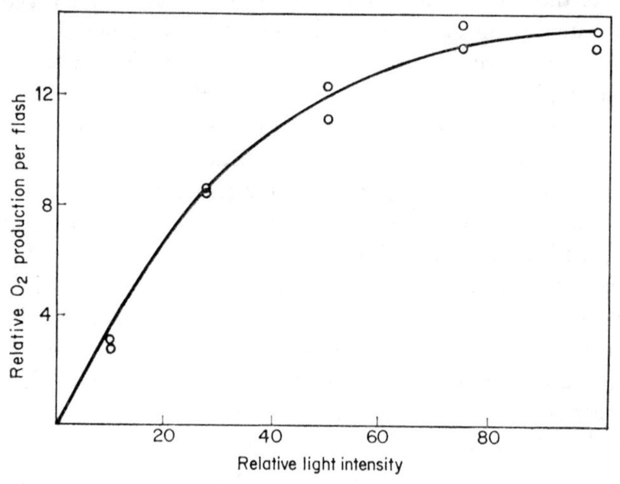

FIG. 9.36 Yield of oxygen per 10 μs flash of neon light, by *Chlorella pyrenoidosa* at 25°C, as a function of light intensity. After Emerson and Arnold [77].

Chlorella vulgaris, *Stichococcus bacillaris*, *Selaginella* sp., *Lemna* sp., *Bryophyllum calycinum* and *Nicotiana langsdorffii* (flowers).

This astonishing result, which at least in *C. pyrenoidosa* could not be attributed to an excess of chlorophyll and limitation by some other factor in view of the linear relation in Fig. 8.6, has received a good deal of subsequent support. Gaffron and Wohl [116] calculated that in thick suspensions of *Chlorella* cells, as used in quantum efficiency determinations, each molecule of chlorophyll absorbed, on the average, a quantum of light once every 8 minutes. Thus a single molecule would need about an hour to collect the 8 quanta necessary to reduce one carbon dioxide molecule. The energy of quanta absorbed anywhere among a very large number of chlorophyll molecules must somehow be transferred to a common reaction centre, for under such conditions photosynthesis begins almost immediately on illumination even with very dim light and shows quantum yields approaching the theoretical maximum.

It seems therefore that the photosynthetic unit must consist of a special reaction centre, where the photochemical reaction actually occurs, associated with a very large number of chlorophyll molecules. When the photochemical reaction has taken place, further quanta absorbed in the unit during a period of about 20 ms at 25°C are ineffective, as the reaction centre is occupied with the dark reactions. Hence the low efficiency of saturating light—we might say that an internal factor (enzyme concentration) was limiting. At 1°C the dark reactions would take about ten times as long (Fig. 9.2) and the rate of photosynthesis (per unit time) would be correspondingly lowered—we should then say the rate was severely temperature-limited.

As we have seen, it is now generally believed that the photochemical reaction is a 2-quantum process resulting in the formation of $NADPH_2$, and ATP by non-cyclic photophosphorylation, with the splitting of one O—H bond in water. This might suggest that the basic unit need consist of only $2,480 \div 4 = 620$ molecules [55]; however, as 8 quanta are necessary for the liberation of one molecule of oxygen it would seem that the energy from all 8 must be concentrated at one reaction centre and that therefore the larger unit is needed. It may be necessary to use the energy from 8 quanta within an exceedingly short space of time (less than 10 μs) and practically in one place (see below). The photosynthetic unit thus seems an admirable mechanism which enables the plant to function in low light intensities by collecting together the energy from relatively scattered quanta.

There is other evidence that the photosynthetic unit may consist of only a few hundred chlorophyll molecules. Thomas *et al.* [289] found that the oxygen evolution from spinach chloroplast fragments of various sizes fell abruptly when their volume was reduced below 10^3 nm^3; they estimated that this corresponded to about 100 chlorophyll molecules and suggested that this was the number per molecule of the rate-limiting enzyme responsible for the dark reaction. They rejected, however, the view that there was any structural unit. Bishop (see ref. [115]) found that an internal concentration of one molecule of the herbicide 3-(4-chlorophenyl)-1-1-dimethylurea (CMU) per 180 chlorophyll molecules was sufficient to prevent oxygen production by the alga *Scenadesmus*. The low concentrations of electron carriers in chloroplasts suggest that these are only found at the reaction centres, for instance cytochrome *f* and P700 in approximately equimolar concentrations and 300–500 times less than that of chlorophyll *a* (195). The granules, first seen on lamellae of disrupted chloroplasts by Frey-Wysseling and Steinmann [109, 282] and later called quantasomes by Calvin, have been discussed in Chapter 1 (page 9); according to Park and Biggins [251], as described there, they contain about 230 molecules of chlorophyll—160 of chlorophyll *a* and 70 of chlorophyll *b*. It has been suggested [267, 252] that these granules are the photosynthetic units, though as noted (page 22) there appears to be no direct evidence that they actually contain chlorophyll. The smallest photosynthetic organelles known are the chromatophores of photosynthetic bacteria—those of *Chromatium* (30 nm diameter) contain some 600 molecules of bacterio-chlorophyll [15] and may perhaps represent single photosynthetic units.

Izawa and Good [179] made a comprehensive study of the number of sites, in isolated spinach chloroplasts, sensitive to inhibitors of oxygen production in photosynthesis. In addition to CMU they used 3-(3,4-dichlorophenyl)-1,1-dimethylurea (DCMU) and 2-chloro-4-(2-propylamino)-6-ethylamino-s-triazine (Atrazine). For all three inhibitors they found that uptake by the chloroplasts was made up of three processes: (1) an irreversible binding of one molecule of inhibitor, for about every 1,000 molecules of chlorophyll, which was then without inhibitory effect; (2) above this threshold a partitioning of the inhibitor between the external medium and the chloroplasts; (3) an absorption which controlled the amount of inhibition and was thought to represent the formation of an enzyme–inhibitor complex. After allowing for (1), two methods of analysis of the results based (2) on the partitioning phenomena or (3) on enzyme kinetics agreed in giving estimates of 2,300 to 2,800 chlorophyll molecules for each inhibitor-sensitive site, regardless of which of the three inhibitors was used. This remarkable agreement with the findings of Emerson and Arnold [77] strongly suggests that 2,500 chlorophyll molecules is indeed the number in the photosynthetic unit—if so it must contain about ten quantasomes of the size given in Table 1.1.

There is no doubt that the inclusion of two pigments in the photochemical scheme greatly complicates the conception of photosynthetic units and their reaction centres. Are there two types of reaction centre, with two different pigments? It would seem that the energy from at least 8 quanta must operate in one place, unless indeed the OH free radicles can migrate so that four can combine to liberate one molecule of oxygen (page 253); therefore the oxygen-evolving system (System II) must be located at the reaction centre, for oxygen evolution starts almost instantaneously on first illumination with very few incident quanta. However, one would think that System I must also be there to reoxidize the electron carriers shown in Fig. 9.35, by removing the electrons and transferring them to ferrodoxin and NADP, otherwise a great excess of H^+ ions would rapidly build up and presumably prevent the system from operating.

There is evidence to suggest that the reaction centre contains P700 [195], the chlorophyll-like pigment absorbing at 700 to 705 nm, with associated and approximately equimolar cytochrome *f*. Light energy absorbed by the long wave length absorbing form of chlorophyll *a* (Chl *a* 683) in the photosynthetic unit might be transferred to P700 at the reaction centre; an electron from P700 might be transferred via ferrodoxin to NADP and be replaced from cytochrome *f*. This picture does not include the locus of oxygen evolution or the pigment responsible. One view is that the two pigment systems are situated at separate reaction centres of different type in a single photosynthetic unit (Fig. 9.37 after Calvin [53]). Short-wave light energy from a quantum $h\nu_2$ is transferred through the chlorophyll of the unit by exciton migration to the System II reaction centre; an electron from the pigment (which is replaced from water) migrates through other chlorophyll molecules and is then transferred by a series of electron carriers from cytochrome b_6 to cytochrome *f*, which is reduced. Another quantum $h\nu_1$ (which may be of far-red light) supplies energy which migrates to the System I reaction centre; here an electron from the pigment P700 is transferred to NADP and the hole migrates back through the chlorophyll of the unit to oxidize the cytochrome *f*. The electron carriers serve to link the two reaction centres and the electron-hole migration separates the two primary products of the reaction; thus back reactions between the oxidation products and reduction products are prevented. Rabinowitch [267] has suggested another way in which this last essential may be brought about, namely that as the pigment layers in chloroplasts are probably monomolecular, all that is needed is for the oxidation and reduction products to be produced on different sides of the layer (see also page 262 below).

Franck [95] and Franck and Rosenburg [97] have put forward hypotheses based entirely on chlorophyll *a*, thus avoiding the necessity for two pigments at the reaction centres. The reaction centre is probably situated at the edge of the photosynthetic unit, where the chlorophyll *a* is supposed to be exposed to water and solutes and therefore photochemically active; it is there complexed with cytochrome *f* and another electron-transport enzyme. Energy is transferred to this active chlorophyll *a* both from accessory pigments and from the main bulk of the photosynthetic unit chlorophyll, which is protected by lipoids

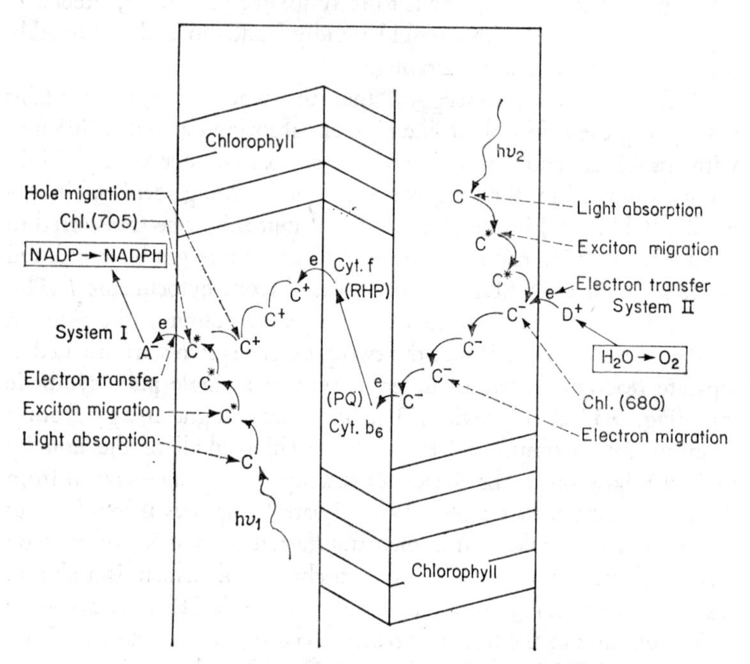

FIG. 9.37 Hypothetical scheme with pigment systems II and I located at separate reaction centres within the same photosynthetic unit. For explanation see text. After Calvin [53].

from contact with water. These pigments deliver energy as excitation of the first excited singlet state (page 19), making the exposed chlorophyll *a* act as System II (Fig. 9.35); however, some of the protected chlorophyll *a* is in a crystalline form absorbing longer wave lengths and this delivers metastable triplet excitation to the exposed chlorophyll, which then behaves as System I (Fig. 9.35). P700 is considered to be a crystalline form of chlorophyll *a* participating in such energy collection from far-red radiation but not essential for the overall process. There is evidence that the crystalline long-wave absorbing chlorophyll is not uniformly distributed throughout the chloroplast but is present in a small proportion only of the photosynthetic units of green plants; on the other hand in red and blue-green algae almost all the units apparently contain crystalline regions. The authors have postulated a 'super-unit', consisting probably of more than 10 photosynthetic units, within which singlet excitation can be transferred from the exposed chlorophyll of one unit to that of another by sensitized fluorescence (this is not possible for metastable triplet excitation). Such transfer could accelerate the succession of the two photochemical steps. If this is correct it may perhaps reconcile the large size of the photosynthetic unit as found by Emerson and Arnold (about 2,500) with the evidence for units of a few 100 molecules mentioned above —the former might be a super-unit including one unit bearing crystalline long-wave absorbing chlorophyll. However, this interpretation would imply that *only* System I, with metastable triplet excitation from crystalline chlorophyll, can reduce NADP and oxidize the electron carriers. Franck and Rosenburg propose that System II with singlet excitation can perform both the photochemical reactions, but System I cannot perform the reduction of the electron carriers and the oxygen-evolving process appropriate to System II with good efficiency. According to them, enhancement involves the migration of singlet excitation from a System II unit to the reaction centre of any unit within the super-unit where oxidized electron carriers await reduction.

In far-red light, System I absorbs most of the incident radiation and as the metastable excitation cannot migrate to the exposed chlorophyll of other units the oxidized cytochrome at the System I reaction centre has a long lifetime. This leads to photoxidation of the exposed chlorophyll; such chlorophyll bleaching may be transferred by hole migration to protected chlorophyll molecules inside the unit before a recovery reaction occurs. (This last is due to reduction of the cytochrome, by transferred singlet excitation, and hence of the oxidized chlorophyll.) Far-red light alone therefore tends to inhibit photosynthesis. This could account for the fall in quantum efficiency in far-red and the negative enhancement found for *Anacystis* with long wave light but not, it would seem, that found with short wave light (see Fig. 9.15 and page 234); also for the consumption of oxygen immediately following far-red found by French and Fork (page 243). Shorter wave length light results in migration of singlet excitation and rapid reduction of the oxidized cytochrome.

Rabinowitch [268] has made suggestions for a two-layered structure with two monomolecular layers of pigment, belonging to System I and System II (Fig. 9.35) respectively. The System I chlorophyll is associated with a lipid layer and therefore non-fluorescent; the System II chlorophyll (Chl *a* 670) is in a hydrophilic protein layer and therefore fluorescent. The reaction centres of photosynthetic units in the two layers are connected by the cytochromes B_6 and *f*. He has also supposed that the chlorophyll layers may be on the surfaces of spherical units (Fig. 9.38), thus combining the evidence from electron micrographs of thin sections, for alternating lamellae of more hydrophilic and more

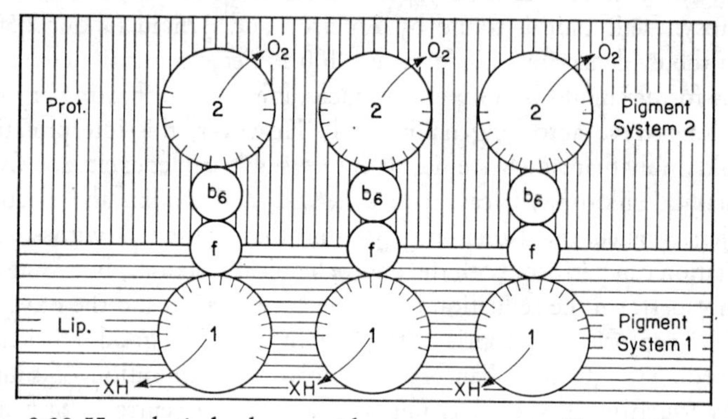

FIG. 9.38 Hypothetical scheme with pigment systems II and I located in protein and lipoid layers respectively. Chlorophyll molecules are supposed arranged in monomolecular layers on spherical units and the two types of photosynthetic unit are linked by cytochromes b_6 and *f*. After Rabinowitch [268].

hydrophobic properties, with that from disrupted chloroplasts for granular structure. Unfortunately, electron micrographs have so far told us nothing of the whereabouts of the chlorophyll. However, Goedheer has optical evidence that it forms separate lamellae less than 1 nm thick, whereas the protein and lipoid layers are about 3–5 nm thick. This seems to support the monomolecular layer suggestion. The highly efficient use of quanta in weak light means that if there are two sorts of photosynthetic unit they must be very effectively coupled. Here the long lifetime of intermediates found by Myers and French [240] is invoked. The separation of the oxidation and reduction products in the two layers should effectively prevent immediate back reactions.

It seems that investigation of the structure and organization of photosynthetic units and their reaction centres will have to remain predominantly in the hands of biophysicists. The function of plant physiology, having indicated the existence of such units, must now be that of testing experimentally the hypotheses they produce. Fortunately there is always the chance that such experiments may also produce some further unexpected discovery.

10: Physiology in Future Work

It may be useful to collect here and enlarge on some of the suggestions and viewpoints that have been put forward in earlier chapters, not so much for their individual value as to indicate the kind of contribution that the plant physiologist may hope to make if he is primarily a botanist and therefore unfortunately knows less of biochemistry, biophysics and mathematics than specialists in those subjects. Such relative ignorance of the more physical sciences and mathematics must be offset by a better knowledge and understanding of the behaviour of intact plants, if the botanical plant physiologist is to carry out his proper functions. The most important of these is to enlarge our knowledge of plant responses to environment conditions. The discovery of new phenomena at the whole plant level not only suggests hypotheses to be tested by experiments at that level but also at the molecular level by biochemists and biophysicists; the hypotheses that they themselves produce in their investigations should in turn be tested at the physiological level. In general, botanical plant physiology, as distinct from the biochemical and biophysical sorts, should attempt to provide a central field of knowledge in which the discoveries of biochemical reactions, fine structure and physical conditions may be co-ordinated with reference to the functioning of the whole plant.

An example of the contribution that can be made is to demonstrate the importance of the treatment that a plant has received previous to the actual experimental determinations of, for instance, rate of photosynthesis. Not only are the long-term effects of such factors as photoperiod, light intensity and temperature important in modifying leaf structure and performance, but less obvious short-term effects can be brought about by variation in conditions during the 24 hours or so preceding the experiment. As described in Chapter 4, Maskell discovered that the rhythmic changes in rate of apparent photosynthesis of cherry laurel leaves under constant illumination could be attributed to a similar rhythm in stomatal movement [213, 214]. It is now known that such stomatal rhythms are found both in light and darkness and are probably of general occurrence; they are apparently started by the change from dark to light or *vice versa* and can be affected by as little as 10 lux of tungsten filament light [219, 210, 147]. Irregularities in illumination of plants to be used for experiments, or determining their photosynthetic rates in the different treatments at various times of day, may therefore lead to misleading results.

A good botanist should be well aware of the complexity of the interactions between environmental factors in their effects on plants, and at the same time should bear constantly in mind the characteristics of the environment in which a given plant flourishes and to which it is presumably adapted. When a plant is subjected to experimental conditions of a kind that it is never exposed to in a state of nature its response is in large measure unpredictable. This response may well reveal new aspects of the internal mechanisms which could not be discovered under normal conditions, but to take advantage of this it is essential to compare the responses under both abnormal and normal ranges of given environmental factors. An example of this need is provided by the almost exclusive use until recently of high concentrations of carbon dioxide in experimental work on photosynthesis. This not only resulted in the CO_2 gush of *Chlorella pyrenoidosa* which caused so much trouble in the quantum efficiency controversy (page 220), but concealed the importance, under natural conditions, of the photosynthetic formation of glycollate. The latter is an intermediate in the formation of the amino-acid glycine and hence in protein synthesis [325]. Moreover, at normal carbon dioxide concentrations *Chlorella* and other algae excrete considerable quantities of glycollate into the water when the light intensity is high. It has been suggested [88] that this glycollate may serve as a reserve carbon source for algal growth when photosynthesis is not possible.

Another example of the difficulties of interpretation that can result from the use of some factor exclusively at a level not usually found in nature is provided by the assumption that as long as some oxygen is present in photosynthesis experiments the concentration is unimportant.

This has resulted in a great deal of experimentation at very low oxygen concentrations because of the demands of convenient or sensitive techniques, while in other investigations ordinary air (or water in equilibrium with it) has been used. The contribution of these differences to the confusion that can arise in comparing the results of one investigator with those of another can be surmised from the results relating to respiration in light discussed in the latter part of Chapter 5. Brown and Weis [44] and Ozbun *et al.* [249] who used very low oxygen concentrations found a partial suppression of carbon dioxide output in light, in mass spectrometer experiments; later, Forrester *et al.* [92] deduced from experiments in which net uptake or output of carbon dioxide was studied that the gross output in light was controlled by oxygen concentration, being less than the dark output below 10 per cent oxygen and more above. It is most desirable that this should be checked by mass spectrometer experiments. In such experiments Hoch *et al.* [168] used oxygen concentrations equivalent to 0·2–0·4 bar and found no effect of oxygen tension (within that range) upon the inhibition at low light intensity of oxygen uptake by *Anacystis*, but they do not mention oxygen concentration effects upon the light-stimulated uptake that occurred at higher intensities nor did they measure carbon dioxide exchange.

It may probably be assumed that oxygen concentrations were subnormal in all mass spectrometer experiments except those with the gas phase eliminated from the reaction vessel (Hoch and Kok [166]), and even lower concentrations are necessary in the Hersch cell (page 88); they must also be very low close to the electrode in a polarograph apparatus. On the other hand in enhancement experiments carried out with Warburg type apparatus it is likely that the cell suspensions were in initial equilibrium with air of 21 per cent oxygen content unless otherwise stated. However, the finding [168] that oxygen uptake is stimulated by near-red light (and for a blue-green alga, inhibited by far-red light) means that the serious study of enhancement phenomena except with isolated chloroplasts now demands a mass spectrometer. Fortunately the modification due to Hoch and Kok makes it possible to include normal oxygen concentration in the range studied and it is to be hoped that the high 'noise' level and other difficulties of their method will be overcome.

It is regrettable that experiments are almost automatically so designed that their results seem to support the hypotheses of the investigator and this tendency is lessened if orthogonal factorial experiments, covering wide ranges of the factors studied, are used. Some of the combination treatments, which have to be included to complete the orthogonal design, have a way of providing unexpected results which can be most damaging to pet hypotheses. Such treatments often provide conditions to which the plant is never normally exposed and, as suggested above, its response to these may be very revealing when compared with the responses to more normal combinations of factors. It is for this reason, as well as to facilitate statistical analysis, that it is important to include in the experiment the complete orthogonal combination of all levels of all factors and not to leave out some of the combinations in the design because they look 'silly'.

Although photosynthesis was the subject of what was perhaps the first large factorial experiment, described by Harder in 1921 [135], the method has been relatively little used in this field—perhaps in part because of the difficulties of many of the techniques employed. A good modern example is provided by the experiments of Fogg and Than-Tun [89] on the interactions of light, temperature, nitrogen and carbon dioxide in controlling nitrogen fixation and photosynthesis in the blue-green alga *Anabaena cylindrinca.*

Factorial experiments with large numbers of levels of several factors in combination lend themselves to regression analysis and here the availability of computers greatly widens the scope. For example, equations such as Maskell's resistance formula and various more modern developments from it are extremely troublesome to fit with a desk calculator but could well be fitted by computer. If mass spectrometry of oxygen and carbon dioxide can be made sufficiently accurate and easy, so that uptake and output of both gases can be measured in really large factorial experiments of the enhancement type with oxygen concentration as one of the factors, it should be possible to present the biochemists with a good deal of new information on the interrelations of photosynthesis and respiration.

Such experiments might also yield explanations of such curious observations as the negative enhancement found when short-wave light is added to far-red (Figs. 9.15 and 9.18)—the suggestion put forward by Hoch and Owens (page 238) would seem only to apply to two long wave lengths. Other intriguing results which might well be further investigated in factorial experiments are the low-temperature effect on the red drop (page 226) and also the effects of beef broth and earth extract (page 239), which Hoch and Owens have suggested may be due to changes of respiration. The adaptation phenomena (page 228) of Brody and Brody should also be re-examined with the separation of respiration and photosynthesis.

It is unfortunate that the mass spectrometer is only suitable for measuring steady-state photosynthesis and respiration. The very intriguing effects of single flashes of light of different wave length, separated by various times and at different temperatures (page 248), will have, therefore, to continue to be investigated in terms of net oxygen output and probably at low oxygen concentration. However, factorial designs can be used for such experiments and information from other investigations with the mass spectrometer should assist greatly in the interpretation of the results. Such assistance in interpretation should also make worth while more extensive investigations, in terms of net gas exchange, of a number of topics including enhancement.

Finally, the comparative physiology of contrasting species provides a fascinating field of work though inevitably the findings are open to many widely differing interpretations. Examples that seem to invite further study are the red and green algae with their differences in enhancement behaviour, or maize that can maintain a zero value of Γ and other plants that cannot. Assimilation by species with contrasting leaf structure is also of interest, for instance *Pelargonium* and *Begonia sanguineum* which have very different distribution of stomata and of chloroplasts (page 132). More attention to effects of differences in size, number and chlorophyll content of chloroplasts might prove rewarding (page 205). In fact, the importance of structure for photosynthesis seems to apply at all levels of organization—not only molecular structure but also the way in which the molecules are packed together to form photosynthetic units, the lamellar structure of the chloroplasts (which only develops fully in light), the size and arrangement of the chloroplasts in the leaf, the characteristics of the intercellular space system, the size, number and disposition of the stomata, the arrangement of leaves on the plant and of the plants in the community or crop.

Investigations in comparative physiology also gain much in relevance and comprehensiveness if the forms are compared in factorial experiments where the ranges of the factors include the natural conditions for each. It may often happen, however, that the ranges of conditions for one species extend into regions where the other cannot survive and though this can sometimes be an interesting finding, death is not usually considered to be a satisfactory response (except perhaps by investigators of weed-killers). In such cases it should be possible by suitable regression analysis (with the aid of a suitably helpful statistician) to study the response surfaces in the two forms for overlapping but not identical ranges of the factors—one of the relatively few cases in which it may be desirable to abandon orthogonality.

It was suggested at the beginning of Chapter 3 that plant physiology grades imperceptibly into ecology on the one hand and into biochemistry and physical chemistry on the other. If this book has done something to encourage and extend the intermingling of plant physiology with these and other sciences, by exciting the interest of actual or potential research workers, I shall be well satisfied that the considerable labour it involved was worth while.

Bibliography

1. Allen, F. L., and Franck, J., 1955. 'Photosynthetic evolution of oxygen by flashes of light': *Arch. Biochem. Biophys.* 58, 124–43.
2. Arnold, W., 1949. 'A calorimetric determination of the quantum yield in photosynthesis': in Franck, J., and Loomis, W. E. (Eds), *Photosynthesis in Plants*: Iowa State Coll. Press, Ames, Iowa, 273–6.
3. Arnold, W., and Kohn, H., 1934. 'The chlorophyll unit in photosynthesis': *J. Gen. Physiol.* 18, 109–12.
4. Arnold, W., and Sherwood, H. K., 1957. 'Are chloroplasts semi-conductors?': *Proc. Nat. Acad. Sci.* 43, 105–14.
5. Arnon, D. I., 1961. 'Cell-free photosynthesis and the energy conversion process': in McElroy, W D., and Glass, B. (Eds), *Light and Life*: Johns Hopkins Press, Baltimore.
6. Arnon, D. I., Allen, M. B., and Whatley, F. R., 1954. 'Photosynthesis by isolated chloroplasts': *Nature* 174, 394–6.
7. Arnon, D. I., Tsujimoto, H. Y., and McSwain, B. D., 1965. 'Photosynthetic phosphorylation and electron transport': *Nature* 207, 1367–72.
8. Audus, L. J., 1940. 'A simple class apparatus for the quantitative determination of oxygen evolution in the photosynthesis of *Elodea canadensis*': *Ann. Bot. Lond.* N.S. 4, 819–24.
9. Audus, L. J., 1953. 'A simplified version of an apparatus for the measurement of oxygen evolution in the photosynthesis of *Elodea*': *School Sci. Rev.* 25, 120.
10. Barer, R., 1955. 'Spectrophotometry of clarified cell suspensions': *Science* 121, 709–15.
11. Bassham, J. A., and Calvin, M., 1957. *The Path of Carbon in Photosynthesis*: Prentice-Hall, New Jersey.
12. Bavel, C. H. M. van, Nakayama, F. S., and Ehrler, W. L., 1965. 'Measuring transpiration resistance of leaves': *Plant Physiol.* 40, 535–40.

13. BEIJERINCK, M. W., 1901. 'Photobacteria as a reactive in the investigation of the chlorophyll function': *Kon. Akad. Wetensch. Amsterdam Proc.* 4, 45–9.
14. BERGERON, J. A., 1963. 'Studies of the localization, physicochemical properties and action of phycocyanin in *Anacystis nidulans*': in Kok, B. (Ed.), *Photosynthetic Mechanisms of Green Plants*: Publ. 1145, Nat. Acad. Sci. Wash., 527–36.
15. BERGERON, J. A., and FULLER, R. C., 1961. 'The submicroscopic basis of bacterial photosynthesis: the chromatophore': in Edds, M. V. (Ed.), *Macromolecular Complexes*: Ronald Press, New York, 179–202.
15*a*. BJÖRKMAN, O., 1968. 'Further studies of the effect of oxygen concentration on photosynthetic CO_2 uptake in higher plants': *Carnegie Inst. Year Bk.* 66, 220–8.
16. BLACK, J. N., 1958. 'Competition between plants of different initial seed sizes in swards of subterranean clover (*Trifolium subterraneum* L.) with particular reference to leaf area and the light microclimate': *Aust. J. Agric. Res.* 9, 299–318.
17. BLACKMAN, F. F., 1895*a*. 'Experimental researches on vegetable assimilation and respiration. I. On a new method for investigating the carbonic acid exchanges of plants': *Phil. Trans. Roy. Soc.* B. 186, 485–502.
18. BLACKMAN, F. F., 1895*b*. 'Experimental researches on vegetable assimilation and respiration. II. On the paths of gaseous exchange between aerial leaves and the atmosphere': *Phil. Trans. Roy. Soc.* B. 186, 503–62.
19. BLACKMAN, F. F., 1905. 'Optima and limiting factors': *Ann. Bot. Lond.* 19, 281–95.
20. BLACKMAN, F. F., and MATTHAEI, G. L. C., 1905. 'Experimental researches in vegetable assimilation and respiration. IV. A quantitative study of carbon-dioxide assimilation and leaf-temperature in natural illumination': *Proc. Roy. Soc.* B. 76, 402–60.
21. BLACKMAN, F. F., and SMITH, A. M., 1911. 'On assimilation in submerged water-plants and its relation to the concentration of carbon dioxide and other factors': *Proc. Roy. Soc.* B. 83, 389–412.
22. BLACKMAN, G. E., and BLACK, J. N., 1959. 'Physiological and ecological studies in the analysis of plant environment XII': *Ann. Bot. Lond.* N.S. 23, 131–45.
23. BLACKMAN, G. E., BLACK, J. N., and MARTIN, R. P., 1953. 'An inexpensive integrating recorder for the measurement of daylight': *Ann. Bot. Lond.* N.S. 17, 529–37.
24. BLACKMAN, G. E., and WILSON, G. L., 1954. 'Adaptive changes in the growth and development of *Helianthus annuus* induced by alteration in light level': *Ann. Bot. Lond.* N.S. 18, 71–94.
25. BLACKMAN, V. H., 1919. 'The compound interest law and plant growth': *Ann. Bot. Lond.* 33, 353–60.

26. BLINKS, L. R., 1957. 'Chromatic transients in photosynthesis of red algae': in Gaffron, H. (Ed.), *Research in Photosynthesis*: Interscience, New York, 444–9.

27. BLINKS, L. R., 1960. 'Relation of photosynthetic transients to respiration': *Science* 131, 1316.

28. BLINKS, L. R., 1964. 'Accessory pigments and photosynthesis': in Giese, A. C. (Ed.), *Photophysiology*, Vol. I: Academic Press, New York, 199–221.

29. BLINKS, L. R., and SKOW, R. K., 1938*a*. 'The time course of photosynthesis as shown by the glass electrode, with anomalies in the acidity changes': *Proc. Nat. Acad. Sci.* 24, 413–19.

30. BLINKS, L. R., and SKOW, R. K., 1938*b*. 'The time course of photosynthesis as shown by a rapid electrode method for oxygen': *Proc. Nat. Acad. Sci.* 24, 420–7.

31. BOLAS, B. D., 1926. 'Methods for the study of assimilation and respiration in closed systems': *New Phytol.* 25, 127–44.

32. BOLAS, B. D., MELVILLE, R., and SELMAN, I. W., 1938. 'The measurement of assimilation and translocation in tomato seedlings under the conditions of glasshouse culture': *Ann. Bot. Lond.* N.S. 2, 717–28.

33. BOYSEN-JENSEN, P., 1932. *Die Stoffproduktion der Pflanzen:* G. Fischer, Jena.

34. BOYSEN-JENSEN, P., and MÜLLER, D., 1929. 'Über die Kohlensäure-assimilation bei Marchantia und Peltigera': *Jb. wiss. Bot.* 70, 503–11.

35. BRANTON, D., 1966. 'Fractured faces of frozen membranes': *Proc. Nat. Acad. Sci.* 55, 1048–56.

36. BRIGGS, G. E., 1920. 'Experimental researches on vegetable assimilation and respiration. XIII. The development of photosynthetic activity during germination': *Proc. Roy. Soc.* B. 91, 249–68.

37. BRIGGS, G. E., 1922. 'Experimental researches on vegetable assimilation and respiration. XV. The development of photosynthetic activity during germination of different types of seeds': *Proc. Roy. Soc.* B. 94, 12–19.

38. BRILLIANT, B., 1924. 'Le teneur en eau dans les feuilles et l'énergie assimilatrice': *C. R. Acad. Sci. Paris* 178, 2122–5.

39. BRODY, M., and BRODY, S. S., 1962. 'Induced changes in the photosynthetic efficiency of *Porphyridium cruentum* II': *Arch. Biochem. Biophys.* 96, 354–9.

40. BRODY, M., and EMERSON, R., 1959*a*. 'The quantum yield of photosynthesis in *Porphyridium cruentum*, and the role of chlorophyll *a* in the photosynthesis of red algae': *J. Gen. Physiol.* 43, 251–64.

41. BRODY, M., and EMERSON, R., 1959*b*. 'The effect of wavelength and intensity of light on the proportion of pigments in *Porphyridium cruentum*': *Amer. J. Bot.* 46, 433–40.

42. BROWN, A. H., 1953. 'The effects of light on respiration using isotopically enriched oxygen': *Amer. J. Bot.* 40, 719–29.

43. BROWN, A. H., NIER, A. O. C., and NORMAN, R. W. VAN, 1952. 'Measurement of metabolic gas exchange with a recording mass spectrometer': *Plant Physiol.* 27, 320–34.

44. BROWN, A. H., and WEIS, D., 1959. 'Relation between respiration and photosynthesis in the green alga, *Ankistrodesmus braunii*': *Plant Physiol.* 34, 224–34.

45. BROWN, A. H., and WHITTINGHAM, C. P., 1955. 'Identification of the carbon dioxide burst in *Chlorella* using the recording mass spectrometer': *Plant Physiol.* 30, 231–7.

46. BROWN, H. T., and ESCOMBE, F., 1900. 'Static diffusion of gases and liquids in relation to the assimilation of carbon and translocation in plants': *Phil. Trans. Roy. Soc.* B. 193, 223–91.

47. BROWN, H. T., and ESCOMBE, F., 1905. 'Researches on some of the physiological processes of green leaves, with special reference to the interchange of energy between the leaf and its surroundings': *Proc. Roy. Soc.* B. 76, 29–111.

48. BROWN, J. S., 1963. 'Forms of chlorophyll *a*': *Photochem. and Photobiol.* (*Chlor. Metabol. Symp.*) 2, 159–73.

49. BROWN, W. H., 1918. 'The theory of limiting factors': *Philippine J. Sci. C. Bot.* 13, 345–50.

50. BURK, D., SCHADE, A. L., HUNTER, J., and WARBURG, O., 1951. 'Three-vessel and one-vessel manometric techniques for measuring CO_2 and O_2 gas exchanges in respiration and photosynthesis': *Symp. Soc. Exp. Biol.* 5, 312–35.

50*a*. BUTLER, W. L., 1964. 'Absorption spectroscopy *in vivo*: theory and application': *Ann. Rev. Plant Physiol.* 15, 451–70.

51. CALVIN, M., 1959*a*. 'From microstructure to macrostructure and function in the photochemical apparatus': *Brookhaven Symp. Biol.* 11, 160–80.

52. CALVIN, M., 1959*b*. 'Free radicals in photosynthetic systems': *Rev. Mod. Phys.* 31, 157–61.

53. CALVIN, M., 1965. 'Energy conversion and the photosynthetic unit: introductory lecture': in Bowen, E. J. (Ed.), *Recent Progress in Photobiology*: Blackwell, Oxford, 225–58.

54. CALVIN, M., and LYNCH, V. H., 1952. 'Grana-like structures of *Synechococcus cedorum*': *Nature* 169, 455.

55. CLAYTON, R. K., 1964. 'Physical aspects of the light reaction in photosynthesis': in Giese, A. C. (Ed.), *Photophysiology* Vol. 1: Academic Press, New York, 155–97.

56. COHEN-BAZIRE, G., and STANIER, R. Y., 1958. 'Specific inhibition of carotenoid synthesis in a photosynthetic bacterium and its physiological consequences': *Nature* 181, 250–2.

57. Decker, J. P., 1955. 'A rapid, postillumination deceleration of respiration in green leaves': *Plant Physiol.* 30, 82–4.

58. Decker, J. P., 1957. 'Further evidence of increased carbon dioxide production accompanying photosynthesis': *J. Solar Energy Sci. and Engng* 1, 30–3.

59 Decker, J. P., 1959. 'Comparative responses of carbon dioxide outburst and uptake in tobacco': *Plant Physiol.* 34, 100–2.

60. Decker, J. P., and Wien, J. D., 1958. 'Carbon dioxide surges in green leaves': *J. Solar Energy Sci. and Engng* 2, 39–41.

61. Dingle, H., and Pryce, A. W., 1940. 'The estimation of small quantities of carbon dioxide in air by the absorption of infra-red radiation': *Proc. Roy. Soc.* B. 129, 468–74.

61*a*. Downton, W. J. S., and Tregunna, E. B., 1968. 'Carbon dioxide compensation—its relation to photosynthetic carboxylation reactions, systematics of the Gramineae, and leaf anatomy': *Canad. J. Bot.* 46, 207–15.

62. Dunning, H. N., 1963. 'Geochemistry of organic pigments': in Breger, I. A. (Ed.), *Organic Geochemistry*: Pergamon Press, Oxford.

63. Dutton, H. J., Manning, W. M., and Duggar, B. M., 1943. 'Chlorophyll fluorescence and energy transfer in the diatom *Nitzschia closterium*': *J. Phys. Chem.* 47, 308–13.

64. Duysens, L. N. M., 1951. 'Transfer of light energy within the pigment systems in photosynthesizing cells': *Nature* 168, 548–50.

65. Duysens, L. N. M., 1952. 'The transfer of excitation energy in photosynthesis': Thesis, Utrecht.

66. Duysens, L. N. M., 1956. 'Energy transformations in photosynthesis': *Ann. Rev. Plant Physiol.* 7, 25–50.

67. Egle, K., and Schenk, W., 1952. 'Untersuchungen über die Reassimilation der Atmungskohlensäure bei der Photosynthese der Pflanzen': *Beitr. Biol. Pflanzen.* 29, 75–105.

68. Egle, K., and Schenk, W., 1953. 'Der Einfluss der Temperatur auf die Lage des CO_2-Kompensationspunktes': *Planta* 43, 83–97.

69. Ehrke, G., 1931. 'Über die Wirkung der Temperatur und des Lichtes auf die Atmung und Assimilation einiger Meeres- und Susswasseralgen': *Planta* 13, 221–310.

69*a*. El-Sharkaway, M. A., Loomis, R. S., and Williams, W. A., 1967. 'Apparent reassimilation of respiratory carbon dioxide by different plant species': *Physiologia Plant.* 20, 171–86.

70. Emerson, R., 1929*a*. 'The relation between maximum rate of photosynthesis and concentration of chlorophyll': *J. Gen. Physiol.* 12, 609–22.

71. Emerson, R., 1929*b*. 'Photosynthesis as a function of light intensity and of temperature with different concentrations of chlorophyll': *J. Gen. Physiol.* 12, 623–39.

72. EMERSON, R., 1935. 'The effect of intense light on the assimilatory mechanism of green plants, and its bearing on the carbon dioxide factor': *Cold Spring Harbor Symp.* 3, 128–37.
73. EMERSON, R., 1957. 'Dependence of yield of photosynthesis in long-wave red on wavelength and intensity of supplementary light': *Science* 125, 746.
74. EMERSON, R., 1958*a*. 'The quantum yield of photosynthesis': *Ann. Rev. Plant Physiol.* 9, 1–24.
75. EMERSON, R., 1958*b*. 'Yield of photosynthesis from simultaneous illumination with pairs of wave-lengths': *Science* 127, 1059–60.
76. EMERSON, R., and ARNOLD, W., 1932*a*. 'A separation of the reactions in photosynthesis by means of intermittent light': *J. Gen. Physiol.* 15, 391–420.
77. EMERSON, R., and ARNOLD, W., 1932*b*. 'The photochemical reaction in photosynthesis': *J. Gen. Physiol.* 16, 191–205.
78. EMERSON, R., and CHALMERS, R. V., 1957. 'On the efficiency of photosynthesis above and below compensation of respiration': in *Research in Photosynthesis*: Interscience, New York, 349–52.
79. EMERSON, R., CHALMERS, R., and CEDERSTRAND, C., 1957. 'Some factors influencing the long-wave limit of photosynthesis': *Proc. Nat. Acad. Sci.* 43, 133–43.
80. EMERSON, R., CHALMERS, R., CEDERSTRAND, C., and BRODY, M., 1956. 'Effect of temperature on the long-wave limit of photosynthesis': *Science* 123, 673.
81. EMERSON, R., and LEWIS, C. M., 1939. 'Factors influencing the efficiency of photosynthesis': *Amer. J. Bot.* 26, 808–22.
82. EMERSON, R., and LEWIS, C. M., 1941. 'Carbon dioxide exchange and the measurement of the quantum yield of photosynthesis': *Amer. J. Bot.* 28, 789–804.
83. EMERSON, R., and LEWIS, C. M., 1942. 'The photosynthetic efficiency of phycocyanin in *Chroococcus*, and the problem of carotenoid participation in photosynthesis': *J. Gen. Physiol.* 25, 579–95.
84. EMERSON, R., and LEWIS, C. M., 1943. 'The dependence of the quantum yield of *Chlorella* photosynthesis on wave length of light': *Amer. J. Bot.* 30, 165–78.
85. EMERSON, R., and RABINOWITCH, E., 1960. 'Red drop and role of auxiliary pigments in photosynthesis': *Plant Physiol.* 35, 477–85.
86. ENGLEMANN, T. W., 1881. 'Neue Methode zur Untersuchung der Sauerstoffausscheidung pflanzlicher und thierischer Organismen': *Bot. Ztg.* 39, 441–8.
87. FLEISCHER, W. E., 1935. 'The relation between chlorophyll content and rate of photosynthesis': *J. Gen. Physiol.* 18, 573–97.
88. FOGG, G. E., and NALEWAJKO, C., 1963. 'The production of glycollate

during photosynthesis in *Chlorella*': discussion of paper by C. P. Whittingham and G. G. Pritchard: *Proc. Roy. Soc.* B. 157, 381–2.

89. FOGG, G. E., and THAN-TUN, 1961. 'Interrelations of photosynthesis and assimilation of elementary nitrogen in a blue-green alga': *Proc. Roy. Soc.* B. 153, 111–27.

90. FORK, D. C., 1963*a*. 'Action spectra for O_2 evolution by chloroplasts with and without added substrate, for regeneration of O_2 evolving ability by far-red and for O_2 uptake': *Plant Physiol.* 38, 323–32.

91. FORK, D. C., 1963*b*. 'Observations on the function of chlorophyll *a* and accessory pigments in photosynthesis': in Kok, B., and Jagendorf, A. T. (Eds), *Photosynthetic Mechanisms of Green Plants*: Publ. 1145. Nat. Acad. Sci. Wash., 352–61.

92. FORRESTER, M. L., KROTKOV, G., and NELSON, C. D., 1966*a*. 'Effect of oxygen on photosynthesis, photorespiration and respiration in detached leaves. I. Soy bean': *Plant Physiol.* 41, 422–7.

93. FORRESTER, M. L., KROTKOV, G., and NELSON, C. D., 1966*b*. 'Effect of oxygen on photosynthesis, photorespiration and respiration in detached leaves. II. Corn and other monocotyledons': *Plant Physiol.* 41, 428–31.

94. FOWLE, F. E., 1929. 'Radiation from a perfect (black-body) radiator': in Washburn, E. W. (Ed.), *Int. Crit. Tables* 5, 238–42.

95. FRANCK, J., 1958. 'Remarks on the long-wave-length limits of photosynthesis and chlorophyll fluorescence': *Proc. Nat. Acad. Sci.* 44, 941–8.

96. FRANCK, J., and FRENCH, C. S., 1941. 'Photoxidation processes in plants': *J. Gen. Physiol.* 25, 309–24.

97. FRANCK, J., and ROSENBERG, J. L., 1964. 'A theory of light utilization in plant photosynthesis': *J. Theoret. Biol.* 7, 276–301.

98. FRENCH, C. S., 1957. 'Derivative spectrophotometry': *Symp. Instrumentation and control.* Instr. Soc. Am. Berkley, Calif. U.S.A., 83.

99. FRENCH, C. S., 1958. 'Various forms of chlorophyll *a* in plants': *Brookhaven Symp. Biol.* 11, 65–73.

100. FRENCH, C. S., 1960. 'The chlorophylls *in vivo* and *in vitro*': in Ruhland, W. (Ed.), *Encycl. Plant Physiol.*: Springer-Verlag, Berlin, 5, 252–97.

101. FRENCH, C. S., 1963*a*. 'The post-illumination survival of photosynthetic O_2 evolution': in Japanese Soc. of Plant Physiologists (Eds), *Studies on Microalgae and Photosynthetic Bacteria*: Univ. Tokyo Press, Tokyo, 271–9.

102. FRENCH, C. S., 1963*b*. 'Experiments with colored light flashes': *Carnegie Inst. Yr Bk* 62, 349–52.

103. FRENCH, C. S., and FORK, D. C., 1961. 'Computor solutions for photosynthesis rates from a two pigment model': *Biophys. J.* 1, 669–81.

104. FRENCH, C. S., and FORK, D. C., 1963. 'Two primary photochemical reactions in photosynthesis driven by different pigments': *Proc. Internat. Congr. Biochem. 5th, Moscow*, 1961, 6, 122–37.

105. FRENCH, C. S., and YOUNG, V. K., 1952. 'The fluorescence spectra of red algae and the transfer of energy from phycoerythrin to phycocyanin and chlorophyll': *J. Gen. Physiol.* 35, 873–90.
106. FRENCH, C. S., and YOUNG, V. M. K., 1956. 'The absorption, action and fluorescence spectra of photosynthetic pigments in living cells and in solutions': in Hollaender, A. (Ed.), *Radiation Biology* Vol. III: McGraw-Hill, New York, 343–91.
107. FREY-WYSSLING, A., 1937. 'Der Aufbau der Chlorophyllkörner': *Protoplasma* 29, 279–99.
108. FREY-WYSSLING, A., 1953. *Submicroscopic Morphology of Protoplasm*: Elsevier, Amsterdam.
109. FREY-WYSSLING, A., 1957. *Macromolecules in Cell Structure*: Harvard Univ. Press, Mass.
110. GAASTRA, P., 1959. 'Photosynthesis of crop plants as influenced by light, carbon dioxide, temperature, and stomatal diffusion resistance': *Med. Landb. Wageningen* 59, 1–68.
111. GABRIELSEN, E. K., 1948*a*. 'Threshold value of carbon dioxide concentration in photosynthesis of foliage leaves': *Nature* 161, 138–9.
112. GABRIELSEN, E. K., 1948*b*. 'Effects of different chlorophyll concentrations on photosynthesis in foliage leaves': *Physiologia Plant.* 1, 5–37.
113. GABRIELSEN, E. K., MADSEN, A., and VEJLBY, K., 1961. 'Induction of photosynthesis in etiolated leaves': *Physiologia Plant.* 14, 98–110.
114. GABRIELSEN, E. K., and VEJLBY, K., 1959. 'On the Kok-phenomenon in photosynthesis of leaves': *Physiologia Plant.* 12, 425–40.
115. GAFFRON, H., 1960. 'Energy storage: photosynthesis': in Steward, F. C. (Ed.), *Plant Physiology*, Vol. 1B: Academic Press, New York, 3–277.
116. GAFFRON, H., and WOHL, K., 1936. 'Zur Theorie der Assimilation': *Naturwissenschaften* 24, 103–7.
117. GIBBS, S. P., 1962. 'The ultrastructure of the chloroplasts of algae': *J. Ultrastruct. Res.* 7, 418–35.
118. GLASSTONE, S., 1950. *The Elements of Physical Chemistry*: MacMillan, London.
119. GLOVER, J., 1959. 'The apparent behaviour of maize and sorghum stomata during and after drought': *J. Agric. Sci.* 53, 412–6.
120. GODNEV, T. N., and KALISHEVICH, S. V., 1940. 'Chlorophyll concentration in chloroplasts of *Mnium medium*': *Compt. rend. Acad. Sci. URSS* 27, 832–3.
121. GOEDHEER, J. C., 1955. 'Orientation of the pigment molecules in the chloroplast': *Biochim. Biophys. Acta* 16, 471–6.
122. GOEDHEER, J. C., 1957. 'Optical properties and *in vivo* orientation of photosynthetic pigments': Thesis, Utrecht, 1–90.
123. GOLDSWORTHY, A., 1966. 'Experiments on the origin of CO_2 released by tobacco leaf segments in the light': *Phytochem.* 5, 1013–19.

124. Good, N. E., and Brown, A. H., 1961. 'The contribution of endogenous oxygen to the respiration of photosynthesizing *Chlorella* cells': *Biochim. Biophys. Acta* 50, 544–54.

125. Goodall, D. W., 1946. 'The distribution of weight change in the young tomato plant. II. Changes in dry weight of separated organs and translocation rates': *Ann. Bot. Lond.* N.S. 10, 305–38.

126. Govindjee, 1963. 'Emerson enhancement effect and two light reactions in photosynthesis': in Kok, B., and Jagendorf, A. T. (Eds), *Photosynthetic Mechanisms of Green Plants:* Publ. 1145. Nat. Acad. Sci. Wash., 318–34.

127. Govindjee and Rabinowitch, E., 1960. 'Action spectrum of the "Second Emerson Effect"': *Biophys. J.* 1, 73–89.

128. Govindjee, R., Thomas, J. B., and Rabinowitch, E., 1960. '"Second Emerson Effect" in the Hill reaction of *Chlorella* cells with quinone as oxidant': *Science* 132, 421.

129. Greenfield, S. S., 1942. 'Inhibitory effects of inorganic compounds on photosynthesis in *Chlorella*': *Amer. J. Bot.* 29, 121–31.

130. Gregory, F. G., 1918. 'Physiological conditions in cucumber houses': *Rep. Exp. and Res. Sta. Cheshunt* 3, 19–28.

131. Gregory, F. G., 1926. 'Effect of climatic conditions on the growth of barley': *Ann. Bot. Lond.* 40, 1–26.

132. Griffiths, M., Sistrom, W. R., Cohen-Bazire, G., Stanier, R. Y., and Calvin, M., 1955. 'Function of carotenoids in photosynthesis': *Nature* 176, 1211–15.

133. Haberlandt, G., 1882. 'Vergleichende Anatomie des assimilatorischen Gewebesystems der Pflanzen': *Jb. wiss. Bot.* 13, 74–188.

134. Haberlandt, G., 1914. Drummond, M. (Trans.), *Physiological Plant Anatomy*: MacMillan, London.

135. Harder, R., 1921. 'Kritische Versuche zu Blackmans Theorie der "begrenzenden Faktoren" bei der Kohlensäureassimilation': *Jb. wiss. Bot.* 60, 531–11.

135*a*. Hatch, M. D., Slack, C. R. and Johnson, H. S., 1967. 'Further studies on a new pathway of photosynthetic carbon dioxide fixation in sugar-cane and its occurrence in other plant species': *Biochem. J.* 102, 417–22.

136. Haxo, F. T., 1960. 'The wavelength dependence of photosynthesis, and the role of accessory pigments': in Allen, M. B. (Ed.), *Comparative Biochemistry of Photoreactive Systems: Symp. Comp. Biol. Kaiser Foundation Res. Inst.* 1, 339–60.

137. Haxo, F. T., and Blinks, L. R., 1950. 'Photosynthetic action spectra of marine algae': *J. Gen. Physiol.* 33, 389–422.

138. Heath, O. V. S., 1939. 'Experimental studies of the relation between carbon assimilation and stomatal movement. I. Apparatus and technique': *Ann. Bot. Lond.* N.S. 3, 469–95.

139. HEATH, O. V. S., 1941. 'Experimental studies of the relation between carbon assimilation and stomatal movement. II. Part I': *Ann. Bot. Lond.* N.S. 5, 455–500.
140. HEATH, O. V. S., 1950. 'Studies in stomatal behaviour. V. The role of carbon dioxide in the light response of stomata. Part I': *J. Exp. Bot.* 1, 29–62.
141. HEATH, O. V. S., 1951. 'Assimilation by green leaves with stomatal control eliminated': *Symp. Soc. Exp. Biol.* 5, 94–114.
142. HEATH, O. V. S., 1959*a*. 'Light and carbon dioxide in stomatal movements': in Ruhland, W. (Ed.), *Encycl. Plant Physiol.:* Springer-Verlag, Berlin, 17(1), 415–64.
143. HEATH, O. V. S., 1959*b*. 'The water relations of stomatal cells and the mechanisms of stomatal movement': in Steward, F. C. (Ed.), *Plant Physiology*, Academic Press; New York, 2, 193–250.
144. HEATH, O. V. S., 1966. 'Light measurements in plant growth investigations': *Nature* 210, 752–3.
145. HEATH, O. V. S., and GREGORY, F. G., 1938. 'The constancy of the mean net assimilation rate and its ecological importance': *Ann. Bot. Lond.* N.S. 2, 811–18.
146. HEATH, O. V. S., and MANSFIELD, T. A., 1962. 'A recording porometer with detachable cups operating on four separate leaves': Proc. Roy. Soc. B. 156, 1–13.
147. HEATH, O. V. S., and MANSFIELD, T. A., 1969. 'The movements of stomata': in Wilkins, M. B. (Ed.), *The Physiology of Plant Growth, Development and Responses*: McGraw-Hill, London.
148. HEATH, O. V. S., and MEIDNER, H., 1961. 'The influence of water strain on the minimum intercellular space carbon dioxide concentration Γ and stomatal movement in wheat leaves': *J. Exp. Bot.* 12, 226–42.
149. HEATH, O. V. S., and MEIDNER, H., 1967. 'Compensation points and carbon dioxide enrichment for lettuce grown under glass in winter': *J. Exp. Bot.* 18, 746–51.
150. HEATH, O. V. S., and ORCHARD, B., 1957. 'Midday closure of stomata. Temperature effects on the minimum intercellular space carbon dioxide concentration "Γ"': *Nature* 180, 180–1.
151. HEATH, O. V. S., and ORCHARD, B., 1968. 'Carbon assimilation at low carbon dioxide levels. II. The processes of apparent assimilation': *J. Exp. Bot.* 19, 176–92.
152. HEATH, O. V. S., and RUSSELL, J., 1951. 'The Wheatstone bridge porometer': *J. Exp. Bot.* 2, 111–16.
153. HEATH, O. V. S., and RUSSELL, J., 1954. 'Studies in stomatal behaviour, VI. An investigation of the light responses of wheat stomata with the attempted elimination of control by the mesophyll. Part 2': *J. Exp. Bot.* 5, 269–92.

154. HEITZ, E., 1936. 'Untersuchungen über den Beau der Plastiden. I. Die gerichteten Chlorophyllscheiben der Chloroplasten': *Planta* 26, 134–63.
155. HERSCH, P., 1952. 'Galvanic determination of traces of oxygen in gases': *Nature* 169, 792–3.
156. HESLOP-HARRISON, J., 1963. 'Structure and morphogenesis of lamellar systems in grana-containing chloroplasts. I. Membrane structure and lamellar architecture': *Planta* 60, 243–60.
157. HESLOP-HARRISON, J., 1966. 'Structural features of the chloroplast': *Sci. Progr.* 54, 519–41.
158. HEWITT, E. J., 1966. *Sand and Water Culture Methods used in the Study of Plant Nutrition*: 2nd Edn, Commonwealth Agric. Bur., Farnham Royal.
159. HILL, R., 1937. 'Oxygen evolved by isolated chloroplasts': *Nature* 139, 881–2.
160. HILL, R., 1965. 'The biochemists' green mansions: the photosynthetic electron-transport chain in plants': in Campbell, P. M., and Greville, G. D. (Eds); *Essays in Biochemistry*, Vol. I: Biochem. Soc., Academic Press.
161. HILL, R., and BENDALL, F., 1960. 'Function of the two cytochrome components in chloroplasts: a working hypothesis': *Nature* 186, 136–7.
162. HILL, R., and WHITTINGHAM, C. P., 1953. 'The induction phase of photosynthesis in *Chlorella* determined by a spectroscopic method': *New Phytol.* 52, 133–48.
163. HILLE, J. C. VAN, 1938. 'The quantitative relation between rate of photosynthesis and chlorophyll content in *Chlorella pyrenoidosa*': *Rec. trav. bot. néerland.* 35, 680–757.
164. HILLER, R. G., and WHITTINGHAM, C. P., 1959. 'Further studies on the carbon dioxide burst in algae': *Plant Physiol.* 34, 219–22.
165. HÖBER, R., 1945. *Physical Chemistry of Cells and Tissues*: Blakiston Philadelphia.
166. HOCH, G., and KOK, B., 1963. 'A mass spectrometer inlet for sampling gases dissolved in liquid phases': *Arch. Biochem. Biophys.* 101, 160–70.
167. HOCH, G., and OWENS, O. v. H., 1963. 'Photoreactions and respiration'; in Kok, B., and Jagendorf, A. T. (Eds), *Photosynthetic Mechanisms of Green Plants*: Publ. 1145. Nat. Acad. Sci. Wash., 409–20.
168. HOCH, G., OWENS, O. v. H., and KOK, B., 1963. 'Photosynthesis and respiration': *Arch. Biochem. Biophys.* 101, 171–80.
169. HODGE, A. J., 1959. 'Fine structure of lamellar systems as illustrated by chloroplasts': *Rev. Mod. Phys.* 31, 331.
170. HODGE, A. J., MCLEAN, J. D., and MERCER, F. V., 1955. 'Ultra structure of the lamellae and grana in the chloroplasts of *Zea mays L.*': *J. Biophys. Biochem. Cytol.* 1, 605–14.
171. HODGSON, G. L., and BLACKMAN, G. E., 1957. 'An analysis of the influence of plant density on the growth of *Vicia faba*. II': *J. Exp. Bot.* 8, 195–219.

172. HUBERT, B., 1935. 'The physical state of chlorophyll in the living plastid': *Rec. trav. bot. néerland.* 32, 323–90.
173. HUDSON, J. P., 1957*a*. 'The study of plant responses to soil moisture': in Hudson, J. P. (Ed.), *Control of the Plant Environment*: Butterworth's, London.
174. HUDSON, J. P., 1957*b*. 'Plants and their water supplies': *Endeavour* 16, 84–9.
175. HUGHES, A. P., and FREEMAN, P. R., 1967. 'Growth analysis using frequent small harvests': *J. Appl. Ecol.* 4, 553–60.
176. HUXLEY, P. A., 1963. 'Some growth characteristics of widely spaced plants grown in full daylight in an equatorial climate': *Proc. Biochem. Soc. Biochem. J.* 89, 76P.
177. ILJIN, W. S., 1923. 'Einfluss des Welkens auf die Atmung der Pflanzen': *Flora*, N.F. 16, 379–403.
178. IRVING, A. A., 1910. 'The beginning of photosynthesis and the development of chlorophyll': *Ann. Bot. Lond.* 24, 807–18.
179. IZAWA, S., and GOOD, N. E., 1965. 'The number of sites sensitive to . . . [DCMU], . . . [CMU] and . . . [Atrazine] in isolated chloroplasts': *Biochim. Biophys. Acta* 102, 20–38.
179*a*. JACKSON, J. E., and SLATER, C. H. W., 1967. 'An integrating photometer for outdoor use particularly in trees': *J. Appl. Ecol.* 4, 421–4.
180. JAMES, W. O., 1928. 'Experimental researches on vegetable assimilation and respiration. XIX.—The effect of variations of carbon dioxide supply upon the rate of assimilation of submerged water plants': *Proc. Roy Soc.* B. 103, 1–42.
181. JEFFREYS, H., 1918. 'Some problems of evaporation': *Phil. Mag.* 35, 270–80.
182. JONES, L. W., and MYERS, J., 1963. 'A common link between photosynthesis and respiration in a blue-green alga': *Nature* 199, 670–2.
183. JONES, L. W., and MYERS, J., 1964. 'Enhancement in the blue-green alga, *Anacystis nidulans*': *Plant Physiol.* 39, 938–46.
184. JOST, L., 1906. 'Über die Reaktionsgeschwindigkeit im Organismus': *Biol. Centr.* 26, 225–44.
185. KAHN, A., and VON WETTSTEIN, D., 1961. 'Macromolecular physiology of plastids. II. Structure of isolated spinach chloroplasts': *J. Ultrastruct. Res.* 5, 557.
186. KAMEN, M. D., 1963. *Primary Processes in Photosynthesis*: Academic Press, New York.
187. KASHA, M., 1959. 'Free radicals in photosynthetic systems': *Rev. Mod. Phys.* 31, 157–69.
188. KENNEDY, S. R., 1940. 'The influence of magnesium deficiency, chlorophyll concentration, and heat treatments on the rate of photosynthesis of *Chlorella*': *Amer. J. Bot.* 27, 68–73.

189. KNIEP, H., 1915. 'Über den Gasaustausch der Wasserpflanzen. Ein Beitrag zur Kritik der Blazenzählmethode': *Jb. wiss. Bot.* 56, 460–509.
190. KOK, B., 1948. 'A critical consideration of the quantum yield of *Chlorella*—photosynthesis': *Enzymologia* 13, 1–56.
191. KOK, B., 1949. 'On the interrelation of respiration and photosynthesis in green plants': *Biochim. Biophys. Acta* 3, 625–31.
192. KOK, B., 1951. 'Photo-induced interactions in metabolism of green plant cells': *Symp. Soc. Exp. Biol.* 5, 209–21.
193. KOK, B., 1957. 'Absorption changes induced by the photochemical reaction of photosynthesis': *Nature* 179, 583–4.
194. KOK, B., 1959. 'Light induced absorption changes in photosynthetic organisms. II. A split-beam difference spectrophotometer': *Plant Physiol.* 34, 184–92.
195. KOK, B., 1961. 'Partial purification and determination of oxidation–reduction potential of the photosynthetic chlorophyll complex absorbing at 700 $m\mu$': *Biochim. Biophys. Acta* 48, 527–33.
196. KRAMER, P. J., 1938. 'Root resistance as a cause of the absorption lag': *Amer. J. Bot.* 25, 110–3.
197. KRAMER, P. J., 1959. 'Transpiration and the water economy of plants': in Steward, F. C. (Ed.), *Plant Physiology*, Vol. II: Academic Press, New York, 607–726.
198. KRASNOVSKY, A. A., 1960. 'The primary processes of photosynthesis in plants': *Ann. Rev. Plant Physiol.* 11, 363–410.
199. KREBS, H. A., 1951. 'The use of "CO_2-buffers" in manometric measurements of cell metabolism': *Symp. Soc. Exp. Biol.* 5, 336–42.
200. LABY, T. H., and NELSON, E. A., 1929. 'Thermal conductivity: gases and vapors': in Washburn, E. W. (Ed.), *Int. Crit. Tables* 5, 213–15: McGraw-Hill, New York.
201. LATIMER, P., 1961. 'Anomalous dispersion of CS_2 and $CHC1_3$ — theoretical predictions': *J. Opt. Soc. Amer.* 51, 116–18.
202. LATIMER, P., 1963. 'Is selective scattering a universal phenomenon?': in Japanese Soc. of Plant Physiologists (Eds), *Studies on Microalgae and Photosynthetic Bacteria*: Univ. Tokyo Press, Tokyo, 213–25.
203. LEECH, R. M., 1964. 'The isolation of structurally intact chloroplasts': *Biochim. Biophys. Acta* 79, 637.
204. LLOYD, F. E., 1908. 'The physiology of stomata': *Carnegie Inst. Wash. Pub.* No. 82, 1–142.
205. MCALISTER, E. D., 1937. 'Time course of photosynthesis for a higher plant': *Smithsonian Inst. Misc. Collections* 95, No. 24, 1–17.
206. MCALISTER, E. D., and MYERS, J., 1940. 'The time course of photosynthesis and fluorescence observed simultaneously': *Smithsonian Inst. Misc. Collections* 99, 1–37.

207. McCree, K. J., 1965. 'Light measurements in plant growth investigations': *Nature* 206, 527–8.
208. McCree, K. J., 1966. 'Light measurements in plant growth investigations': *Nature* 210, 753.
209. Magee, J. L., Witt, T. W. de, Smith, E. C., and Daniels, F., 1939. 'A photo-calorimeter. The quantum efficiency of photosynthesis in algae': *J. Amer. Chem. Soc.* 61, 3529–33.
210. Mansfield, T. A., and Heath, O. V. S., 1964. 'Studies in stomatal behaviour X. An investigation of responses to low-intensity illumination and temperature in *Xanthium pennsylvanicum*': *J. Exp. Bot.* 15, 114–24.
211. Martin, A. E., Reid, A. M., and Smart, J., 1958. 'Infra-red gas analysers for plant control': *Research, Lond.* 11, 258–65.
212. Maskell, E. J., 1927. 'Field observations on starch production in the leaves of the potato': *Ann. Bot. Lond.* 41, 327–44.
213. Maskell, E. J., 1928*a*. 'Experimental researches on vegetable assimilation and respiration. XVII. The diurnal rhythm of assimilation in leaves of cherry laurel at "limiting" concentrations of carbon dioxide': *Proc. Roy. Soc.* B. 102, 467–587.
214. Maskell, E. J., 'Experimental researches on vegetable assimilation and respiration. XVIII. The relation between stomatal opening and assimilation: *Proc. Roy. Soc. B.* 102, 488–533.
215. Matthaei, G. L. C., 1905. 'Experimental researches on vegetable assimilation and respiration. III. On the effect of temperature on carbon-dioxide assimilation': *Phil. Trans. Roy. Soc.* B. 197, 47–105.
216. Meidner, H., 1961. 'The minimum intercellular-space carbon-dioxide concentration in leaves of the palm *Phoenix reclinata*': *J. Exp. Bot.* 12, 409–13.
217. Meidner, H., 1962. 'The minimum intercellular-space CO_2 concentration (Γ) of maize leaves and its influence on stomatal movements': *J. Exp. Bot.* 13, 284–93.
218. Meidner, H., 1967. 'Further observations on the minimum intercellular space carbon dioxide concentration (Γ) of maize leaves and the postulated role of "photorespiration" and glycollate metabolism': *J. Exp. Bot.* 18, 177–85.
219. Meidner, H., and Mansfield, T. A., 1965. 'Stomatal responses to illumination': *Biol. Rev.* 40, 483–509.
220. Meidner, H., and Spanner, D. C., 1959. 'The differential transpiration porometer': *J. Exp. Bot.* 10, 190–205.
221. Menke, W., 1938. 'Über den Feinbau der Chloroplasten': *Kolloid-Z.* 85, 256–9.
222. Menke, W., 1939. 'Direkter Nachweis des Lamellaren Feinbaues der Chloroplasten': *Naturwissenschaften* 27, 29.

223. MENKE, W., 1961. 'Über das Lamellarsystem des Chromatoplasmas von Cyanophyceen': *Z. Naturforsch.* 16*b*, 543.
224. METZNER, H., 1952. 'Die Reduktion wässriger Silbernitratlösungen durch Chloroplasten und andere Zellbestandteile': *Protoplasma* 41, 129–67.
225. MEYER, A., 1883. *Das Chlorophyllkorn*: Felix, Leipzig, 1–91.
226. MILLER, E. S., and BURR, G. O., 1935. 'Carbon dioxide balance at high light intensities': *Plant Physiol.* 10, 93–114.
227. MILLER, J., 1914. 'A field method for determining dissolved oxygen in water': *J. Soc. Chem. Ind.* 33, 185–6.
228. MILTHORPE, F. L., 1961. 'Plant factors involved in transpiration': in *Plant–Water Relationships in Arid and Semi-Arid Conditions*: Proc. Madrid Symp (1959), Unesco, 107–13.
229. MITSCHERLICH, E. A., DÜHRING, F., and SAUCKEN, S. VON, 1921. 'Das Wirkungsgesetz der Wachstumsfaktoren': *Landw. Jahrb.* 56, 71–93.
230. MONSI, M., and SAEKI, T., 1953. 'Über den Lichtfaktor in den Pflanzengesellschaften und seine Bedeutung für die Stoffproduktion': *Jap. J. Bot.* 14, 22–52.
231. MONTEITH, J. L., 1963. 'Gas exchange in plant communities': in Evans, L. T. (Ed.), *Environmental Control of Plant Growth*: Proc. Canberra Symp. (1962), Academic Press, New York, 95–111.
232. MONTEITH, J. L., 1965. 'Light distribution and photosynthesis of field crops': *Ann. Bot. Lond.* N.S. 29, 17–38.
233. MONTEITH, J. L., and SZEICZ, G., 1960. 'The carbon dioxide flux over a field of sugar beet': *Quart. J. Roy. Meteorol. Soc.* 86, 205–14.
234. MOOR, H., 1964. 'Die Gefrier-Fixation lebender Zellen und ihre Anwendung in der Elektronenmikroskopie': *Z. Zellforschung* 62, 546.
235. MOSS, D. N., 1962. 'The limiting carbon dioxide concentration for photosynthesis': *Nature*, 193, 587.
236. MOSS, D. N., 1966. 'Respiration of leaves in light and darkness': *Crop. Sci.*, 6, 351–4.
237. MOSS, R. A., and LOOMIS, W. E., 1952. 'Absorption spectra of leaves. I. The visible spectrum': *Plant Physiol.* 27, 370–91.
238. MYERS, J., 1963. 'Enhancement': in Kok, B., and Jagendorf, A. T. (Eds), *Photosynthetic Mechanisms in Green Plants*: Publ. 1145. Nat. Acad. Sci. Wash., 301–7.
239. MYERS, J., and FRENCH, C. S., 1960*a*. 'Evidences from action spectra for a specific participation of chlorophyll *b* in photosynthesis': *J. Gen. Physiol.* 43, 723–36.
240. MYERS, J., and FRENCH, C. S., 1960*b*. 'Relationships between time course, chromatic transient, and enhancement phenomena of photosynthesis': *Plant Physiol.* 35, 963–9.
241. MYERS, J., and GRAHAM, J. R., 1963. 'Further improvements in stationary platinum electrode of Haxo and Blinks': *Plant Physiol.* 38, 1–5.

242. NEWTON, R. G., 1935. 'An improved electrical conductivity method for the estimation of carbon dioxide and other reactive gases': *Ann. Bot. Lond.* 49, 381–98.

243. NIR, I., and POLJAKOFF-MAYBER, A., 1967. 'Effect of water stress on the photochemical activity of chloroplasts': *Nature* 213, 418–19.

244. NISHIMURA, M. S., WHITTINGHAM, C. P., and EMERSON, R., 1951. 'The maximum efficiency of photosynthesis': *Symp. Soc. Exp. Biol.* 5, 176–210.

245. NORMAN, R. W. VAN, and BROWN, A. H., 1952. 'The relative rates of photosynthetic assimilation of isotopic forms of carbon dioxide': *Plant Physiol.* 27, 691–709.

246. OLSON, R. A., and ENGEL, E. R., 1959. 'Chlorophyll absorption microscopy of *in vivo*, cell-free and fragmented Chlorella chloroplasts': in *The Photochemical Apparatus: Its Structure and Function*: Brookhaven Symp. in Biol. No. 11, Upton, New York, 303–9.

247. ORCHARD, B., 1956. 'Studies in carbon assimilation': Ph.D. Thesis Univ. Lond.

248. ORCHARD, B., and HEATH, O. V. S., 1964. 'Carbon assimilation at low carbon dioxide levels. I. Apparatus and technique': *J. Exp. Bot.* 15, 314–30.

249. OZBUN, J. L., VOLK, R. J., and JACKSON, W. A., 1964. 'Effects of light and darkness on gaseous exchange of bean leaves': *Plant Physiol.* 39, 523–7.

250. PARK, R. B., 1965. 'Substructure of the chloroplast lamellae': *J. Cell. Biol.* 27, 151–61.

251. PARK, R. B., and BIGGINS, J., 1964. 'Quantasome: size and composition': *Science* 144, 1009–11.

252. PARK, R. B., and PON, N. G., 1961. 'Correlation of structure with function in *Spinacea oleracea* chloroplasts': *J. Mol. Biol.* 3, 1–10.

253. PARK, R. B., and PON, N. G., 1963. 'Chemical composition and substructure of lamellae isolated from *Spinacea oleracea* chloroplasts': *J. Mol. Biol.* 6, 105–14.

254. PENMAN, H. L., and LONG, I. F., 1960. 'Weather in wheat: an essay in micro-meteorology': *Quart. J. Roy. Meteorol. Soc.* 86, 16–50.

255. PENMAN, H. L., and SCHOFIELD, R. K., 1951. 'Some physical aspects of assimilation and transpiration': *Symp. Soc. Exp. Biol.* 5, 115–29.

256. PETERING, H. G., DUGGAR, B. M., and DANIELS, F., 1939. 'Quantum efficiency of photosynthesis in *Chlorella*. II': *J. Amer. Chem. Soc.* 61, 3525–9.

257. PFEFFER, W., 1900. *The Physiology of Plants* (Ewart, A. J., Trans. and Ed.): Vol. I, Oxford.

257*a*. PICKETT, J. M., and FRENCH, C. S., 1968. 'Some essential considerations in the measurement and interpretation of absorption spectra of heterogeneous samples': *Carnegie Inst. Year Bk.* 66, 171–5.

258. PIRIE, N. W., 1958. 'Unconventional production of foodstuffs': in Yapp, W. B., and Watson, D. J. (Eds), *The Biological Productivity of Britain*: Inst. Biol., London, 115–23.

258*a*. PLEIJEL, G., and LONGMORE, J., 1952. 'A method of correcting the cosine error of selenium rectifier photocells': *J. Sci. Instr.* 29, 137–8.

259. PORTER, G., 1963. 'Primary photoprocesses of chlorophyll and related molecules': *Proc. Roy. Soc.* B. 157, 293–9.

260. PORTER, H. K., PAL, N., and MARTIN, R. V., 1950. 'Physiological studies in plant nutrition. XV. Assimilation of carbon by the ear of barley and its relation to the accumulation of dry matter in the grain': *Ann. Bot. Lond.* N.S. 14, 55–68.

261. PORTSMOUTH, G. B., 1949. 'The effect of manganese on carbon assimilation in the potato plant as determined by a modified half-leaf method': *Ann. Bot. Lond.* N.S. 13, 113–33.

262. POWELL, M. C., and HEATH, O. V. S., 1964. 'A simple and inexpensive integrating photometer': *J. Exp. Bot.* 15, 187–91.

263. POWELL, R. W., 1940. 'Further experiments on the evaporation of water from saturated surfaces': *Trans. Inst. Chem. Engnrs* 18, 36–50.

264. RABINOWITCH, E. I., 1945. *Photosynthesis and Related Processes.* Vol. I: Interscience Publishers, New York.

265. RABINOWITCH, E. I., 1951. *Photosynthesis and Related Processes.* Vol. II, Part I: Interscience Publishers, New York.

266. RABINOWITCH, E. I., 1956. *Photosynthesis and Related Processes.* Vol. II, Part 2: Interscience Publishers, New York.

267. RABINOWITCH, E., 1959. 'Primary photochemical and photophysical processes in photosynthesis': *Plant Physiol.* 34, 213–18.

268. RABINOWITCH, E., 1963. 'The mechanism of photosynthesis': in Kok, B., and Jagendorf, A. T. (Eds), *Photosynthetic Mechanisms of Green Plants*: Publ. 1145. Nat. Acad. Sci. Wash, 112–21.

269. RICHARDS, F. J., 1941. 'The diagrammatic representation of the results of physiological and other experiments designed factorially': *Ann. Bot. Lond.* N.S. 5, 249–61.

270. RICHARDS, F. J., and TEMPLEMAN, W. G., 1936. 'Physiological studies in plant nutrition. IV. Nitrogen metabolism in relation to nutrient deficiency and age in leaves of barley': *Ann. Bot. Lond.* 50, 367–402.

271. RYLE, G. J. A., 1967. 'Growth rates in *Lolium temulentum* as influenced by previous regimes of light energy': *Nature* 213, 309–11.

272. SCARTH, G. W., LOEWY, A., and SHAW, M., 1948. 'Use of the infra-red total absorption method for estimating the time course of photosynthesis and transpiration': *Canad. J. Res.* C26, 94–107.

273. SCARTH, G. W., and SHAW, M., 1951. 'Stomatal movement and photosynthesis in Pelargonium. II': *Plant Physiol.* 26, 581–97.

274. SCHIMPER, A. F. W., 1885. 'Untersuchungen über die Chlorophyll-körper und die ihnen homologen Gebilde': *J. wiss. Bot.* 16, 1–247.

275. SHIBATA, K., 1958. 'Spectrophotometry of intact biological materials': *J. Biochem. (Tokyo)* 45, 599–623.

276. SHIBATA, K., BENSON, A. A., and CALVIN, M., 1954. 'The absorption spectra of suspensions of living organisms': *Biochim. Biophys. Acta* 15, 461–70.

277. SMITH, J. H. C., 1954. 'The development of chlorophyll and oxygen-evolving power in etiolated barley leaves when illuminated': *Plant Physiol.* 29, 143–8.

278. SMITH, J. H. C., DURHAM, L. J., and WURSTER, C. F., 1959. 'Formation and bleaching of chlorophyll in albino corn seedlings': *Plant Physiol.* 34, 340–5.

279. SPANNER, D. C., 1953. 'On a new method for measuring the stomatal aperture of leaves': *J. Exp. Bot.* 4, 283–95.

280. SPANNER, D. C., 1967. Appendix to Heath, O. V. S., and Meidner. H. 'Compensation points and carbon dioxide enrichment for lettuce grown under glass in winter': *J. Exp. Bot.* 18, 750–1.

281. SPIERINGS, F. H., HARRIS, G. P., and WASSINK, E. C., 1952. 'Applications of the diaferometer technique to studies on the gas exchange and the carbon dioxide content of potato tubers': *Med. Landb. Wageningen* 52, 93–104.

282. STEINMANN, E., 1952. 'An electron microscope study of the lamellar structure of chloroplasts': *Exp. Cell. Res.* 3, 367–72.

283. STEINMANN, E., and SJÖSTRAND, F. S., 1955. 'The ultra-structure of chloroplasts': *Exp. Cell. Res.* 8, 15.

284. STILES, W., and LEACH, W., 1931. 'On the use of the katharometer for the measurement of respiration': *Ann. Bot. Lond.* 45, 461–88.

285. STREHLER, B. L., and ARNOLD, W., 1951. 'Light production by green plants': *J. Gen. Physiol.* 34, 809–20.

286. SVEDBERG, T., and KATSURAI, T., 1929. 'The molecular weights of phycocyan and of phycoerythrin from *Porphyra tenera* and of phycocyan from *Aphanizomenon flos aquae*': *J. Amer. Chem. Soc.* 51, 3573–83.

287. TAGAWA, K., and ARNON, D. I., 1962. 'Ferrodoxins as electron carriers in photosynthesis and in the biological production and consumption of hydrogen gas': *Nature* 195, 537–43.

288. TANADA, T., 1951. 'The photosynthetic efficiency of carotenoid pigments in *Navicula minima*': *Amer. J. Bot.* 38, 276–83.

289. THOMAS, J. B., BLAAUW, O. H., and DUYSENS, L. N. M., 1953. 'On the relation between size and photochemical activity of fragments of spinach grana': *Biochim. Biophys. Acta* 10, 230–40.

290. THOMAS, J. B., MINNAERT, K., and ELBERS, P. F., 1956. 'Chlorophyll concentrations in plastids of different groups of plants': *Acta bot. néerland.* 5, 315–21.

291. THOMAS, J. B., POST, L. C., and VERTREGT, N., 1954. 'Localisation of chlorophyll within the chloroplast': *Biochim. Biophys. Acta* 13, 20–30.

292. THOMAS, M. D., HENDRICKS, R. H., and HILL, G. R., 1944. 'Apparent equilibrium between photosynthesis and respiration in an unrenewed atmosphere': *Plant Physiol.* 19, 370–6.

293. THOMAS, M. D., and HILL, G. R., 1937. 'The continuous measurement of photosynthesis, respiration and transpiration of alfalfa and wheat growing under field conditions': *Plant Physiol.* 12, 285–307.

294. TONNELAT, J., 1944. 'Mesure calorimétrique du rendement de la photosynthèse': *Compt. Rend.* 218, 430–2.

295. TREADWELL, F. P., 1935. Hall, W. T. (Trans.), *Analytical Chemistry*. Vol. II. *Quantitative Analysis*: Wiley, Chapman & Hall, New York, 8th Edn 1935.

296. TREBST, A. V., TSUJIMOTO, H. Y., and ARNON, D. I., 1958. 'Separation of light and dark phases in the photosynthesis of isolated chloroplasts': *Nature* 182, 351–5.

296*a*. TREGUNNA, E. B., and DOWNTON, J., 1967. 'Carbon dioxide compensation in some members of the Amaranthaceae and some related families': *Canad. J. Bot.* 45, 2385–7.

297. TREGUNNA, E. B., KROTKOV, G., and NELSON, C. D., 1961. 'Evolution of carbon dioxide by tobacco leaves during the dark period following illumination with light of different intensities': *Canad. J. Bot.* 39, 1045–56.

298. TREGUNNA, E. B., KROTKOV, G., and NELSON, C. D., 1964. 'Further evidence on the effects of light on respiration during photosynthesis': *Canad. J. Bot.* 42, 989–97.

299. TREGUNNA, E. B., KROTKOV, G., and NELSON, C. D., 1966. 'Effect of oxygen on the rate of photorespiration in detached tobacco leaves': *Physiologia Plant.* 19, 723–33.

300. UMBREIT, W. W., BURRIS, R. H., and STAUFFER, J. F., 1957. *Manometric Techniques*: 3rd Edn: Burgess Publ., Minneapolis, Minnesota.

301. VEEN, R. VAN DER, 1949. 'Induction phenomena in photosynthesis. I': *Physiologia Plant.* 2, 217–34.

302. VEJLBY, K., 1958. 'Induction phenomena in photosynthesis. Experiments with *Polytrichum attenuatum*': *Physiologia Plant.* 11, 158–69.

303. VEJLBY, K., 1959. 'Induction phenomena in photosynthesis. Simultaneous measurements of CO_2 and O_2 exchange': *Physiologia Plant.* 12, 162–72.

304. VIRGIN, H. I., 1954. 'The distortion of fluorescence spectra in leaves by

light scattering and its reduction by infiltration': *Physiologia Plant.* 7, 560–70.

305. VIRGIN, H. I., 1956. 'Some notes on the fluorescence spectra of plants *in vivo*': *Physiologia Plant.* 9, 674–81.
306. WARBURG, O., 1919. 'Über die Geschwindigkeit der photochemischen Kohlensäurezersetzung in lebenden Zellen. I': *Biochem. Z.* 100, 230–70.
307. WARBURG, O., 1920. 'Über die Geschwindigkeit der photochemischen Kohlensäurezersetzung in lebenden Zellen. II': *Biochem. Z.* 103, 188–217.
308. WARBURG, O., and KRIPPAHL, G., 1960. 'Notwendigkeit der Kohlensäure für die Chinon und Ferricyanid-Reaktionen in grünen Grana': *Z. Naturforsch.* 15b, 367–9.
309. WARBURG, O., and NEGELEIN, E., 1923. 'Über den Einfluss der Wellenlänge auf den Energieumsatz bei der Kohlensäureassimilation': *Z. phys. Chem.* 106, 191–218.
310. WARREN-WILSON, J., 1966. 'High net assimilation rates of sunflower plants in an arid climate': *Ann. Bot. Lond.* N.S. 30, 745–51.
311. WATSON, D. J., 1952. 'The physiological basis of variation in yield': in *Advances in Agronomy*, Vol. 4: Academic Press, New York, 101–45.
312. WATSON, D. J., 1958. 'The dependence of net assimilation rate on leaf-area index': *Ann. Bot. Lond.* N.S. 22, 37–54.
313. WATSON, D. J., and BAPTISTE, E. C. D., 1938. 'A comparative physiological study of sugar-beet and mangold with respect to growth and sugar accumulation. I': *Ann. Bot. Lond.* N.S. 2, 437–80.
314. WATSON, W. F., and LIVINGSTON, R., 1950. 'Self-quenching and sensitization of fluorescence of chlorophyll solutions': *J. Chem. Phys.* 18, 802–9.
315. WEBER, C., 1882. 'Ueber specifische Assimilationsenergie': *Arb. bot. Inst. Würzburg* 2, 346–52.

315*a*. WEIER, T. E., STOCKING, C. R., and SHUMWAY, L. K., 1967. 'The photosynthetic apparatus in chloroplasts of higher plants': in *Energy Conversion by the Photosynthetic Apparatus*: Brookhaven Symp. in Biol. No. 19, Upton, New York, 353–74.

316. WEIS, D., and BROWN, A. H., 1959. 'Kinetic relationships between photosynthesis and respiration in the algal flagellate, *Ochromonas malhamensis*': *Plant Physiol.* 34, 235–9.
317. WELLS, D. A., and SOFFE, R., 1962. 'A bench method for the automatic watering by capillarity of plants grown in pots': *J. Agric. Engng Res.* 7, 42–6.
318. WEST, C., BRIGGS, G. E., and KIDD, F. L., 1920. 'Methods and significant relations in the quantitative analysis of plant growth': *New Phytol.* 19, 200–7.
319. WETTSTEIN, D. VON, 1960. 'Multiple allelism in induced chlorophyll

mutants. II. Error in the aggregation of the lamellar discs in the chloroplast': *Hereditas* 46, 700.

320. WHATLEY, F. R., and LOSADA, M., 1964. 'The photochemical reactions of photosynthesis': in Giese, A. C. (Ed.), *Photophysiology*, Vol. I: Academic Press, New York, 111–54.
321. WHITTINGHAM, C. P., 1954. 'Photosynthesis in *Chlorella* during intermittent illumination of long periodicity': *Plant Physiol.* 29, 473–7.
322. WHITTINGHAM, C. P., and BISHOP, P. M., 1961. 'Thermal reaction between two light reactions in photosynthesis': *Nature* 192, 426–7.
323. WHITTINGHAM, C. P., and BISHOP, P. M., 1963. 'Studies with flash illumination on the enhancement effect in chloroplasts': in Kok, B., and Jagendorf, A. T. (Eds), *Photosynthetic Mechanisms of Green Plants*: Publ. 1145. Nat. Acad. Sci. Wash., 371–80.
324. WHITTINGHAM, C. P., and BROWN, A. H., 1958. 'Oxygen evolution from algae illuminated by short and long flashes of light': *J. Exp. Bot.* 9, 311–19.
325. WHITTINGHAM, C. P., and PRITCHARD, G. G., 1963. 'The production of glycollate during photosynthesis in *Chlorella*': *Proc. Roy. Soc.* B. 157, 366–80.
326. WIESNER, J., 1907. *Der Lichtgenuss der Pflanze*: Liepzig.
327. WILLIAMS, R. F., 1939. 'Physiological ontogeny in plants and its relation to nutrition. 6. Analysis of the unit leaf rate': *Australian J. Exp. Biol. and Med. Sci.* 17, 123–32.
328. WILLSTÄTTER, R., and STOLL, A., 1913. *Untersuchungen über Chlorophyll*: Berlin.
329. WILLSTÄTTER, R., and STOLL, A., 1918. *Untersuchungen über die Assimilation der Kohlensäure*: Berlin, 1918.
330. WILMOTT, A. J., 1921. 'Experimental researches on vegetable assimilation and respiration. XIV.—Assimilation by submerged plants in dilute solutions of bicarbonates and of acids: an improved bubble counting technique': *Proc. Roy. Soc.* B. 92, 304–27.
331. WINTER, E. J., 1963. 'Automatic watering for amateurs': *Gdnr's Chron.* Oct. 26, 304–5.
332. WIT, C. T. DE, 1965. 'Photosynthesis of leaf canopies': *Agric. Res. Rep. Wageningen* No. 663, 1–57.
333. WITHROW, R. B., and WITHROW, A. P., 1956. 'Generation, control and measurement of visible and near-visible radiant energy': in Hollaender, A. (Ed.), *Radiation Biology*, Vol. III: McGraw-Hill, New York, 125–258.
334. WOLKEN, J. J., and SCHWERTZ, F. A., 1953. 'Chlorophyll monolayers in chloroplasts': *J. Gen. Physiol.* 37, 111–21.

334a. WOLKEN, J. J., and STROTHER, G. K., 1963. 'Microspectrophotometry': *Appl. Optics* 2, 899–907.

335. WRIGHT, W. D., 1949. *Photometry and the Eye*: Hatton Press, London.
336. ZELITCH, I., 1964. 'Organic acids and respiration in photosynthetic tissues': *Ann. Rev. Plant Physiol.* 15, 121–42.
337. ZSCHEILE, F. P., and COMAR, C. L., 1941. 'Influence of preparative procedure on the purity of chlorophyll components as shown by absorption spectra': *Bot. Gaz.* 102, 463–81.
338. ZSCHEILE, F. P., WHITE, J. W., BEADLE, B. W., and ROACH, J. R., 1942. 'The preparation and absorption spectra of five pure carotenoid pigments': *Plant Physiol.* 17, 331–46.

Author Index

Pages in the text are shown in lightface type and pages in the Bibliography in boldface. Reference numbers are shown in square brackets to enable the reference to be traced where the author's name is not given in the text.

Subject Index

The main text references are given in bold type.